Seeing the Unseen

Mount Wilson's role in high angular resolution astronomy

AAS Editor in Chief

Ethan Vishniac, Johns Hopkins University, Maryland, USA

About the program:

AAS-IOP Astronomy ebooks is the official book program of the American Astronomical Society (AAS), and aims to share in depth the most fascinating areas of astronomy, astrophysics, solar physics and planetary science. The program includes publications in the following topics:

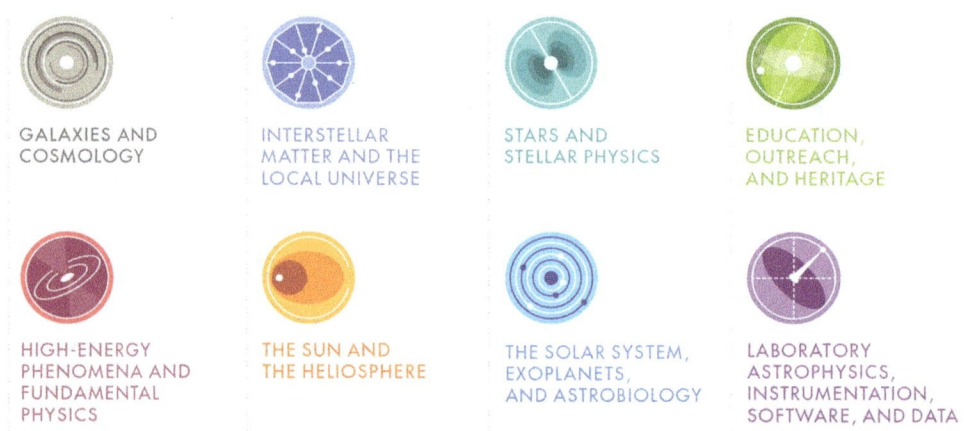

GALAXIES AND COSMOLOGY

INTERSTELLAR MATTER AND THE LOCAL UNIVERSE

STARS AND STELLAR PHYSICS

EDUCATION, OUTREACH, AND HERITAGE

HIGH-ENERGY PHENOMENA AND FUNDAMENTAL PHYSICS

THE SUN AND THE HELIOSPHERE

THE SOLAR SYSTEM, EXOPLANETS, AND ASTROBIOLOGY

LABORATORY ASTROPHYSICS, INSTRUMENTATION, SOFTWARE, AND DATA

Books in the program range in level from short introductory texts on fast-moving areas, graduate and upper-level undergraduate textbooks, research monographs and practical handbooks.

For a complete list of published and forthcoming titles, please visit iopscience.org/books/aas.

About the American Astronomical Society

The American Astronomical Society (aas.org), established 1899, is the major organization of professional astronomers in North America. The membership (~7,000) also includes physicists, mathematicians, geologists, engineers and others whose research interests lie within the broad spectrum of subjects now comprising the contemporary astronomical sciences. The mission of the Society is to enhance and share humanity's scientific understanding of the universe.

Seeing the Unseen

Mount Wilson's role in high angular resolution astronomy

Harold A McAlister

*Regents' Professor Emeritus of Astronomy, Georgia State University,
Atlanta, GA 30302, USA*

IOP Publishing, Bristol, UK

ISBN   978-0-7503-2208-9 (ebook)
ISBN   978-0-7503-2206-5 (print)
ISBN   978-0-7503-2209-6 (myPrint)
ISBN   978-0-7503-2207-2 (mobi)

DOI   10.1088/2514-3433/abb4de

Version: 20201101

AAS–IOP Astronomy
ISSN 2514-3433 (online)
ISSN 2515-141X (print)

British Library Cataloguing-in-Publication Data: A catalogue record for this book is available from the British Library.

Published by IOP Publishing, wholly owned by The Institute of Physics, London

IOP Publishing, Temple Circus, Temple Way, Bristol, BS1 6HG, UK

US Office: IOP Publishing, Inc., 190 North Independence Mall West, Suite 601, Philadelphia, PA 19106, USA

Contents

Preface

I first became aware of Mount Wilson Observatory as a fourth grader shortly after the stunning and world-changing launch of Sputnik I. Just a few blocks from my home in Chattanooga, Tennessee was the Clarence T. Jones Observatory—its director was Karel Hujer who was traveling in the Americas in 1938–1939 when the Nazis occupied his native Czechoslovakia. He never lived there again and settled in the U.S. with his academic trajectory taking him in 1946 to the University of Chattanooga. Karel had first visited Mount Wilson during his three-year appointment as a pre-doctoral student at Yerkes Observatory from Prague's Charles University and spoke often of that great observatory and its extraordinary founder George Ellery Hale. By the time I was in the sixth grade, I was what amounted to a junior docent at the Jones Observatory where one of my duties was to explain the wall-mounted, back-illuminated transparencies to groups of visitors during the Friday-night open houses. Among those transparencies was an aerial view of the Mount Wilson Observatory, a photo of the 100 inch dome, and images of the 150 ft solar tower and the 60 and 100 inch telescopes. Under Karel's mentorship, I would become an astronomer, and my engagement with Mount Wilson would ultimately extend to overseeing the construction of a large interferometer there as well as becoming the Observatory's director from 2002–2014. Writing this book has been a sentimental journey for me.

Twenty-six-year-old Karel Hujer on the footbridge from the Galley to the 100 inch dome on Mount Wilson in June 1928. (Photo inherited from Karel's estate by the author.)

Interferometry at Mount Wilson was predestined at its 1904 founding. George Ellery Hale had attended Albert Abraham Michelson's lecture at the 1888 Cleveland, Ohio meeting the American Association for the Advancement of Science. Three years later, Michelson would observe Jupiter's moons interferometrically, and the year after that the two would be among the first faculty recruited for the newly established University of Chicago. In 1907, Hale would support Michelson's consideration by the Nobel Committee. Hale's obsession with expanding classical astronomy—preoccupied with determining positions and motions of stars—into knowing the physical properties and processes powering the stars themselves compelled him to develop the telescopic and instrumentation means of achieving those goals. Interferometry could provide the physical dimensions of stars and their surface temperatures. It could resolve binary stars to reveal masses, distances, and luminosities. It was inevitable that Hale would develop interferometry as a tool for the new science of observational astrophysics, and he had a friend and colleague with unrivaled expertise in the field. How could Hale not have ensured that the 100 inch telescope be equipped from the outset with the tool for resolving the largest suspected stars?

This book is the story of how stellar interferometry came to be in the early years of the 19th century and how it subsequently and specifically developed on Mount Wilson—first by Hale and his staff and later by individuals inspired to continue elsewhere what John Anderson and Francis Pease had started on Hale's mountain. Ultimately, groups from other institutions wishing to exploit astronomical seeing conditions not found at other North American sites would build stand-alone interferometers on Mount Wilson.

It is as much the story of those who rose to interferometry's challenge as it is of how they met it. It is not an exhaustive compilation of all astronomical interferometers everywhere and shamelessly ignores the development of interferometry at radio wavelengths. It is not a highly technical overview of the field, a subject area presently well served by other books. My goal has been to provide detail of sufficient depth to see how an instrument works generally and explore the scientific results it produced. Extensive references are given for those wishing to dig deeper. And, as a kind of primer on interferometry, I've put together an appendix and frequently refer to relevant sections therein for those who want to know a little more on the spot.

I am not a historian of science as was my mentor Karel Hujer. My lack of such training will be painfully apparent to many. Rather, I have spent my career pursuing high angular resolution astronomy along a path that led to a site famous for birthing our modern understanding of the universe. Along this path, I have gained unbounded respect for the enormous gift that George Ellery Hale gave to the world by what he created on Mount Wilson.

Foreword

For most of human history, observational astronomy has been an art form, in which we humans have developed ever-more ingenious means of capturing and analyzing starlight in order to describe and understand the universe surrounding our planet home. We began with unaided eyesight to evolve mostly fanciful ideas of what those mostly stationary pinpoints of light in the night sky might be, as well as the Sun, the Moon, and a few planetary wanderers. The development of lenses and telescopes revealed that some of those pinpoints had structure and coloring, while photography provided permanence of observations and the opportunity for deeper analysis. The parallel development of allied sciences like physics and chemistry expanded the context and intellectual tools that informed those analyses. The directions of the starlight gave us ideas of the structure of our surroundings and the first rungs on the cosmic distance ladder. From the colors, and eventually the detailed spectra, we inferred a vast range of information about the chemistry, structure, and lifecycles of stars and other objects. In recent times, this science originally derived from visible starlight has been augmented by similar efforts using other parts of the electromagnetic spectrum from radio waves to gamma rays, by other "cosmic messengers" from the physicists' particle zoo—and now even from gravitational radiation.

Stellar interferometry, the subject of much of *Seeing the Unseen*, is a prime example of astronomers' efforts to tease fascinating information out of the continuous trickles of light coming to us from colossally distant objects. Visible starlight has been the mainstay of astronomy, but it suffers from the presence of the earth's atmosphere, which makes it easy for astronomers to breathe but blurs the images formed by big telescopes. We can build large telescopes with excellent optics to produce sharp images down to about 0.01 arcsec (~3 millionths of a degree of angle), but the atmosphere smears them out by at least 100 times that ideal. This puts direct measurements of the diameters of even nearby stars (not counting the Sun) out of reach. The origin and development by clever astronomers of how to overcome this limitation by exploiting the subtle wave properties of the light itself makes a compelling story, as we shall see.

Science and science history are multifaceted topics. There are the goals, methods and results of any particular discipline, plus the theories that provide the motivation and validation of the observations. But it is also valuable in many cases to understand the places where the work was done and the discoveries made. Mount Wilson Observatory is a worthy site to be celebrated, not only for its key role in stellar interferometry, but also for its history as the first modern observatory—chosen for sake of its astronomical possibilities rather than for the convenience of astronomers. Its 60 inch and 100 inch telescopes made it the home of the world's largest telescope from 1908 to 1949, the longest reign of any observatory. Those telescopes enabled legendary astronomers like Edwin Hubble to revolutionize our ideas of the size and history of the universe. This reminds us of a third facet of a good science story: the scientists themselves—their personalities, skills, and motivations. Knowing about these people and their lives humanizes the history, helps a reader grasp the context and importance of the discoveries, and may inspire those readers to pursue scientific inquiry themselves.

In *Seeing the Unseen*, Harold McAlister, himself one of the key innovators and drivers of stellar interferometry, has given us a nicely multifaceted history of that discipline—the what, the where, and the who. Growing up in the Los Angeles suburbs, I was aware of Mount Wilson only as the place where our TV signals originated. Even during my later professional visits to the observatory, I was still mostly oblivious to the storied general history of the site and its telescopes. Reading Hal's manuscript was a delightful education for me, and made me admire and appreciate the varied cast of characters who have created the Mount Wilson interferometry story. I believe that all readers will share these sentiments.

William C. Wickes
Corvallis, Oregon

Acknowledgments

In researching this book, I have had the great pleasure of meeting and interviewing a few of the individuals who are part of this chronicle, and I am grateful to them for their time and enthusiasm in recollecting portions of their careers and insights from their own research as well as for previewing relevant sections of the book. And so I am indebted to Willet Beavers, Manfred Bester, Theo ten Brummelaar, Doug Currie, Walt Fitelson, Bill Hartkopf, Don Hutter, John Monnier, Rick Wasson, Bill Wickes, and Ed Wishnow for their enrichment of this book. If only I could time travel back to 1930 to spend a few hours with Francis Pease.

This book could not have been written were it not for the resources available from the Caltech Archive's wonderfully curated George Ellery Hale Papers collection. Now that the book is done, I see many hours ahead of reading that remarkable man's letters to and from others of his era. The same can be said of the Huntington Library's digital collection of images in the Observatories of the Carnegie Institution for Science Collection (COPC). Lowell Observatory Historian Kevin Schindler and Archivist Lauren Amundson kindly provided the 1962–67 correspondence between Lowell Director John Hall and Willet Beavers.

I am grateful to Daniel Bonneau, Mark Colavita, David Hale, James Lequeux, Peter Tuthill, Mike Shao, Andrei Tokovinin, and Denis Weaire for their respective replies to inquiries I sent their ways in researching this project.

I thank the following individuals for their courtesy in granting permission for the use of many photographs and other illustrations that embellish the book's layout: America Association of Variable Star Observers for access to photometric data for Betelgeuse, Willet Beavers, Manfred Bester (ISI), Peter Collopy (Caltech Archives and Special Collections), Doug Currie (Univ. of Maryland), Reed Estrada (speckle interferometry), Bill Finsen (who gave me permission for his images nearly 25 years ago), Thierry Forveille (Astronomy and Astrophysics), Steve Golden (CHARA), David Hale (ISI), Bill Hartkopf (for calculating orbit residuals), Don Hutter (Mark III), Brian Kloppenborg (CHARA), Kathryn Lester (CHARA), Susan McAlister (CHARA), Daniel Meyer (Univ. of Chicago Special Collections Research Center), John Monnier (Univ. of Michigan), Bill Pounds (artist), Rachael Roettenbacher (Yale Univ.), Jessica Rose (NOIRLab), Laszlo Sturmann (CHARA), Bill Wickes, Kit Whitten (Carnegie Institution of Science), and Lily Wilmes (Living Editions Publishers).

I owe a special debt of gratitude to two friends and colleagues who patiently read the entire manuscript. Larry Webster, whose long and continuing association with Mount Wilson Observatory has saved me from embarrassing errors of fact I would have endured had it not been for his alertness to such gaffes. We've spent many hours during my regular visits to Mount Wilson over the years talking about the Observatory's history and its people, and I look forward to continuing those visits in the years ahead. Bill Wickes and I first met at Doug Currie's 1978 interferometry conference at the University of Maryland. I think I remember talking with Bill at the meeting, but he doesn't recall meeting me. As we are standing next to each other in the group photo (Figure 8.12), I'm convinced we did indeed chat at College Park. In any event, while interviewing Bill in preparation for writing about his work on the mountain in the 1970s, he very kindly volunteered to read the drafts of each chapter in a critical manner for style and readability (as well as for my physics and astronomy). Bill's technical corrections were almost invariably on the mark, and his frequent deconvolution of my sentence structure has made for a far more readable book.

As I submit this book to IOP in preparation for its publishing, I want to acknowledge the kind and patient guidance given to me during its writing and editing by Leigh Jenkins, Poppy Emerson, Sarah Armstrong, and Robert Trevelyan. IOP's Production Department did a wonderful job with the layout, and I thank them for their devotion to detail. That I embarked on this effort at all is the result of the encouragement by my Georgia State University astronomy colleague Piet Martens, who serves on the AAS–IOP Astronomy editorial advisory board.

Author Biography

Harold A McAlister

 Harold A. "Hal" McAlister is Regents Professor Emeritus of Astronomy at Georgia State University where he founded the Center for High Angular Resolution Astronomy (CHARA) in 1983 and served as its director until he retired in 2015. As principal investigator for the CHARA Array, he led the team that turned a dream into a reality when ground was broken on Mount Wilson on 1996 July 13. During 2002–2014 he served pro-bono as CEO of the Mount Wilson Institute and Director of Mount Wilson Observatory. McAlister received a B.A. in physics from the University of Tennessee at Chattanooga in 1971, where he was subsequently honored as UTC's commencement speaker in 2001. He earned M.A. and Ph.D. degrees in astronomy from the University of Virginia in 1974 and 1975. He then spent two years as a postdoctoral research associate at the Kitt Peak National Observatory during 1975–1977 after which he joined the faculty of the Department of Physics and Astronomy at GSU.

He has authored or co-authored more than 400 scientific papers in the areas of binary star speckle interferometry and long-baseline optical/near-infrared interferometry. In his retirement, McAlister has taken to writing. His other books now include the novel *Sunward Passage* (2013), *Policing Greene* (2018, non-fiction in collaboration with Thomas C. Lewis), *The 2009 Station Fire Threat to Mount Wilson Observatory* (2019, with co-author Susan J. McAlister), *Photographs of a Traveling Astronomer: Karel Hujer's 1937–38 Tour of the Americas* (2019, with co-author Tor N. G. Westin), *The Unfolding Universe: Karel Hujer's Reflections on Cosmic Consciousness* (2019), and *Mount Wilson Observatory: A Self-Guided Walking Tour* (2019).

Seeing the Unseen
Mount Wilson's role in high angular resolution astronomy
Harold A McAlister

Chapter 1

The Birth of Stellar Interferometry

1.1 It All Started with the Last Man Who Knew Everything

With only modest exaggeration, Andrew Robinson's biography of Thomas Young (1773–1829) is entitled *The Last Man Who Knew Everything* (Robinson 2006). Young, who hailed from a small village in southwest England, was trained as a physician in Edinburgh and Göttingen, but his intellectual and experimental attention was drawn to a variety of topics—including investigations into vision, music, language, Egyptology, and even life insurance—a diversity sufficient to tag him as a polymath. A portrait of him at age 49 appears in Figure 1.1. Fearful of damaging his professional reputation in medicine, Young published his earliest papers anonymously. Within a few years, though, he was widely noted for his brilliance and was openly and prolifically publishing in the *Philosophical Transactions of the Royal Society of London*. His lecture to the Society on 1803 November 24 on "Experiments and Calculations Relative to Physical Optics" (Young 1804) elegantly described work he began two years earlier regarding the colored fringes produced when sunlight admitted through a small aperture into a darkened room passes around the opposite edges of a paper card and onto a screen. The clincher in his deduction of interference in support of the wave theory of light of Christiaan Huygens (1629–1695) is that these fringes disappear if a second card blocks light from striking one or other edge of the occulting card before being allowed to combine with light from the other edge. I elaborate on this episode in the history of optics not only because of the fascination inherent in such a person as Young, but also because the simple and famous Young's double slit experiment laid out the beginnings of basic principles that would, after substantial theoretical buttressing by Augustin-Jean Fresnel (1788–1827), eventually enable interferometry's employment in stellar astronomy. For more about the basic principles and terminology of stellar interferometry, see Appendix B.

The suggestion that interference fringes might be analyzed so as to reveal the diameters of stars is widely attributed to the French physicist Armand Hippolyte

Figure 1.1. Thomas Young, engraving by G. Adcock of an 1822 portrait by Thomas Lawrence (Courtesy of Smithsonian Institution).

Louis Fizeau (1819–1896), shown in Figure 1.2 as he appeared later in life (Lawson 2000). In 1848, Fizeau had independently discovered the Doppler effect—sometimes referred to in France as the "Effet Doppler–Fizeau"—albeit six years after Christian Doppler's (1803–1853) postulation of this phenomenon. The following year Fizeau measured the speed of light to ±5% accuracy. He was clearly at his peak by 1850.

Fizeau's connection to measuring star diameters is oddly buried in a report from a commission of the French Academy of Sciences charged with devising criteria for the award of the 1867 Prix Bordin in Mathematics (Fizeau 1868). Fizeau was appointed as the "*rapporteur*" (designated reporter) of the deliberations of this commission to the Academy. The other commissioners were Jean-Marie Duhamel (1797–1872), Claude Pouillet (1790–1868), Henri Victor Regnault (1810–1878), Joseph Louis François Bertrand (1822–1900), and either Alexandre Edmond Becquerel (1820–1891) or his father Antoine Cesar Becquerel (1788–1878). All were distinguished physicists except for Bertrand, primarily a mathematician who also did important work in thermodynamics.

The final of the topics desired by these luminaries to be treated by the prize winner was described in the following translation from the original specification statement. "*For most of the phenomena of interference, such as the fringes of Young, those of Fresnel mirrors and those which are the result of the scintillation of stars according to Arago, there is a remarkable and necessary relationship between the size of the fringes*

Figure 1.2. A. H. L. Fizeau photographed by C. H. Reutlinger (Courtesy of Smithsonian Institution).

and that of the light source, so that extremely tenuous fringes can only arise when the light source has angular dimensions almost insensitive; hence, to put it in passing, it is perhaps possible to hope that by relying on this principle and by forming interference fringes, for example, by means of two large widely spaced slits at large instruments to observe the stars, it will become possible to obtain some new data on the angular diameters of these stars.[1]"

One naturally wonders how this particular prize criterion came to be. Was it purely the brainchild of Fizeau or was he merely reporting a goal set by another member? In the middle ground between these extremes, did Fizeau throw this out to the other committee members who participated actively in refining his idea to its final statement? Fizeau mentioned the scintillation theory of François Arago (1786–1853) in which the distorted image of a star arises from interference and cancellation of portions of the wavefront diverted by atmospheric irregularities rather than variations of refractive index of turbulent cells of air as would be correctly proposed later by others (see Rayleigh 1893; Danjon 1955). Arago had been an important figure in advancing interference phenomena in astronomy, and so it was natural that he would be cited by Fizeau.

As all the commission members lived in Paris or its suburbs, it seems likely they did meet together to carry out their work for the French Academy. Regrettably,

---

[1] Translated using Google Translate, from Fizeau (1868).

there was no fly on the wall of their committee room. In any event, there appears to have been no Prix Bordin awarded that cycle for someone's success in satisfying the commission's challenge.

As for the issue of Fizeau's ownership of the concept of measuring stellar diameters interferometrically, the matter was settled by Paris Observatory astronomer James Lequeux and described in his 2014 book *Hippolyte Fizeau, Physicien de la lumière* (Lequeux 2014a). Lequeux discovered Firzeau's original working notes that had been donated by his family to the Museum of Urban and Social History of Suresnes, France and to the archives of the French Academy of Sciences. These document clearly show that since 1851 June 22—smack in the middle of his most productive years—Fizeau had realized that the apparent diameter of a star could be measured by interferometry (Lequeux 2014b). But, he would leave that task to others. What better way to encourage someone to action, when the opportunity arose, than baiting a cash prize with stellar interferometry?

Five years would pass before Édouard Stephan (1837–1923), director of the Marseille Observatory, reported his successful observation of fringes from a stellar interferometer he devised for the Marseille 80 cm Foucault Reflector telescope (Figure 1.4). The account that appeared in the weekly bulletin for 1873 April 27 of the Scientific Association of France was labeled as *"from a letter from M. Stephan, addressed to M. Fizeau"* (Stephan 1873). Stephan, shown in Figure 1.3, began by reiterating Fizeau's 1868 awards document, quoting the passage translated above, and noting that no one to his knowledge had yet taken up Fizeau's commission's task of turning a telescope into an interferometer—until now, that is. Stephan went on to report that he had succeeded in obtaining fringes on the bright star Sirius after installing a mask with double slits separated by 15 cm over the 80 cm objective mirror of the Marseille telescope. The following night, he again attempted to see fringes but this time used a mask with slits separated by 50 cm. No fringes were to be seen, regardless of the magnification used at the telescope's eyepiece whereas bright stars in nearby Orion, which were a little higher in the sky than Sirius, did show fringes. Succeeding nights were troubled by a *"poor state of the sky,"* but Stephan was anxious to report his observations and went out on a limb by saying that while he was *"very far from presenting this result as definitive; but, by the way the fringes persist, regardless of the waviness of the images, I bow very strongly to think that the disappearance of the fringes of Sirius is not just an atmospheric influence."* Stephan also published a similar report in the more formal *Comptes Rendus* of the French Academy of Sciences ending with the statement (translated from the original French) *"I very much incline to think that the disappearance of the fringes of Sirius does not depend solely on an atmospheric influence. I have the firm hope that later experiments will show, with evidence, that the diameter of this star is not insensitive, and will make it possible to obtain some rough estimate"* (Stephan 1874a).

This hope that the fringe behavior was representative of the diameter of Sirius rather than the result of instrumental or atmospheric effects was something that Stephan intended to explore with additional observations of stars. It also required a mathematically-based analysis of the observing parameters and the geometrical

Figure 1.3. Édouard J-M Stephan (Wikimedia Commons).

configuration of his aperture mask, which employed large openings required to admit starlight sufficient in brightness to see fringes.

With this work completed sometime during late 1873 to early 1874, Stephan returned to the *Comptes Rendus* to summarize his efforts and conclusions (Stephan 1874b). The paper's title—"On the Extreme Smallness of the Apparent Diameters of the Fixed Stars"—is a tipoff that Sirius had, in reality, not been resolved. Stephan's very thorough geometrical and optical analysis of the problem showed that his large apertures had little effect and, more importantly, that the fringe behavior was indicative of angular diameters well below the threshold of resolution with the 80 cm Foucault telescope at Marseilles. The case was thus closed, and Stephan blamed the disappearing fringes of Sirius on atmospheric conditions. He never returned to the problem of measuring stellar diameters.

Lequeux's discovery of Fizeau's papers at Suresnes included a fascinating personal letter from Stephan to Fizeau dated 1874 February 1. Stephan set out to describe his efforts with an apologetic preamble stating that he had not stopped worrying about the problem since they were last together, indicating that the two had personally discussed the observing program. Stephan had by then observed most stars down to third magnitude and some a magnitude fainter. He had also

Figure 1.4. The 80 cm Foucault Telescope of the Marseilles Observatory. (Adapted from the article "Le Plus Grands Telescopes du Monde," by Camille Flammarion, La Nature, 1873 Nov 15, p. 370.)

adopted a modified mask laid directly on the mirror rather than mounted to the top of the tube to avoid path-length differences arising from flexure of the telescope's wooden tube, which might explain his earlier false positive. He reported anxiously awaiting the seasonal return of Sirius to the sky and then his great disappointment in seeing Sirius' fringes behave as they did for the other stars implying that it was far from resolved. He deduced, correctly of course, that all stars are much smaller in angular size than might have been thought. Stellar parallax data were in hand by 1874, and the distance to Sirius was known with reasonable accuracy. It does not appear that he had estimated Sirius' diameter from its distance and adopting the size of the Sun as an approximation that would have shown him the unlikelihood of its resolution with this modest aperture.

Of course, no one then had much of a notion as to the physical diameters of stars other than by analogy to the Sun—that was necessarily a few decades away pending the development of the theory of stellar evolution and its anticipation of giant and supergiant stars. Perhaps Sirius was the largest star because it was the brightest and relatively nearby. With the maximum aperture spacing on the refractor's mask of 65 cm used in the second round of observations, the angular diameter that would have corresponded with fringes that just washed out at that spacing would be 160 milliarcseconds (mas). The actual angular diameter of Sirius was first determined more than three-quarters of a century after Stephan's attempts to be 7.1 ± 0.6 mas, by Robert Hanbury Brown and Richard Twiss (Hanbury Brown & Twiss 1956, 1958) using the University of Sydney's Intensity Interferometer at Narrabri, New South Wales, Australia. This value is more than 20 times too small to have been resolved by the Marseilles reflector acting as an interferometer. So, Stephan was indeed carried away by his original optimism as to the cause of Sirius' fringe disappearance at the wider slit spacings.

Stellar interferometry at optical wavelengths entered a hiatus at this point but would re-emerge in fits and starts in the coming century and a half. We can mark the brief period from 1868 to 1874 as the birth of the field. The problem was that there was no straightforward way to obtain the resolution that Stephan reckoned would be necessary to measure stellar angular diameters. They required sampling two apertures far wider than permitted by simply masking any existing telescope. Thus, Stephan set star diameters and interferometry aside and continued his work at Meudon on "nebulae," virtually all of which were distant galaxies.

However, Fizeau had already envisioned the means to achieve an extended baseline by 1851. In his Suresnes notes, there is a rough sketch of mirrors on a beam mounted perpendicular to the optical axis of a refracting telescope (Lequeux 2014b, p. 5). This exact configuration would be fulfilled 70 years later at the Mount Wilson 100 inch telescope, although there is no indication that its implementers were aware of Fizeau's originality in their concept. In the meantime, interferometry went to sleep for more than 15 years until it reawakened in America in the hands of Albert Michelson.

1.2 The Master of Light

Albert Abraham Michelson (1852–1931), (Figure 1.5), is justly celebrated for his experimental determinations of an accurate value of the speed of light—a quest he began in earnest as early as 1878 when he was a young naval officer and instructor of physics at Annapolis. His experimental apparatus derived from the rotating-mirror method used by the French physicist Leon Foucault (1819–1868) in 1850 who had adopted the approach originally suggested by the intellectually relentless Fizeau. This French connection points back to Michelson's odd lack of citation of Fizeau's precedence in stellar interferometry. In his first scientific paper (Michelson 1878)—essentially an abstract—the future Nobel laureate described how his approach *"dispenses with Foucault's concave reflector, and permits to use of any distance."* This important modification allows for a potentially very long lever arm of sorts to improve measurement precision, which was a goal inherent in Michelson's focus on this problem for decades. Foucault is not cited here other than by name, and Fizeau is not mentioned at all. A few years later, Michelson detailed his experiments while in the Navy in a beautifully-illustrated, fifty-four page paper published by the U.S. Naval Observatory (Michelson 1882). Again, Foucault's work was referred to without its ties to Fizeau.

During many years of mentoring graduate astronomy students, I was frequently struck by their lack of historical interest in digging back in time to the progenitor of

Figure 1.5. Albert Abraham Michelson (Courtesy of Smithsonian Institution).

their subfield or technique. They would merely cite the most recent work before them as sufficient with no concern for crediting the person who had the first spark of brilliance that enabled their particular dissertation topic. While doing my best to dispel them of this inclination, I would ask how they would feel if their own great ideas appeared to be assigned to a successor. Perhaps Michelson, who was only 26 in 1878, had this same disregard for the achievements of his scientific ancestors. Others have pondered Michelson's failure to acknowledge Fizeau, (see DeVorkin 1975; Lawson 2000, p. 326), but it is what it is.

In Michelson's 1903 book *Light Waves and Their Uses*, Fizeau's name appears in four places, none of which pertain to astronomical applications, a topic addressed in the 20 pages of the book's final lecture on "Applications of Interference methods to Astronomy." That chapter is largely descriptive of the theory and applications at an introductory level with the exception of a description of Michelson's work in the summer of 1891 measuring the diameters of the Galilean satellites of Jupiter with the Lick Observatory's 12 inch refractor.

In her biography of her father (Livingston 1973), Dorothy Michelson Livingston describes how Michelson's interest turned in 1890 toward astronomical applications of the interferometer, stating that he "*discovered that the interferometer could be adapted so as to measure the diameters of very small and distant sources of light such as planetoids and satellites.*" After setting up an artificial double star in his laboratory at Clark University, where he had been appointed to its inaugural faculty as professor of physics, Michelson was inspired to make actual measurements of astronomical sources, settling on Jupiter's four bright Galilean satellites. During this time frame, he also fully developed the relevant theory to support his goal of obtaining data at a telescope. He published that theory in a mathematically detailed presentation in the *Philosophical Magazine* for 1890 July (Michelson 1890). It is in this paper where he first published his famous definition of fringe visibility V in the form of:

$$V = (I_{max} - I_{min})/(I_{max} + I_{min}),$$

where I_{max} and I_{min} are the maximum and minimum brightnesses of a fringe. Visibility is essentially a measure of fringe contrast and ranges from 0 to 1. He goes on to show that V is related to the wavelength at which the observation is made, the linear separation (or baseline) between the wavefront sampling mirrors or telescopes in an interferometer, and the angular diameter of the source. Visibility remains the basic observable of astronomical interferometers (see Appendix B.2.3). Michelson being Michelson, he closed out this thorough presentation with an equally complete discussion of the experimental apparatus he had developed.

Enthusiastic from his theoretical exploration of making practical astronomical measurements, Michelson contacted Harvard College Observatory Director Edward C. Pickering (1846–1919) that same summer seeking access to HCO's 15 inch refractor in order to observe the bright Jovian moons. Receiving a positive response, Michelson quickly set out for Cambridge where through a combination of delays, bad weather, and poor "seeing" (the motion and blurring of astronomical images by atmospheric turbulence; see Appendix B.6)—a triple threat familiar to most astronomers—he was unable to obtain measurements of the diameters of these ostensibly low-hanging fruit. That stay at Harvard, at least gave Michelson the opportunity to spend some time with George Ellery Hale (1868–1938)—Mount Wilson Observatory's future founder—who was there working with Pickering. Michelson and Hale had first met at the 1888 meeting of the American Association for the Advancement of Science in Cleveland (Livingston 1973, p. 140).

Michelson then turned to Lick Observatory and its director Edward S. Holden (1846–1914), a famously explosive individual (Osterbrock 1984), but with whom Michelson would have no difficulties. Holden took to the idea of measuring the Jovian moons as there was scant agreement among the existing micrometer measurements made by several observers skilled at observing closely separated "double stars" (gravitationally bound pairs of stars aka "binary stars") in varying seeing conditions. In 1891 August, Michelson mounted his variable slit mask device (Figure 1.6), on Lick's 12 inch refractor to obtain diameters for each satellite on four nights. The seeing on only one of those nights was judged as "good," the remaining three being "poor." In this effort, Michelson had observing assistance from William Wallace Campbell (1862–1939)—later president of the University of California.

Michelson published his results in the 1891 October issue of the *Publications of the Astronomical Society of the Pacific* (Michelson 1891a) followed more or less verbatim in the 1891 December 17 issue of *Nature* (Michelson 1891b). The duplicate publishing was presumably due to his wanting to attract European attention to

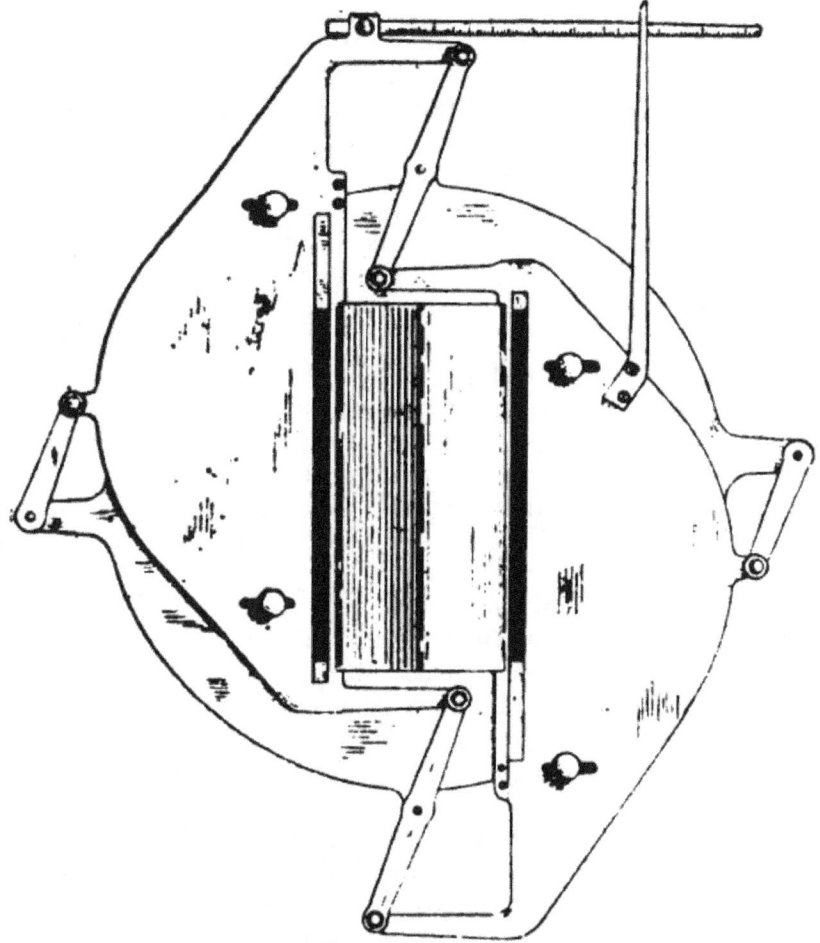

Figure 1.6. The aperture mask used by Michelson for his measurements of the Galilean satellites at the Lick 12 inch refractor. (Printed by permission from Macmillan Publishers Ltd: Michelson 1891b, Copyright (2020).)

the work. In order to judge the quality of his results, Michelson compared them with micrometer measures made by Rudolf Engelmann (1841–1888) of Leipzig Observatory, Friedrich G. W. Struve (1793–1864) of Dorpat (later Tartu) Observatory in Estonia, George W. Hough (1836–1909) of Dearborn Observatory, and Lick Observatory's Sherburne W. Burnham (1838–1921). Burnham's micrometer measurements had been made at the Lick 36 inch telescope on Michelson's last night at the 12 inch. After he transformed all measurements to represent an angular diameter subtended at the Sun from Jupiter's mean distance, Michelson could then compare them.

A few years later, these same data would draw the attention of another famous astronomer. With the goal of determining the physical diameters of these satellites, Edward Emerson Barnard (1857–1923) published an analysis in 1895 of their angular diameter data going as far back as the 1693 eclipse timings of Giovanni Domenico Cassini (1625–1712) (Barnard 1895). Barnard settled upon using "later measures" starting with those from 1829 of F. G. W. Struve, a founder of double star astronomy. Barnard's sample, shown in Table 1.1, included Michelson's along with results from four additional observers.

Unavailable to Michelson and Barnard, of course, were the actual physical diameters of the moons based upon spacecraft measurements, which today give us an absolute level of intercomparison of the historical results. Inspection of the last column of Table 1.1 shows that Michelson more or less tied with F .G. W. Struve's 1829 visual micrometer observations made at Tartu. Struve used the "Great Dorpat Refractor," a 9 inch instrument built by Fraunhofer, that was smaller in aperture than the instrument Michelson used at Mount Hamilton. Comprising the second tier are Engelmann and Barnard whose residuals from the true diameters are very slightly systematic. The results of the other observers are substantially inferior and exhibit real systematic errors. Note that the large outlier is Pickering's photometric estimate, which apparently relied on a common albedo assumption about which these moons would a century later show many surprises.

What can we deduce from this? First, it has long been known that visual micrometry is a skill susceptible to systematic effects most likely related to an observer's ability to deal with astronomical seeing conditions. Some observers are simply better than others and have what has been dubbed the "double star eye," as elaborated upon in the next paragraph. Struve and Engelmann had that rare skill. A few of the others in the table apparently did not. On the other hand, measuring diameters of disks is not the same problem as measuring the separations between two unresolved disks as is the case with a double star. Interferometry, as practiced visually, then relied upon the disappearance of fringes at the first null in visibility at which point the angular diameter is given by the interferometer slit spacing B at which the visibility goes to 0 and the wavelength of the observation through the simple relation $1.22\lambda/B$ ($V = 0$). This disappearance is less susceptible to atmospheric disturbances than is the far more complex act of comparing an image with the separation of micrometer threads. Michelson was, of course, a remarkably skilled instrumentalist so that his apparatus was very well calibrated. See Appendix B.3.2 for more on micrometry and visual interferometry.

Michelson closed out his summary of the Galilean satellites by noting, without elaboration, that Engelmann's measures were *"probably more reliable than the succeeding ones."* Was that because they were in better agreement with his results? He also claimed that a 6 inch telescope equipped with adjustable slits was *"fully equal to the largest telescopes now used without them."* Of course, he was unaware of Struve's 1829 Dorpat measurements with a small refractor. Finally, he expressed hope that the Lick 36 inch telescope, which would afford higher resolution from wider slit spacings, would be suitably equipped for *"definitive measurement of the satellites of Jupiter and Saturn and such of the asteroids as may come within the range of the instrument."* Alas, no one rose to that challenge.

Table 1.1. Angular Diameter Measurements of the Galilean Satellites (in Seconds of Arc as Seen from Jupiter's Mean Solar Distance)

Observer—Date	I. Io	II. Europa	III. Ganymede	IV. Callisto	(Obs. – Actual) Mean ± Std. Dev.
F. G. W. Struve—1829 *Micrometry*	1.015	0.911	1.488	1.273	0.052 ± 0.043
Father Secchi—1856 *Micrometry*	0.985	1.054	1.609	1.496	0.166 ± 0.101
J. H. von Mädler—1863 *Micrometry*	1.200	1.132	1.519	1.300	0.168 ± 0.124
R. Engelmann—1871 *Micrometry*	1.081	0.910	1.537	1.282	0.083 ± 0.059
C. Pickering—1879 *Photometry*	0.924	0.866	1.096	0.651	−0.236 ± 0.300
G. W. Hough—1880 *Micrometry*	1.114	0.980	1.778	1.457	0.212 ± 0.111
S. W. Burnham—1891 *Micrometry*	1.112	1.002	1.783	1.609	0.257 ± 0.118
A. A. Michelson—1891 *Interferometry*	1.02	0.94	1.37	1.31	0.040 ± 0.058
E. E. Barnard—1894 *Micrometry*	1.048	0.874	1.521	1.430	0.098 ± 0.047
Actual—spacecraft based	0.968	0.828	1.395	1.278	—

Michelson's interest in astronomical applications turned next to spectroscopy, and he would only come back to astrometric measurements by interferometry thirty years later in collaboration with George Ellery Hale, when a superb new telescope would allow slit spacings and resolutions far beyond what he had utilized in 1891. However, there was still more to come in the 19th century, including a set of high-quality double star measurements employing an innovative use of interferometry by one of the most brilliant astronomers of all time.

1.3 Along the Way to the Event Horizon

Nearly lost among his other monumental achievements are the results of a brief 1895 venture into interferometry by Karl Schwarzschild (1873–1916), seen in his prime in Figure 1.7. In 1895, he was a young graduate student working under Professor Hugo von Seeliger (1849–1924) in Munich. His dissertation was on the theory of rotating stars and had nothing to do with interferometric observations other than mentioning that one day interferometry would image the oblate spheroids of the most extreme of these spinning stars, a prospect likely unimaginable in the 1890s (see Suhendro 2008; Hertzsprung 1917).

It seems that his foray into this field was at Seeliger's introduction to him of Michelson's work from a few years earlier. However, Schwarzschild chose to measure double stars rather than redo Jupiter's moons. And, rather than applying Michelson's method of measuring the disappearance of fringes at the first null in visibility, Schwarzschild's device acted more like a coarse, full-aperture diffraction grating. His mask, shown in Figure 1.8, employed a clever means of varying the slit spacing so that the angular separation of the low-frequency fringes it yields for each component of the double star could be adjusted to reproduce the angular separation

Figure 1.7. Karl Schwarzschild. Courtesy of The Leibniz Institute for Astrophysics Potsdam (AIP)

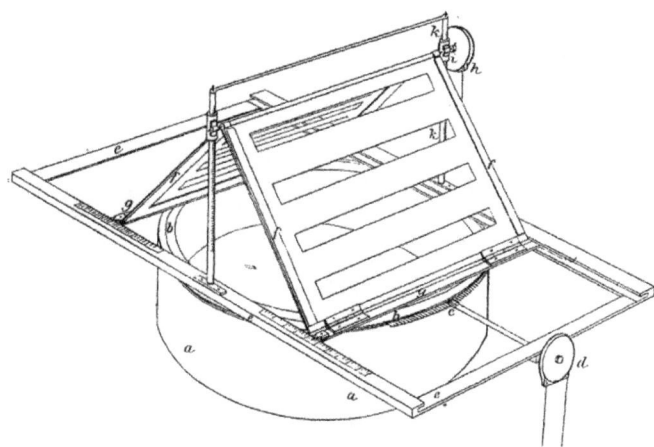

Figure 1.8. The aperture mask devised by Schwarzschild for his double star measurements made at the Munich Observatory's 10 inch refractor. (Schwarzschild 1896. John Wiley & Sons. Copyright © 1896 WILEY-VCH Verlag GmbH & Co. KGaA, Weinheim.)

on the sky of the components themselves. Rotation of the mask about the optical axis would align the projection of the fringes onto the vector separating the stellar components. While still involving the interference phenomenon, this is a far more direct way of getting at the angular separation of a double star than by attempting to match it to the separation of micrometer threads. We shall see that the instrument worked quite well for bright double stars on the 10 inch refractor of Munich University's observatory (Schwarzschild 1896).

The mask was designed and fabricated with the assistance of *"Herrn Mechaniker Sendtner,"* undoubtedly the observatory's instrument maker, whom Schwarzschild so kindly thanked in his publication. He was assisted in the observing itself by Walther A. Villiger (1872–1938), who would not be generally remembered for his contribution to this history,[2] although his observations are separately attributed to him in the Washington Double Star (WDS) Catalog.[3] Schwarzschild and Villiger observed 13 star systems over 18 nights during May through 1895 August, accumulating a total of 46 measurements of bright binaries mostly discovered by F. G. W. Struve. The angular separations ranged from 0.86 to 4.25 arcsec.

To illustrate the quality of their data among the visual micrometry carried out in the same period by a number of international observers, I show in Figure 1.9 mean values of the interferometric measures by Schwarzschild and Villiger for the 59.9 year binary star system ξ Ursae Majoris—in reality at least a quadruple star system —surrounded by similarly calculated means for the contemporary micrometry. The mean residuals and their standard deviations in arcseconds from the orbit referred to in the figure caption in X and Y (i.e., right ascension and declination) for the

---

[2] Regrettably, there appears to be no authoritative or even brief biography of Villiger except for what is found online. His work with Zeiss is mentioned in numerous books and articles on planetaria.

[3] *The Washington Double Star Catalog*, maintained by the U.S. Naval Observatory, is available at https://www.usno.navy.mil/USNO/astrometry/optical-IR-prod/wds/WDS.

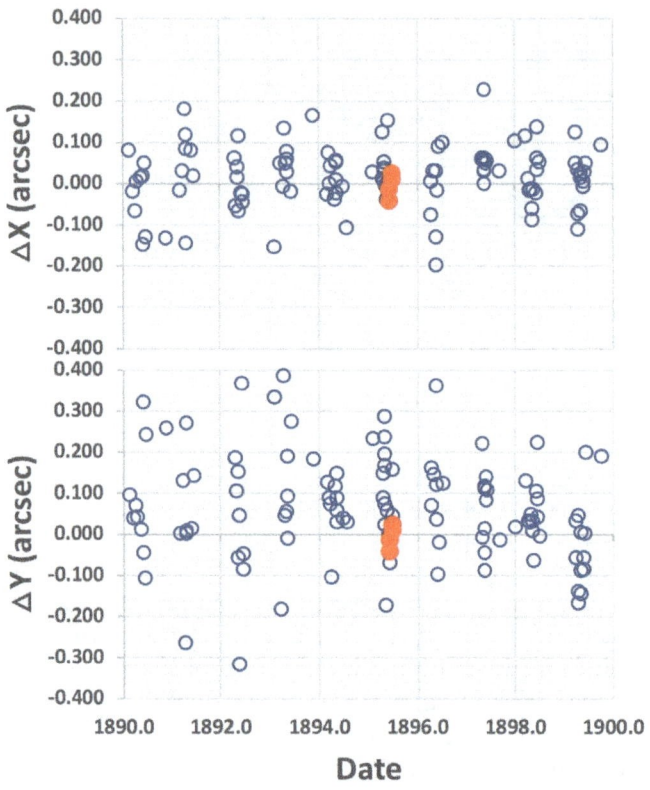

Figure 1.9. The residuals to the modern orbit of ξ Ursae Majoris (by Mason et al. Astron. Journal 109, 332, 1995) in the X (North–South) and Y (East–West) directions are shown above for the decade surrounding the epochs of Schwarzschild's measurements. His mean values are shown as filled red circles against mean values of micrometer measures as open blue circles. The quality of the interferometry is apparent. This multiple star system has a period for the outer binary components of 59.878 yr. Over the decade of the 1890s, the separation was approximately 2 arcsec. (Author's illustration, using residuals kindly calculated by William I. Hartkopf, retired from the U.S. Naval Observatory.)

micrometry are 0.018 ± 0.073 and 0.077 ± 0.147 compared with 0.007 ± 0.029 and 0.000 ± 0.050 for the interferometry. The latter clearly wins this competition. As with the results of Michelson, however, if we went through and selected only the "good" micrometrists, it is highly likely that their means would be comparable with those of the young Munich interferometrists.

Neither Schwarzschild nor Villiger would continue pursuing this technique. Schwarzschild, who was destined to take Einstein's General Relativity Theory to another level at the end of his mere 43 yr of life, finished his doctorate at Munich and left there for a position in Vienna. He was appointed director of the Astrophysical Observatory in Potsdam in 1909, from which post he volunteered for the German Army in 1914, serving as a junior artillery officer. Villiger remained in Munich for another decade before joining Zeiss as an engineer, wherein he developed the Mark II Zeiss planetarium projector. In 1906 and 1907, Villiger was the second author on three papers with his old Munich colleague. Schwarzschild's last paper, entitled *About the Gravitational Field of a Sphere of Incompressible Liquid According to Einstein's Theory*, appeared in 1916 March. Two months later, he died from an illness contracted while serving Germany on the Eastern Front in World War I, a tragic loss for science.

The last 19th century manifestation of interferometry in observational astronomy came from Maurice Hamy (1831–1936) at the Paris Observatory. Hamy was working in the Observatory's meridian circle department where positions on the sky of stars, minor planets, etc were measured with the greatest possible precision. In 1895, he wrote a paper (Hamy 1895) describing how *"the use of M. Fizeau's fringes considerably simplifies the control of the trunnions of the meridian instrument."* Notice that he cites Fizeau, while Michelson is nowhere mentioned in the paper. The following year, Hamy continued experimenting with an interferometer attached to the meridian circle, this time using it to measure displacements produced by heating the instrument in order to see how the thermal radiation of the human observer can result in displacement errors in the results. He found it to be a non-trivial effect— Michelson would have liked this (Hamy 1896). *Nature* also liked it and published a quarter-column note describing how Hamy found the effect to reduce as the square of the distance of the heat source from the instrument. The human influence arose because observers traditionally lie supine below the meridian telescope, heating it differentially from below. Hamy concluded that illumination using gas flames with chimneys is to be *"studiously avoided"* and that high conductivity metals such as copper are desirable to reduce the preferential heating of one side and not the other of the telescope tube.[4] Hamy was thus a careful observer who sought to reduce the influence of external factors on the quality of observational data prior to securing them.

Hamy next undertook interferometric observations of Jupiter's bright moons and published his results in 1899 (Hamy 1899). As the bright minor planet Vesta was then in the sky, Hamy threw that into the mix as well. This paper also includes a very detailed mathematical development of the interferometric technique, and this time he does cite Michelson as his new observations could be compared with the American's results. As described in the modern article by James Lequeux about *The Coudé Equatorials* put into service during 1884–1892 in France, Algeria, and Austria (Lequeux 2011). Hamy used the Great Equatorial Coudé telescope at Paris. On this 60 cm aperture instrument, Hamy installed an interferometric mask, whose specific design I have been unable to find. With it, he obtained values for the four satellite diameters, which compared with the now known physical diameters, show a mean residual in terms of fractional radius of -0.019 ± 0.076 compared with those for Michelson of 0.051 ± 0.063. Thus, both men achieved similar levels of accuracy in their measurements. Hamy's value for the diameter of Vesta was almost a third smaller than reality.

---

[4] From Our Astronomical Column (1896).

These results were more demonstrative of the new technique of interferometry than they were useful in shedding new light on these solar system bodies. They also more or less exhausted the accessible targets in regard to brightness and angular size. And so, interferometry again dozed off in the closing years of the 19th century.

One can summarize these activities as having developed the needed mathematical tools and observational techniques to explore the feasibility of interferometry as an observational methodology—with mixed results. Apparently, satellite diameters could be no better determined interferometrically than those measured by classical micrometers employed by the best observers. On the other hand, the double star measures of Schwarzschild were more promising. Astronomy is a small field wherein individuals can make a big difference. Had the right person come along in the 1880s and 1890s, double star interferometry might have supplanted the use of visual micrometry, a fate that speckle interferometry would ultimately deal to that last visual technique.

References

Barnard, E. E. 1895, MNRAS, 55, 383

Danjon, A. 1955, VA, 1, 377

DeVorkin, D. H. 1975, JHA, 6, 1

Fizeau, H. 1868, C. R. Hebd. Seances Acad. Sci., 66, 932

Hamy, M. 1895, BuAsI, 12, 49

Hamy, M. 1896, BuAsI, 13, 178

Hamy, M. 1899, BuAsI, 16, 257

Hanbury Brown, R., & Twiss, R. Q. 1956, Natur, 178, 1046

Hanbury Brown, R., & Twiss, R. Q. 1958, RSPSA, 248, 222

Hertzsprung, E. 1917, ApJ, 45, 285

Lawson, P. 2000, in Course Notes from the 1999 Michelson Summer School, ed. P. Lawson (Pasadena, CA: JPL), 325

Lequeux, J. 2011, JAHH, 14, 191

Lequeux, J. 2014a, Hippolyte Fizeau, Physicien de la lumière (Paris: EDP Sciences), 148 pp

Lequeux, J. 2014b, Bibnum Physique, http://journals.openedition.org/bibnum/687

Livingston, D. 1973, The Master of Light (New York: Charles Scribner's Sons), 376

Michelson, A. A. 1878, AmJS, 15, 394

Michelson, A. A. 1882, in Astronomical Papers Prepared for the Use of the American Ephemeris and Nautical Almanac, Vol. 1, ed S. Newcomb (Washington, DC: US Nautical Almanac Office), 109

Michelson, A. A. 1890, Lond. Edinb. Dubl. Philos. Mag. J. Sci., 30, 1

Michelson, A. A. 1891a, PASP, 3, 27

Michelson, A. A. 1891b, Natur, 45, 160

Osterbrock, D. E. 1984, JHA, 15, 81

Our Astronomical Column, 1896, Natur, 54, 84

Rayleigh, Lord 1893, PMag, 36, 129

Robinson, A. 2006, The Last Man Who Knew Everything (New York: Pi Press), 288 pp

Schwarzschild, K. 1896, AN, 139, 353

Stephan, E. 1873, Bull. Hebd. Assoc. Scient. Fr., 12, 68
Stephan, E. 1874a, C. R. Hebd. Seances Acad. Sci., 76, 1008
Stephan, E. 1874b, C. R. Hebd. Seances Acad. Sci., 78, 1008
Suhendro, I. 2008, Abraham Zelmanov J., 1, 14
Young, T. 1804, RSPT, 94, 1

Chapter 2

The Mount Wilson Opportunity

2.1 Go West, Young Man

George Ellery Hale's name must be high on the list of the most-important but least-known of American scientists (Figure 2.1). While he made seminal advances in solar physics, including the discovery of magnetic fields in sunspots and the development of the spectroheliograph, he was a consummate organizer and creator of transcendent facilities. He was to astronomy as Eisenhower was to D-Day. Hale's accomplishments include the founding of *The Astrophysical Journal*, the International Union for Cooperation in Solar Research—progenitor of the International Astronomical Union, the National Research Council, Yerkes Observatory, Mount Wilson Observatory, and Mount Palomar Observatory. He was the primary co-founder of Caltech, instigator of the Huntington Library and Art Gallery in San Marino, California, and leader of the development of the Pasadena Civic Center District. He was a man totally controlled by the irresistible urge to build the bigger and the better. Mount Wilson Observatory, the primary axis of this book, was a major result of that compulsion, which would ultimately swing its yearning gaze toward Mount Palomar. Most incredibly, all that Hale created was achieved in spite of episodes of debilitating mental distress (Adams 1938).

Briefly told, Hale launched Mount Wilson Observatory—the second of what would be his three forefront astronomical facilities—in 1904, only seven years after founding Yerkes Observatory and becoming its first director. Located at Williams Bay on the shores of Lake Geneva in extreme southeast Wisconsin within reasonable distance of its parent institution the University of Chicago, Yerkes was long on instrumentation and costly but irrelevant architectural magnificence (see Figure 2.2) but short on conditions ideal for astronomical observations. Its geographical location deprived it of the appropriate weather and stable atmospheric conditions that led to good astronomical seeing.

2-1

Figure 2.1. George Ellery Hale (1868–1938) in 1910 at age 42 (left), in 1921 (middle) when interferometry was thriving on Mount Wilson, and as drawn by Stephen Seymour Thomas in 1929 December (right). (Images courtesy of the Observatories of the Carnegie Institution for Science Collection at the Huntington Library, San Marino, California.)

Figure 2.2. The domes of the Yerkes Observatory 40 inch refractor (left) from 1897 and the 1912 Mount Wilson 60 inch reflecting telescope (right) show how Hale, in order to optimize nighttime performance, forsook ornate and costly masonry structures, which absorb and retain heat, in favor of double-skinned metal exteriors with highly reflective paints. This very significantly reduces image-degrading dome seeing that results from daytime solar heating and requires hours of post-sunset cooling to eliminate. The larger-diameter reflector, with its folded optical path, is significantly shorter than the great refractor, enabling it to be housed in a 58 ft diameter dome compared with Yerkes' 90 ft dome. Hale thus put Mount Wilson Observatory on a very different price–performance curve in comparison with most major astronomical facilities of the 19th century. (Author's photos.)

Hale was driven to create the most powerful telescopes and, equally importantly, advanced instrumentation to attach to them. Those telescopes and their accompanying tools, when placed in the hands of the cleverest scientists Hale could find, would transform our understanding of the stars and the universe. In considering the scope and magnitude of the scientific advances from his 100 inch telescope, that instrument is at or near the top of the list of the most important telescopes of all times.

Hale's new brand of *astrophysics* would tilt well away from traditional astronomy's emphasis on the *where* of the stars—their positions, distances, and motions—to *what* they are. The astrophysical questions were many and varied. What is the source of the radiant energy of stars? How long will this source last? What are their internal temperatures, densities, and pressures? How do their properties change over time? What are their chemical compositions, surface temperatures, masses, and diameters?

It was, of course, the need to know this last astrophysical parameter that led Hale to Albert Michelson. From simple trigonometry, if we know the distance D to a star, then its angular diameter Θ can be converted to a physical diameter d by the relation $d = D \cdot \Theta$. An increasing number of distance measurements to the nearer stars, obtained through the method of trigonometric parallax, had been growing into the early 20th century, but even if a star's distance is not known directly its angular diameter is still of great use. For example, a star's surface temperature can be determined from $T = (4F/\sigma\Theta^2)^{1/4}$, where σ is the Stefan–Boltzmann constant and F is the flux of light we receive from a star (over all wavelengths), measured with traditional techniques. The physical size and the surface temperature of a star derived from angular-diameter measurements together provide significant penetration into the mysteries of stellar astrophysics. Interferometry would be one key to Hale's new kingdom, but we are getting ahead of the storyline here.

After Yerkes, Hale wanted to build an observatory emphasizing solar research, in accordance with his belief that in order to understand the stars generally we must first understand our own star in detail. Realizing his mistake in locating Yerkes under frequently cloudy skies and a turbulent, star-blurring atmosphere, Hale sought sites based on their observing conditions rather than their convenience. He looked westward toward the mountains of Southern California.

Mount Wilson, overlooking the Los Angeles Basin, had been explored as an astronomical site as early as 1889 when Edward C. Pickering himself, who we met in the previous chapter, came to Southern California in search of a much better site for his astronomers than was being provided in Cambridge, Massachusetts. Pickering liked what he saw and quickly launched an evaluation process that would employ a 13 inch telescope of sufficiently massive design to require six weeks to lug it up the primitive trail to the summit. The University of Southern California was also interested in the site, although they did not actively participate in its evaluation. While Harvard's seeing tests were encouraging, they abandoned Mount Wilson in late 1890 noting harsh winters, scarcity of water, and surplus of rattlesnakes as the basis for their decision. More than likely, friction between them and the USC astronomers was a deciding impediment (Robinson 1991).

There was yet another telescope installed in the San Gabriels well before Hale's first instrument. A 16 inch aperture refractor, produced by the preeminent American telescope makers Alvan Clark & Sons, was situated on Echo Mountain some 6 km southwest of Mount Wilson. This remarkable facility was a component of the fabulous resort of Thaddeus S. C. Lowe (1832–1913) and others that existed in the front range of the San Gabriels from 1893 until 1938. In addition to an incline railway up the mountain and elegant hotel facilities, Lowe, who had been appointed Chief Aeronaut in the Union Army Balloon Corps by President Lincoln, wanted a first-rate telescope as the basis for a scientific institute. Fate would lead it to serve primarily as a tourist enticement to the resort facilities. But, an even worse destiny befell the Lowe Observatory when a windstorm blew off the dome in 1928 February, fortunately without destroying the telescope or killing the astronomer who was in the dome at the time (Seims 1976). By that date, Mount Wilson had already become home to the world's greatest astronomical observatory.

In 1903 June, Hale first set foot on Mount Wilson, where he would shortly redirect his ambitions away from Yerkes and, taking some of that observatory's finest staff with him, would seek the fulfillment of his science imperatives (Wright 1994). Yerkes Observatory, and its telescope featuring the largest-ever-fabricated objective lens, represents the pinnacle of refractor-based technology, but Hale would equip his new facility with reflecting telescopes with their many advantages over refractors, including the practical aspect of cost. Most importantly, these instruments and his astronomers would enjoy a site where the prevailing weather pattern coming in off the deep and cool Pacific Ocean created a nearly laminar airflow over the San Gabriel Mountains. The seeing from Mount Wilson would be far superior to that in Williams Bay. (See Appendices A and B.6 for justification of this statement.) Unsentimentally, Hale headed west, never looking back toward Williams Bay, and would reside in Pasadena the rest of his life.

The story of the establishment of the Mount Wilson Solar Observatory through Hale's courtship of Andrew Carnegie and his new Carnegie Institution of Washington is well known (Wright 1994, pp. 179–196). This courtship resulted in marriage on 1904 December 20, while Hale was resting at Martin's Camp in the saddle separating "Wilson's Peak" from Mount Harvard. A telephone wire had been draped up the mountain, and Hale received a call on that fateful Tuesday notifying him of a telegram that had just arrived from the Carnegie Institution. He was informed that *the Executive Committee has appropriated $150,000 a year for two years and has authorized the immediate execution of the larger plan.* Hale had been dipping into his personal funds to support his activities on the mountain, and so this must have been music to his ears. More than likely, though, Hale rather confidently anticipated this award, which amounts to nearly nine million of today's dollars.

The Carnegie Institution had been established on 1902 January 29, and Hale was appointed at the outset to a five-man Advisory Committee on Astronomy chaired by the ubiquitous E. C. Pickering. The committee was charged with advising the Institution as to how it could significantly advance astronomical sciences. Two of the committee members were classical positional astronomers—Lewis Boss (1846–1912), director of the Dudley Observatory in Schenectady, New York, and Simon Newcomb (1835–1909), shown in Figure 2.3, director of the U.S. Naval

Figure 2.3. Professor Simon Newcomb ca. 1905–1909. (Courtesy of the Library of Congress, Harris and Ewing Collection.)

Observatory's Nautical Almanac Office and Professor of Mathematics and Rear Admiral in the U.S. Navy.[1]

The fourth member was Smithsonian Institution Secretary Samuel P. Langley (1834–1906) who had wide-ranging interests with solar studies being his particular astronomical niche. At Hale's urging, Langley wrote a sub-appendix to the committee's report entitled *Proposal for a Distinctly Solar Observatory*. After

---

[1] Newcomb is often tarred with an 1888 statement that *"We are probably nearing the limit of all we can know about astronomy."* Google this and you'll find endless repetition of this famously ill-advised declaration. I was initially unable to find the original source of this quote and wondered if it might be apocryphal. In any case, by the time he was serving the Carnegie Institution, Newcomb (1903) expressed a conviction that astronomy was a burgeoning field with unlimited prospects, *"the existence of which was scarcely suspected ten years ago."* This apparent contradiction was resolved by further digging, revealing that the "limit" comment was extracted—inaccurately in its wording—from a lecture Newcomb gave at the dedication of the new observatory of the University of Syracuse on 1887 November 18. The lecture was subsequently published serially in *The Sidereal Messenger*, 1888, Vol. 7, No. 1, pp. 14–20 and No. 2, pp. 65–73 as "the Place of Astronomy Among the Sciences." On p. 69, Newcomb (1888) states: *"It would be too much to say with confidence that the age of great discoveries in any branch of science has passed by: yet, so far as astronomy is concerned, it must be confessed that we do appear to be fast reaching the limits of our knowledge."* An article by Newcomb's biographers Bill Carter and Merri Sue Carter in the 2010 November issue of *Physics Today* (p. 8) entitled "Newcomb Looked to Astronomy's Future" points out that—just two paragraphs following the apparently notorious remark—Newcomb makes clear that he was talking about positional astronomy as opposed to "the new science of physical astronomy." In particular, what was on his mind was the spectroscopic study of stars—astrophysics' launching pad. He then recommended that undertaking with Syracuse's new 8 inch Alvin Clark refractor. George Ellery Hale would have nodded in complete agreement had he been in Newcomb's audience that November Friday in Syracuse. Can you imagine a dedication address for an observatory with the guest of honor telling the audience there was little remaining undone for the new facility!

defining the science goals for such a facility, Langley set the equipment cost to fulfill them at $150,000. The seed for the Mount Wilson Solar Observatory was herewith planted. Hale's own contribution to the committee's report was an appendix on *Programs and Present State of Certain Departments of Astrophysical Research*. It was the longest of the appendices, longer even than the main body of the report. Hale used the words "*astrophysics*" or "*astrophysical*" 16 times.

Hale was just 34 years old in 1902. His four colleagues were a generation older than him and either at the zenith—or sliding over it—of their careers. Hale was the fair-haired lad on this important advisory committee and in a position to capitalize on their recommendations while reinventing the ancient science of astronomy. In Andrew Carnegie's statement creating the Institution's trust (Carnegie 1902), he wrote "*the chief purpose of the Founder* [i.e., himself] *being to secure if possible for the United States of America leadership in the domain of discovery and the utilization of new forces for the benefit of man.*" What better person to achieve U.S. leadership in the new science of astrophysics than this brilliant and already remarkably accomplished young man—George Ellery Hale.

The 1903 Carnegie *Year Book* reported that Hale received the Institution's Grant No. 13 of $4000 ($117,000 today) to support photographic work at the Yerkes refractor. Additionally, the astronomy advisory committee had morphed into the Committee on Southern and Solar Observatories, reduced in number to three members—Boss as chair, Hale, and Lick Observatory director William Wallace Campbell (1862–1938). The latter two were each given $5000 to further investigate Langley's call the year before for a "distinctly solar observatory" as well as for a facility in the neglected southern hemisphere. Theirs was a serious report with 165 pages of details including 64 pages of letters of support from the world's preeminent astronomers. A detailed discussion of the attributes of several solar sites—with a clear focus on Mount Wilson, which was regarded as having been selected provisionally—was contributed at the committee's request by Lick Observatory astronomer William Joseph Hussey (1862–1926) (Hussey 1904).

My point here is that Hale was not your typical supplicant to a funding agency, throwing himself on the mercies of anonymous reviewers. He used his many talents to more or less prearrange funding for building a successor to Yerkes Observatory. Indeed, in 1904, the Institution gave him funds to relocate the Snow Solar Telescope from Williams Bay to Mount Wilson. There would be no stopping this extraordinary visionary from achieving his goals.

Only four years after its founding, the adornment of the Mount Wilson Solar Observatory with telescopes of revolutionary design and unrivaled performance was well underway. True to his ambition to establish a solar observatory, The Snow telescope was joined by a 60 foot tower atop which was a coelostat optical system that fed the Sun's image to a spectrograph. The advantage of the tower was that it captured the sunlight above the turbulence in the air close to the ground, providing a sharper image. Hale immediately set about searching for evidence of magnetic fields in the Sun with an emphasis on studying sunspots at high resolution. The sharp images revealed structures resembling iron filings around a magnet—highly suggestive of solar magnetism. Hale promptly confirmed this to be the case by using the tower's spectrograph to photograph the telltale Zeeman splitting of certain spectral features. Not satisfied with the capability of the 60 foot, he soon had plans for a higher tower with a larger aperture coelostat to pursue solar magnetism with better resolution and greater sensitivity (Harvey 1999). Those improvements would arrive in 1910 with the completion of the 150 foot solar tower telescope adjacent to the 60 foot. This impatience for larger and better telescopes would characterize Hale for the rest of his life.

Recalling that studying the Sun was the doorway to understanding the stars, we would fully expect Hale to build nighttime telescopes on his mountain. And so, the busy year of 1908 also saw the installation of a 60 inch telescope for stellar astrophysics. This instrument was remarkable for the quality of its optics but also for the precision of its electrical control systems for tracking stars during long photographic exposures. It signaled a new generation of optically and mechanically precise giant instruments created to fulfill Hale's goals for observational astrophysics, and telescope design would thereafter emphasize both of these attributes.

True to his nature, Hale wanted a bigger telescope than the 60 inch. But that ambition would take more than the two year increment between the solar tower telescopes—time required for fund raising and for obtaining and then polishing the largest glass disk ever made. In 1919, Hale's 100 inch telescope would complete the retinue of magnificent instruments for studying the Sun by day and the stars by night. The 100 inch would be the world's largest telescope for three decades until Hale's compulsion resulted in the 200 inch telescope on Mount Palomar. The saga of George Ellery Hale's realization of his telescopic ambitions is well told elsewhere (see Florence 2011; Wright 1944, Wright et al. 1972).

2.2 Michelson and Hale

As already noted, the Master of Light and the Master of Organization first met in Cleveland in 1888. Hale was a youngster of 20 as compared with Michelson's 36 years. Hale was a sophomore at MIT and had obtained a position at Harvard College Observatory, just up the Charles River, as a volunteer assistant to Pickering. He was already an extraordinarily advanced amateur astronomer who was just then building his own research laboratory next to his wealthy family's home on Drexel Boulevard in the Kenwood section of Chicago. Hale's Kenwood Physical Laboratory would grow to become Kenwood Astrophysical Observatory when a 12 inch refracting telescope was subsequently added. There, Hale—still an MIT undergraduate—would develop the spectroheliograph, an instrument that enables imaging of the Sun at narrow and tunable wavelength bands (Wright 1994, pp. 48–96).

Michelson was a member of the physics faculty at the Case School of Applied Science in Cleveland at the time of the 1888 August meeting of the AAAS in that city. His prominence as a scientist, justifiably earned with an accurate measurement of the speed of light and the first iteration of the Michelson–Morley

Experiment already under his belt, was evidenced by his appointment as Vice President for the AAAS Section B, Physics. In that capacity, he gave the introductory address for the Section B meeting in Cleveland. One can imagine Hale grimacing when scanning the program for Section A, Mathematics and Astronomy, and then cheering up when he read the Section B program—although Asaph Hall's astronomical paper "On the Supposed Canals on the Surface of the Planet Mars" might have earned a check mark (Putnam 1889).

The full text of Michelson's Cleveland address, entitled "A Plea for Light Waves," was printed in the conference proceedings (Putnam 1889, pp. 67–78). Following a flowery introduction regarding the accomplishments of optical instrumentation, Michelson plunged into a discussion of *"a limited number of problems"* whose solutions *"all depend upon the phenomenon of interference of light waves."* In particular, they would involve an apparatus with an *"inconvenient"* name of *"interferential refractometer,"* as the device was apparently originally dubbed by Arago in 1854 (Fulton 1854). Michelson went on to describe how the instrument can be used to measure such things as absolute wavelengths, the flatness of plane surfaces, and the coefficients of expansion, elasticity, and refraction. Although astronomical applications were omitted, one can imagine Hale's mind spinning around those possibilities. It seems very likely that Hale resolved to follow Michelson's work into the future as a result of that conference. And, as described here in Section 1.2, almost exactly three years after the Cleveland gathering, Michelson would briefly resurrect astronomical interferometry with his 1891 observations of the Galilean satellites.

Michelson's and Hale's paths converged the following year when both were recruited to join the inaugural faculty of the new University of Chicago by its founding president William Rainey Harper (1856–1906). Michelson would complete his career there, while Hale resigned from his position in 1905 in favor of his new observatory in California (Wright 1994, p. 92). Although no longer faculty colleagues at Harper's great new university, they remained close friends, and Hale even became a customer of Michelson's by occasionally purchasing diffraction gratings for Mount Wilson spectrographs that Michelson's staff fabricated at UC's Ryerson's Physical Laboratory. Michelson's fame spiked in 1907 when he became America's first Nobel Laureate in a scientific field by winning the physics prize *"for his optical precision instruments and the spectroscopic and metrological investigations carried out with their aid."* (Livingston 1973).

Hale had a hand in that high recognition of Michelson as a preeminent physicist. In a letter to *"Professor Dr B. Hasselberg"* dated 1907 August 8, Hale was apparently continuing a correspondence with this member of the Nobel Prize committee for physics, regarding a request for background information on Michelson. Hale begged off from the task while promising that he would have the University of Chicago Press send the professor a copy of Michelson's 1902 book *Light Waves and Their Uses*, which includes a chapter on Interference Methods in Astronomy. Hale told Hasselberg that the book *"in spite of its popular character, might contain the additional information you desire in preparing an account of Michelson's work."* He referred to an enclosed list of *"the chief events in*

Michelson's life" along with a list of the candidate's papers. The letter closed with: *"It will be a great satisfaction to me if you succeed in obtaining the Nobel prize for Professor Michelson, as I am quite sure that no one deserves it more richly than he. I shall of course regard your letter as confidential."* So, here was the 39 year-old George Ellery Hale supporting the candidacy of his significantly older friend and colleague, Albert Abraham Michelson, for the Nobel Prize. This correspondence, and much more to follow below, was mined from the extensive and wonderfully-curated George Ellery Hale Papers collection of the Caltech Archives.[2]

Hale's superb new 60 inch telescope would become operational the year following Michelson's Nobel award. Nine years later, it would be joined by the 100 inch Hooker Telescope where interferometry would see a 20th century resurgence. It was obvious to Hale that a fruitful collaboration with Michelson aimed at achieving remarkable advances in the resolution of stars was in the offing.

The question naturally arises as to when these two gentlemen first started talking about using the Mount Wilson telescopes for interferometry. The available correspondence between Hale and Michelson does not clearly indicate a genesis date for this venture. It is likely that their arrangement resulted from an invitation from Hale some time in 1918 to carry out interferometry on the nearly completed 100 inch telescope (Pease 1931). That must have appealed to Michelson who also saw Mount Wilson as a site from which to conduct other projects that were high on his to-do list. Their subject matter during much of 1919 largely involved an Earth-tide experiment Michelson wished to place on the mountain. The subject of stellar interferometry must have been under discussion for some time but first appeared in the Michelson letters among the Caltech archives in a letter of 1919 May 8 to Carnegie Institution President Robert S. Woodward (1849–1924) expressing gratitude for the offer of a Research Associate[3] appointment engineered by Hale. While Michelson was confident of the feasibility of fabricating an instrument *"for an attempt to measure the diameters of the stars,"* the uncertainty existed that obstacles imposed by atmospheric seeing *"may not be an insuperable difficulty."* He intended to carry out preliminary experiments at Yerkes Observatory before inaugurating the work at Mount Wilson. In an accompanying letter to Hale in which the stellar diameter project was only referred to as *"the proposed work,"* Michelson made it clear he was nervous that the atmosphere would slam the door on their efforts and wanted to postpone the Carnegie appointment until local experiments yielded

---

[2] See the California Institute of Technology Archives and Special Collections George Ellery Hale Papers at https://hale.archives.caltech.edu//. The subsection "Director's Files of the Mount Wilson Observatory, 1900–1926" is a particularly rich resource for historians at https://collections.archives.caltech.edu/repositories/2/archival_objects/57837.

[3] Far from being the junior positions that the title implies in today's scientific world, these were rare, long-term appointments to distinguished astronomers with strong connections to Mount Wilson. Michelson's 1919 appointment was the second following that of Jacobus Kapteyn (1851–1922) a decade earlier. Only two more Research Associates would be designated—Henry Norris Russell (1877–1957) in 1921 and Joel Stebbins (1878–1966) in 1931. The program would be retired in 1946 with Russell and Stebbins still holding those positions.

encouragement to their expectation that seeing would not be an obstacle. Hale expressed his agreement on 1919 May 18.

The subject next appeared when Hale closed his letter to Michelson of 1919 June 9 with "*I shall be glad to hear how your tests of the interference apparatus are getting on, and what the probability is that you will come out here before autumn.*" Michelson responded on June 15 describing preliminary experiments with a small effective aperture as "*coming out very well.*" The next step would be "*an interferometer with equatorial mounting and an effective aperture of sixty inches to be tried on a star. This apparatus is nearly half done and I hope to have it finished in about three weeks.*" He was here referring to the optics to be used at the Mount Wilson 60 inch telescope. Michelson went on to state that the feasibility of starting on "*the larger instrument (fifteen feet effective "aperture") will depend on the results with the five foot* [i.e., the 60 inch telescope]." Michelson also felt it advisable to postpone implementation of interferometry as well as the Earth-tide installation on the mountain until 1920. On 1919 June 23, Hale acknowledged the advisability of proceeding conservatively.

And then, seemingly out of the blue—success! On 1919 September 3, Michelson sent Hale a telegram reporting fringes at the Yerkes refractor in spite of mediocre seeing. That must have been quite a relief, but then Hale got cold feet when he telegraphed Michelson on 1919 September 13 that he was concerned that the refrigeration system to maintain the temperature of the 100 inch mirror was not yet installed. Without it, thermal cycles in the dome would likely result in a slightly warped mirror surface providing highly aberrated imagery that might swamp fringes. He advised waiting until the next summer so that system could be installed and made operational. In turn, and on that very same day, Michelson sent Hale a telegram to where he thought Hale was staying in Santa Barbara saying he was coming out that very evening and may stop by Santa Barbara to "*join you for a day or two.*" Hale replied immediately by regular mail reiterating his concerns about the mirror figure and advising that "*it would really be wiser to wait until your chances here are likely to be the very best.*"

Regrettably, their next round of exchanges are not found in the Caltech archives. But, fortunately as it turns out, Michelson did indeed travel to Pasadena, and successful interferometric testing was achieved at both the 60 and 100 inch telescopes on the night of 1919 September 18. In a brief paper in the 1919 October issue of the *Publications of the Astronomical Society of the Pacific*, Hale closed out a perform-ance comparison of the two nighttime telescopes on Mount Wilson with a single paragraph reporting Michelson's success at both telescopes where fringes were "*beautifully seen*" for the star Altair in even worse seeing—"*2 to 3 on a scale of 10*"— than Michelson experienced during his initial tests at Yerkes. Hale expressed enthusiasm for the production of fringes from mirrors "*placed as much as 25 feet*" atop the telescope and reported that Professor Michelson was presently designing such an apparatus (unaware that this very idea had been conceived nearly 70 years earlier by Fizeau). In the meantime, the existing capability would be used for measuring close binary stars and the diameters of novae and planetary nebulae. The paragraph closed optimistically with: "*this method will be of great importance in future work.*" (Hale 1919).

Frustratingly, this historic first feasibility testing is not mentioned in their remaining letters, nor is it described in the Carnegie *Year Book* for 1919 except for a cursory note that "*Professor A. A. Michelson, recently appointed Research Associate of the Carnegie Institution, is preparing apparatus for use on Mount Wilson in measuring the diameter of stars by interference methods.*" Although an observing log for the 100 inch had been initiated on 1919 September 1 to record the now regularly assigned use of the telescope, the log does not mention Michelson's crucial test. That night had been assigned to Harlow Shapley for spectroscopic observations. The record does show that conditions were indeed poor as Michelson had reported. The seeing on the Mount Wilson scale of 1–10, where higher numbers indicate better seeing, was recorded as "*1 to 2,*" likely a result of the "high wind" also noted—even worse than Hale's published report of "*2 to 3 on a scale of 10.*" (See Appendices A and B.6 for more on astronomical seeing and the Mount Wilson scale.)

Nevertheless, 1919 September 18 must be regarded as the quiet embarkation of Mount Wilson Observatory's long and continuing journey into high angular resolution astronomy. The failure to report this milestone in the 1919 *Year Book* is puzzling in view of Hale's published optimism. However, the interferometry achievement came at a time when Hale had a lot of plates spinning (Wright 1994, pp 330–332). First, he was dealing with his brilliant but excruciatingly difficult and increasingly erratic and insubordinate optical designer George W. Ritchey (1864–1945), who he finally terminated that October. He was also involved in the selection of the new Carnegie Institution president while resisting the desire of the Carnegie board chairman Elihu Root, who had served in the cabinets of Presidents Roosevelt and McKinley, that Hale accept the presidency. To top these off, Hale was in a period of illness wherein he could hardly cope with anything. Perhaps these circumstances contributed to the lack of fanfare given Michelson's success on Mount Wilson.

In any event, Michelson described these first interferometric observations from Mount Wilson in a paper that appeared in the summer of 1920, saying "*On invitation from Dr George E. Hale the test was applied (1919 September 18) to the 60 inch reflector of the Mount Wilson Observatory and then to the 100 inch reflector, and in both cases the experience at the Yerkes Observatory was confirmed.*

"*In the case of the 60 inch telescope the apertures were applied, as in the case of the 40 inch, to the objective; while for the 100 inch, it was found quite as effective and far more convenient to use a small screen with two apertures near the eyepiece, the distance and orientation being thus much more readily controlled while the effective size and separation of the two interfering pencils remain the same.*" (Michelson 1920).

A year later it would be a different story altogether. Hale kicked off the report for his observatory in the *Year Book* for 1920 with confidence that there was nothing serious to worry about in terms of atmospheric turbulence obliterating interference fringes. He graciously credited the expertise of Albert Michelson and his compelling demonstration decades earlier with the Jovian satellites. He puzzled with the "*fact that no astronomical applications of the method have since been made is not easily explained"* while stating that Michelson's interferometer *"promises to play an important part in the future of sidereal astronomy."*

Also, by then the binary system comprising the bright star Capella had been resolved interferometrically, and the 20 foot interferometer beam was already at work on reaching the holy grail of stellar astrophysics—measuring the diameter of a star. Before we continue with that landmark effort, though, we turn to the work of other Mount Wilson astronomers who were also engaged in this latest reawakening of interferometry.

2.3 Anderson and Merrill

During the war years and for periods thereafter, Hale spent much of his time in Washington, DC in his role of president of the National Research Council, which he had founded in 1916. This had double-edged consequences, of course, but one positive result for posterity is that he communicated with his staff routinely by mail. Those letters are preserved in the Caltech Archives.

Among the important players in this new, key episode of interferometry was John August Anderson (1876–1959). Anderson, shown in Figure 2.4, received a doctorate from Johns Hopkins in 1907 and was appointed the following year to a professorship there in astronomy. His specialty was the ruling of diffraction gratings with the institution's Rowland ruling machines, and the Hopkins concave gratings were a prized specialty for the laboratory's customers. The Hale Papers contain extensive exchanges with Anderson starting in 1911 when Mount Wilson began ordering its gratings from Anderson's laboratory. In 1916, Hale recruited Anderson to join the observatory staff in Pasadena, which he did in spite of the move being lateral in salary (Bowen 1962).

At the Pasadena headquarters of the Mount Wilson Observatory on Santa Barbara Street, Anderson assumed the responsibilities for the observatory's own gratings lab and experimented with the Stark Effect, seeking specific lines that would show the tell-tale broadening that electric fields might produce in the Sun. Anderson began a letter to Hale, dated 1920 April 23, with a progress report on the ruling machine for which the *"diamond that is in now gives a really fine ruling—very brilliant spectra so it will be possible to test the grating on the higher orders, which I particularly wish to do."* That must have been good news to Hale.

The very next paragraph described the "extra-focal interferometer" that Anderson has designed to pursue the "Jupiter–Einstein test." Among his other interests was the development of an interferometer for measuring the subtle displacement of a background star closely aligned on the sky with Jupiter as predicted by Einstein's General Relativity Theory. His accompanying sketch of the compact instrument, which would reside in a telescope's converging beam, employs small lenses and prisms to separately collect the light of Jupiter and a bright star as far as a few arc-minutes from the planet to be super-positioned on the entrance slits of an interferometer. In principle, the apparent angular separation change resulting from General Relativity might be detectable as Jupiter moved with respect to the star. Anderson expressed a healthy skepticism for its success in noting that the apparatus *"should do other things besides the Jupiter–Einstein experiment, of course, provided it works at all."* Because the star

Figure 2.4. John A. Anderson. (Image courtesy of the Observatories of the Carnegie Institution for Science Collection at the Huntington Library, San Marino, California.)

and Jupiter were sufficiently far apart to have different paths through the atmosphere —what today would be called non-isoplanatism (see Appendix B.7)—the relative motion arising from seeing might swamp the GR signal. Anderson proposed "*to build the apparatus anyway—and test it out. If it works, one could practice it on other things*" while awaiting for a fortuitous near occultation by Jupiter of a bright star. Apparently, that's what Anderson did—or at least planned to do—as machine shop drawings, dated 1920 October, exist for the extra-focal interferometer front end.

There is no indication that the device was ever used at the telescope. The 1920 *Year Book* describes the interferometric approach to measuring the Jupiter–Einstein effect and reports that "*Professor Michelson has designed several forms of interferometer for this purpose, and one of them will soon be tested with the Hooker telescope. It is feared, however, that atmospheric disturbances, which would differ along the optical paths of the stars several minutes apart, may prevent the fringes from being observed.*" And that is all that exists for this concept, but this was likely the device that Pease mentioned in a 1931 review paper as being unsuccessfully applied to stellar parallax and proper motion studies (Pease 1931, p. 85). No further mention of the experiment appears in subsequent yearbooks.

Yet another application of interferometry went forward—this time with great success. With the help of Francis Gladheim Pease (1881–1938), who will play the lead role in this story, Anderson mounted a simple, experimental interferometer in the converging beam with a slit pair with variable separation at a position about 2 ft from focus where the beam diameter from the 100 inch telescope's primary mirror was 73 mm. A trial run was made on 1919 December 30. For a given slit spacing, the instrument was rotated with respect to the position angle vector of the two components of a resolvable double star, the fringe visibility would be modulated by the presence of two stars, passing through four minimum visibilities. When the known close binary Algol was observed, fringes maintained constant visibility at all spacings —Algol was unresolved. Capella, however, exhibited very significant variations in fringe contrast indicative of resolution. In his brief published report of this night, Anderson concluded that interferometry did indeed double the resolving power of a given telescope aperture, was relatively immune to seeing conditions, and could be used to obtain the orientation and separation of the two components in a close double-star system just as accurately as those of a wide binary. This simple experiment, which followed Michelson's fringe demonstration a few months earlier, yielded the first science observation from interferometry in the 20th century.

It convinced Anderson that additional unique measurements of binaries could be harvested with an interferometer. As a result, work began immediately on the fabrication of an appropriately designed instrument—the Anderson Double Star Interferometer—that exists to this day (Anderson 1920a). Figure 2.5 shows how the device functioned on the 100 inch along with a photo of the interferometer itself. Details of its components are identified in Figure 2.6.

Six weeks after that first resolution of Capella, Anderson was back at the telescope with this newly completed instrument. Well known from spectroscopy to be a binary-star system with an orbital period of 104 days, Capella (α Aurigae) would continue as his primary target. Measurements were taken on the nights of 1919 February 13, 14, 15, March 15, and April 23 with the results from the last date compromised by daylight. Anderson folded the four acceptable data sets in with the spectroscopically-determined orbital elements to solve for the parameters of the "visual" orbit as projected on the sky. His approach led to values for the angular semimajor axis of the projected, elliptical orbit as well as its inclination to the plane of the sky. Combining these with other spectroscopic orbital elements, Anderson found the distance to the system as well as the component masses (see Appendix B.3.4). Those values were 16.7 parsecs and 4.62 and 3.65 solar masses. He also adjusted the orbital period in the best-fit solution to 104.066 days. The angular separation residuals from the orbit solution were astonishingly and, as will be seen below, artificially small, ranging from 0 to 4 in the fifth decimal place (Anderson 1920b).

Anderson's interferometry work would be short lived. His final foray was a fine paper on the importance of knowing the effective wavelength at which interferometer observations are being made as that parameter enters into the problem linearly just as does the slit spacing. In his discussion of this complex topic, Anderson showed that this wavelength is dependent upon the star's effective temperature (which, of course, parameterizes its blackbody spectrum), as well as the transmission curve for the atmosphere (which is itself dependent upon zenith angle), and the sensitivity curve for the eye. He developed a laboratory setup for measuring the required quantity and the mathematics needed to calculate each term. We would today add to Anderson's list of parameters pertaining to effective wavelength the transmission curves for any glass in the beam, reflectivity of non-transmissive surfaces, and the wavelength response of the detector, i.e. a solid-state device rather than the human retina (Anderson 1922).

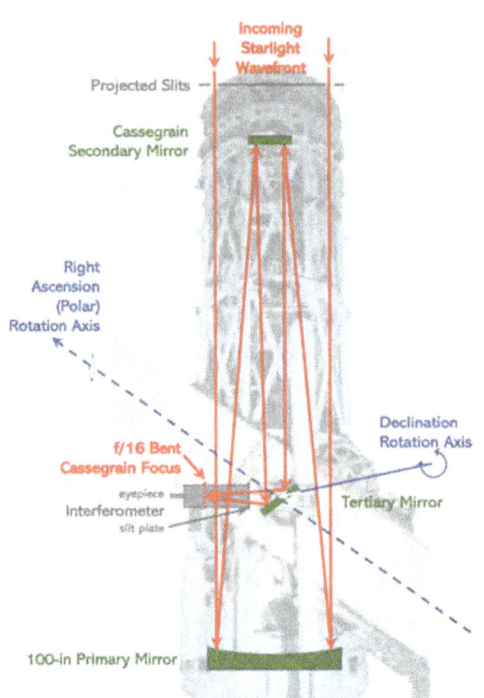

Figure 2.5. (Left) The Anderson Interferometer, shown mounted at the north Cassegrain observing port of the 100 inch telescope, could potentially resolve binary stars with angular separations down to about 0.03 arcsec = 30 milliarcseconds (mas). (Right) CHARA Array Site Manager Larry Webster, who has rescued many old Mount Wilson instruments, books, and archival materials, inspects the Anderson Interferometer outside the CHARA Exhibit Hall where it is displayed with the 20 foot Michelson–Pease Interferometer beam. The interferometer is mounted in the standard circular instrument adaptor used at the 100 inch Cassegrain foci, with the slit assembly facing the viewer and the eyepiece facing Webster. (Author's diagram and photo.)

With that effort completed, Anderson bid interferometry farewell and returned to his work in laboratory spectroscopy, except, as we shall see in the next chapter, when he backed up a colleague with an experienced eye for interferometry. Later in his career he would be preoccupied with realizing Hale's goal of a 200 inch telescope.

Just two months after Anderson's final observations of Capella, the same interferometer was reintroduced to the same star by another Mount Wilson astronomer—Paul Willard Merrill (1887–1961), whose photograph appears in Figure 2.7. After receiving his PhD at the University of California in 1913, Merrill had stints at the University of Michigan and the National Bureau of Standards before joining the Mount Wilson staff in 1919.[4] In scanning his long list of scientific publications, one finds paper after paper dealing with spectroscopy

---

[4] Merrill had inquired about a job at MWO as early as June 1914. Hale responded that there were no openings at present. (See Letters dated 1914 May 30 and 1914 June 13 from the George Ellery Hale papers of the Caltech Archives in the Paul W. Merrill File.)

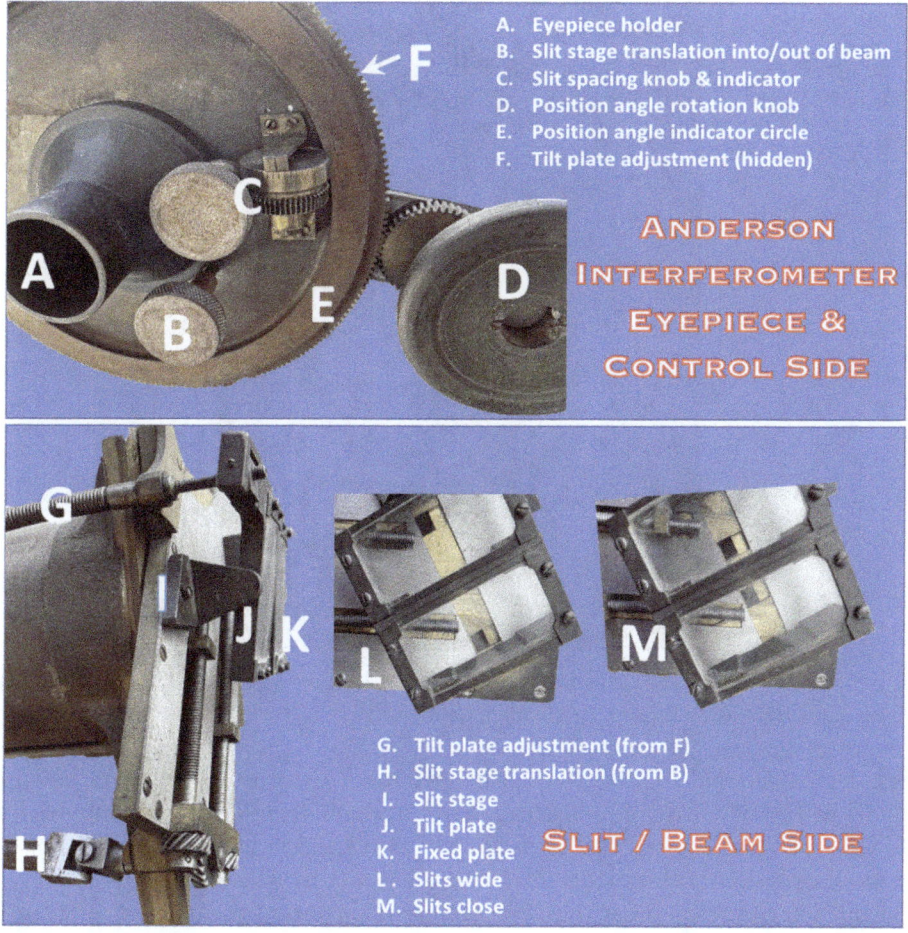

A. Eyepiece holder
B. Slit stage translation into/out of beam
C. Slit spacing knob & indicator
D. Position angle rotation knob
E. Position angle indicator circle
F. Tilt plate adjustment (hidden)

ANDERSON INTERFEROMETER EYEPIECE & CONTROL SIDE

G. Tilt plate adjustment (from F)
H. Slit stage translation (from B)
I. Slit stage
J. Tilt plate
K. Fixed plate
L. Slits wide
M. Slits close

SLIT / BEAM SIDE

Figure 2.6. Details of the eyepiece and slit assembly ends of the Anderson Double Star Interferometer are shown here. (Author's diagram and photos.)

of peculiar stars, especially variable stars. But in 1922, there appears the anomalous "Interferometer Observations of Double Stars." Merrill supplemented Anderson's observations of Capella with 30 measurements on ten nights scattered between 1920 October 5 and 1921 April 1. He combined these new data with Anderson's more limited sample to recalculate Capella's orbit while again adopting certain spectroscopic elements. The revised masses were about 9% smaller than Anderson's. The mean residuals were +0.41° ± 1.34° in position angle and −0.1° ± 8.4° milliarcseconds (mas) in angular separation (Merrill 1922).

I appreciated the quality of Anderson's and Merrill's data firsthand in solving for Capella's orbit incorporating the Mount Wilson visual interferometry along with speckle interferometric measurements accumulated between 1968 and 1980. The speckle technique, to be described later in more detail, is not dependent on the human eye and was generally employed at telescopes with apertures larger than 100

Figure 2.7. Paul W. Merrill. (Image courtesy of the Observatories of the Carnegie Institution for Science Collection at the Huntington Library, San Marino, California.)

inches. I was startled to find that those Mount Wilson measurements had mean residuals to my new orbit of +0.16° ± 1.61° and −0.4° ± 0.6° mas compared with +0.67° ± 1.59° and −1.4° ± 1.2° mas for my own speckle observations. While mine were more or less comparable to the 1920–1921 data in position angle accuracy, the old visual interferometry had half the standard deviation in angular separation! I addressed this then with the comment *"It is interesting, and perhaps opposite to expectations, to note that the sample of observations with the smallest dispersion in the residuals is the original set of visual interferometer measures from Mt. Wilson during 1920–1921, whereas the sample with the largest dispersions is the 13.8 meter baseline interferometer measures from CERGA."* Interesting indeed. MWO data had originally appeared somewhat poorer in quality than it actually was due to the imprecision of the orbit they employed (McAlister 1981). John Anderson and Paul Merrill knew what they were doing!

Further insight into the quality of the early Capella data may be had from calculating their residuals from the best modern orbit of the system (Torres et al. 2015). That analysis is based upon spectroscopic and interferometric data, the latter including speckle interferometry as well as the superbly accurate long-baseline interferometry obtained some 60 years after Anderson and Merrill from the same mountaintop by the Mark III Interferometer, to be explored in Section 7.2. Figure 2.8 shows the 16 measurements obtained by Anderson and Merrill with predicted positions (open circles) for the epochs of observation along with the measured positions (+ signs).

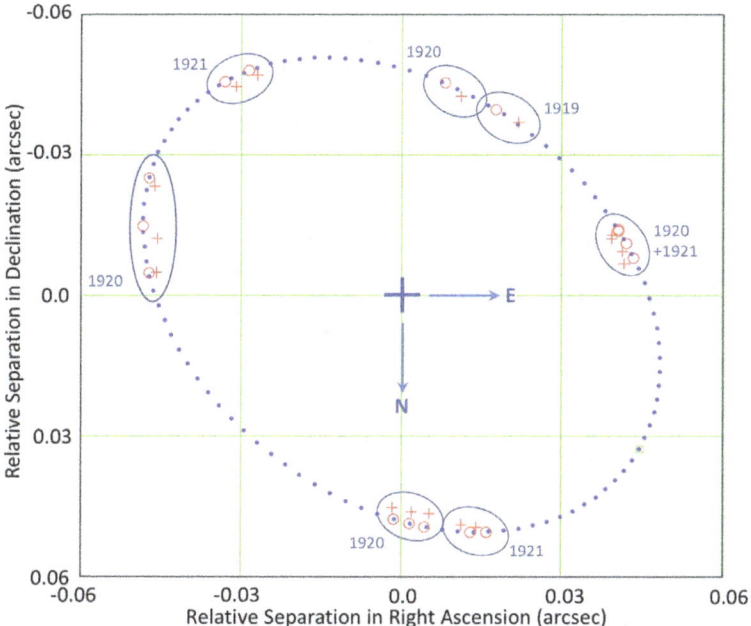

Figure 2.8. 100 inch Anderson interferometer observations of the binary star Capella (α Aurigae) made by Anderson and Merrill during 1919–1921 are plotted against the 2015 orbit by Torres et al., which includes highly accurately measurements from long-baseline interferometers. Open circles indicate predicted positions while the red + symbols are the actual measurements. In observations of this type, the brighter component is positioned as stationary at the blue + symbol, while the angular separation and position angle of the secondary star indicate the motion within the system. In reality, both stars revolve around a common center of mass. (Author's diagram.)

Table 2.1 contains the residuals of these pioneering measurements to the modern orbit (Pease 1931, p. 85), which was determined to have a period of 104.02128 ± 0.00016 days—only about 64 min longer than that found by Anderson and Merrill. Although Merrill, whose explanation of the observing methodology was particularly lucid, chose to ignore the position angle of the one 1919 result on the basis of Anderson labeling it as a rough estimate, I show it here for completeness and note that while it has the largest residual in position angle, it also has the smallest in angular separation. The means and standard deviations of the residuals for the two observers are also included. They indicate modest systematic errors in the position angles and, especially, in the angular separations. Those latter data are about 4% smaller than the separations predicted by the orbit, an effect likely attributable to an uncertainty at the same percentage level of the effective wavelength adopted by Merrill. We will see that a value of 575 nm was used in the culminating interferometry carried out at the 100 inch.

Merrill also published relative positions of the κ Ursae Majoris system with an orbital period of just over 13,000 days. The star was noted as being quadruple when Merrill was assisting Anderson in testing the apparatus the latter devised for measuring effective wavelengths employing artificially produced double stars.

Table 2.1. MWO Observations of Capella (Residuals from Orbit of Torres et al. 2015)

Epoch	$\theta(°)$	ρ(mas)	$\Delta\theta(°)$	$\Delta\rho$(mas)
	Observations by J. A. Anderson			
1919.9963	[149.0] (estimate)	42.8	[−6.38]	−0.5
1920.1177	5.6	46.9	1.37	−2.7
1920.1204	1.6	46.2	0.67	−2.5
1920.1232	357.0	45.4	-0.39	−2.4
1920.2025	242.6	51.7	1.01	−1.7
	$\overline{\Delta\theta} = -0.67 \pm 0.75$ $\overline{\Delta\rho} = -1.96 \pm 0.90$			
	Observations by P. W. Merrill			
1920.3093	107.6	41.2	−0.42	−1.6
1920.7621	254.5	47.4	1.99	−3.1
1920.8413	165.0	43.9	−4.39	−2.0
1920.0378	263.2	46.1	−0.59	−1.4
1921.0842	214.4	54.4	−1.12	−2.0
1921.0897	209.5	54.2	−0.86	−1.6
1921.1635	106.5	40.9	−1.85	−1.8
1921.1662	102.2	42.1	−1.84	−1.2
1921.1690	98.7	42.0	−1.00	−2.0
1921.2457	15.1	51.4	−1.67	−1.6
1921.2485	12.1	50.2	−1.69	−2.0
	$\overline{\Delta\theta} = -1.22 \pm 1.51$ $\overline{\Delta\rho} = -1.85 \pm 0.50$			

Instead of the visibility going through the number of minima expected for their artificial binary, they experienced the signature of a quadruple system indicating that their presumed single star was itself a double. They subsequently found that the star's duplicity had been discovered by the Robert Grant Aitken (1864–1951), one of the great visual double-star micrometrists, and that their measurements showed the system's separation had closed to 0.08 arcsec, nearly half what Aitken had measured in 1919. Merrill's five measurements, shown with the modern orbit in Figure 2.9, exhibit mean residuals to the orbit of $\overline{\Delta\theta} = 0.47° \pm 0.30°$ and $\overline{\Delta\rho} = -1.7 \pm 0.6$ mas (Muterspaugh et al. 2010). With a mean angular separation of 83 mas over Merrill's observation epoch, the residuals are again systematically small, although this time only at 2% with a 3σ significance level.

Aitken had recommended the nine year period binary star system ν^2 Boötis, which he suspected would be below 0.1 arcsec in separation. This star showed variable visibility although insufficient to extract other than a "most probable result" at 61 mas. Some 73 stars down to 5th magnitude were inspected for variable visibility without exhibiting it. Scanning down the list, I recognize several of these as having subsequently been resolved as binaries by speckle interferometry (e.g., β Per, η Vir, β CrB, β Cap, η Peg, and o And).

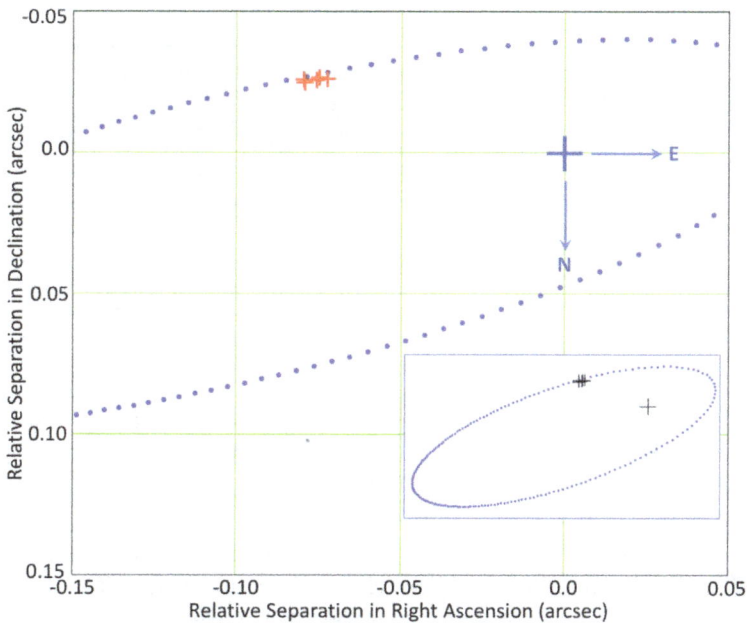

Figure 2.9. 100 inch Anderson Interferometer observations of the binary star κ UMa made by Paul Merrill in 1921 are shown plotted against the best modern orbit for the system. (Author's diagram.)

It is likely that they were missed by Merrill because of any combination of several factors—large magnitude differences of the components, which decrease the amplitude of the visibility curve; wide angular separation; and/or separation below their resolution limit. They strongly suspected duplicity in the additional five stars: δ Cancri, 10 Leo Minoris, o Leonis, ε Ursae Majoris, and υ Sagittarii. Results from modern speckle inspection of these stars (Hartkopf et al. 2016) are summarized in Table 2.2. Perhaps their suspicions were due to some observational factor. In any event, they expressed hope that others will one day inspect these *"as the double star work at the Mount Wilson Observatory has been discontinued."*

The interferometric programs of Anderson and Merrill, largely undertaken without collaboration between the two young scientists, were isolated events in both of their careers that were abruptly terminated in spite of the fact that either could have productively continued indefinitely. Why did they venture into interferometry only to move on to other scientific tasks? While the answer to this question is by no means clear, the possible explanation relates to their junior status on the Mount Wilson staff. Perhaps Hale asked each of them to do this work merely to explore the method's potential. Observations of binary stars are a component of classical astronomy, an arena that Hale preferred leaving to other observatories. Thus, once these explorations were complete, Anderson and Merrill returned to their respective fields, both of which fed into what Hale saw as the astrophysics mission for his observatory.

Table 2.2 Speckle Interferometry of Merrill's Suspected Binaries

δ Cnc	Found single by several observers but double at <0.1 arcsec by two
10 LMi	SI shows no close companion
o Leo	SI shows no close companion, but HIPPARCOS astrometry satellite found one >1 arcsec
ε UMa	One SI shows companion at 0.05 arcsec, but many show no secondary. Probably single
υ Sgr	Many inspections by SI show no companion

Their success did infect, although with considerable delay in its outbreak, at least one more Mount Wilson astronomer. We have already described the fate of the Jupiter–Einstein effect instrument. A somewhat analogous device was proposed by MWO staff astronomer Gustave Strömberg (1882–1962) a decade after the original outburst of interferometry activity. He designed an interferometer with complicated front-end optics and a second slit set to superimpose a beam pair from binaries with angular separations out to 5 arcsec and use the fringes to obtain the relevant astrometry. The Stromberg interferometer is apparently only mentioned in Carnegie Year Book numbers 31 and 32 as I am unable to find any other reference to it. As a result, there is no explanation as to how it functioned. Year Book No. 32 reported that it was tested and improvements made sometime during the reporting period of 1932 July 1 to 1933 June 30, but *"it is as yet impossible to judge of the performance of the instrument under actual observing conditions."* The ensuing silence likely indicates the outcome of that judgment.

Much was happening on Mount Wilson in the interferometry arena in the early 1920s, and it is worth taking a pause to summarize the "who" and "what" was going on during 1919–1922 and thereafter. Table 2.3 is a cheat sheet of sorts to keep these matters straight. Although there were at least two other interferometers fabricated or planned for Mount Wilson, the two working science instruments were the Anderson Double Star Interferometer and the Michelson–Pease 20 foot Interferometer, with the latter the primary subject of the next chapter.

By 1921, the Roaring Twenties of Interferometry were underway on Hale's mountaintop as several princes had kissed the sleeping beauty of interferometry. She was now only reawakening, and would slip into her hibernation again, but not until well after Albert Michelson had successfully converted the 100 inch telescope into an interferometer with a 240 inch baseline. His colleague, Francis Pease, would optimistically endeavor to more than double that achievement. Then, for reasons we will explore in Chapter 4, somnolence would set in yet again. But, the value of interferometric access to astrophysical parameters beyond the reach of classical imaging would not allow the promise to be forgotten.

Table 2.3. MWO 100 inch Telescope Interferometry Activity During the Early 1920s

Dates	Observer(s)	Instrument	Science Goals	Publication
1919 Sep 18	Michelson & Hale	Sit mask	Fringes at 60 & 100 in telescopes.	Hale (1919)
1919 Dec 30	Anderson & Pease	Prototype double star interferometer	Search for variable fringe visibility on Algol & Capella	Anderson (1920a)
1920 Feb 13–15	Anderson	Anderson Double Star Interferometer (aka Anderson Int.)	Measurement of Capella (α Aur) binary star system	Anderson (1920b)
1920 Mar 15	Anderson	Anderson Int.	"	"
1920 Apr 23	Anderson	Anderson Int.	"	"
1920 Aug 6–8	Pease	Michelson–Pease 20-ft Interferometer (aka Mich.–Pease Int.)	Testing instrument. Search for fringes on Vega	Hale (1920), p. 251
1920 Oct 15	Merrill	Anderson Int.	Double star measurements	Merrill (1922)
1920 Nov 3	Merrill	Anderson Int.	"	"
1920 Dec 11–20	Michelson & Pease	Mich.–Pease Int.	Diameter of α Orionis	Michelson & Pease (1921)
1921 Jan 14	Merrill	Anderson Int.	Double star measurements	Merrill (1922)
1921 Jan 31	Merrill	Anderson Int.	"	"
1921 Feb 2	Merrill	Anderson Int.	"	"
1921 Mar 1–3	Merrill	Anderson Int.	"	"
1921 Mar 31	Merrill	Anderson Int.	"	"
1921 Apr 1	Merrill	Anderson Int.	"	"
1921 Apr 30	Merrill	Anderson Int.	"	"
1921 June 14–17	Merrill	Anderson Int.	"	"
1921 Sep 24–29	Pease	Mich.–Pease Int.	Further observations of α Ori for diameter variability	Pease (1925a)
1921 Nov 23–26	Pease	Mich.–Pease Int.	"	"
1922 Oct 14–15	Pease	Mich.–Pease Int.	"	"
1925 Apr 15–19	Pease	Mich.–Pease Int.	Mizar binary star system	Pease (1925b)
1927 May 31–Jun 3	Pease	Mich.–Pease Int.	"	Pease (1927)

References

Adams, W. S. 1938, ApJ, 87, 369 (an excellent early memorial and biography by Hale's successor as director of Mount Wilson Observatory)

Anderson, J. A. 1920a, PASP, 32, 58

Anderson, J. A. 1920b, ApJ, 51, 263

Anderson, J. A. 1922, ApJ, 55, 48

Bowen, I. S. 1962, John August Anderson 1876–1959 Biographical Memoir (Washington, DC: National Academy of Sciences), 18 pp

Carnegie, A. 1902, Carnegie Institution of Washington Year Book No 1 (Washington, DC: Carnegie Institution of Washington), XI

Florence, R. 2011, The Perfect Machine (New York: Harper Collins), 451 pp

Fulton, 1854, FrInJ, 57, 204

Hale, G. E. 1919, PASP, 31, 257

Hale, G. E. 1920, Carnegie Institution of Washington Year Book No 19 (Washington, DC: Carnegie Institution of Washington), 251

Hartkopf, W. I., Mason, B. D., & McAlister, H. A. 2016, Fourth Catalog of Interferometric Measurements of Binary Stars (Washington, DC: US Naval Observatory), available online at https://www.usno.navy.mil/USNO/astrometry/optical-IR-prod/wds/int4

Harvey, J. 1999, ApJ, 525, 60

Hussey, W. J. 1904, Appendix A to Report of Committee on Observatories, Carnegie Institution of Washington Year Book No. 2 (Washington, DC: Carnegie Institution of Washington), 71

Livingston, D. 1973, The Master of Light (New York: Charles Scribner's Sons), 232

McAlister, H. A. 1981, AJ, 86, 795

Merrill, P. W. 1922, ApJ, 56, 40

Michelson, A. A. 1920, ApJ, 51, 257

Michelson, A. A., & Pease, F. G. 1921, ApJ, 53, 24

Muterspaugh, M. M., Hartkopf, W. I., Lane, B. F., et al. 2010, AJ, 140, 1623

Newcomb, S 1903, Sci, 17, 121

Newcomb, S 1888, The Sidereal Messenger, 7, 65

Pease, F. G. 1925a, PASP, 34, 346

Pease, F. G. 1925b, PASP, 37, 155

Pease, F. G. 1927, PASP, 39, 313

Pease, F. G. 1931, ErNW, 10, 84

Putnam, F. W. (ed) 1889, in Proc. of the American Assoc. for the Advancement of Science (Salem, MA: AAAS), 47

Robinson, J. W. 1991, The San Gabriels (Arcadia: Big Santa Anita Historical Society), 121

Seims, C. 1976, Mount Lowe: The Railway in the Clouds (San Marino, CA: Golden West Books), 238 pp

Torres, G., Claret, A., Pavlovski, K., & Dotter, A. 2015, ApJ, 807, 26

Wright, H. 1994, Explorer of the Universe (Woodbury, NY: American Institute of Physics), 166 (the full biography of George Ellery Hale)

Wright, H., Warnow, J. N., & Weiner, C. 1972, The Legacy of George Ellery Hale (Cambridge, MA: MIT Press), 293 pp

Chapter 3

The Pièce de Résistance

3.1 The Stellar Diameter Imperative

The Copernican world view plucked the stars from their crystalline sphere and scattered these lights in the night sky as Suns at distances that could be scaled from their brightnesses. Galileo sought to find the zero point of this relationship by measuring stellar diameters telescopically. In his famous "*Dialogue Concerning the Two Chief World Systems* (see Drake 2001)," Galileo asserted that first magnitude stars had diameters of five arcseconds while those of the sixth magnitude were six-times smaller and six-times more distant at 2160 au. Ignorant of the wave nature of light and diffraction theory, the turbulence theory of atmospherically-induced "seeing," the huge intrinsic variation among stellar sizes, and any real idea of interstellar distances, Galileo was at least correct that more distant stars would have smaller angular sizes. At the beginning of the 20th century, diffraction theory was in hand, seeing was blamable on the atmosphere, and the curtain was about to be lifted on the intrinsic brightness range and distance scales of the stellar family. Most significantly, actual measurements of star diameters were in the offing.

The angular diameter of the Sun as seen from the distance of the nearest star is no more than about 7 milliarcseconds (mas)—the size of a quarter at 600 km. The famous Rayleigh criterion (see Appendix B.7) of $\theta_{min} = 1.22\lambda/D$, where D is the telescope aperture and λ is nominally 550 nm for visible observations, results in a diffraction limit of about 55 mas for a 100 inch telescope. Even in the rare instance of a ground-based telescope having diffraction-limited optics, the atmosphere always steps in and blurs stellar images beyond that limit by an order of magnitude and more. The Rayleigh criterion is reachable and even surpassed if that same aperture is employed as the baseline of an interferometer, where two small apertures produce an image crossed by interference fringes characteristic of the apertures' separation ($\lesssim D$). And yet, a 100 inch baseline still misses the 7 mas mark by a factor of eight. In its simplest application, where there is no accurate means for measuring fringe visibility, one only needs to find that baseline at which fringes vanish. If the visibility as a function of baseline can be accurately measured, one might need only, say, $D/4$ to calculate the diameter, as Michelson explained in words and a sketch he sent to Hale on 1919 November 27. An extract of that letter is shown in Figure 3.1. Michelson's approach as sketched in his diagram is commonly practiced at long-baseline interferometers today in measuring stellar diameters well below the formal limit of λ/D. See the primer on interferometry in Appendix B for more on the practical aspects of the field.

In the years before the outbreak of World War I, as the connections between spectral classification schemes and temperature were emerging, observational techniques were developed to provide estimates of stellar surface brightness. It then became evident that there were two classes of red stars. Indeed, the recognition of the existence of two luminosity regimes among the red stars was one of the great achievements of early 20th century stellar astronomy. The subsequent reconciliation of that fact was a major impetus toward a correct theory of stellar evolution.

Astronomer and historian David DeVorkin has written of the confluence of Michelson's interferometer and the challenge presented by the realization that the brightest red stars in the night sky might have measurable angular diameters (DeVorkin 1975). DeVorkin's separate and complementary discussion of the revelation of the Hertzsprung–Russell Diagram also plays into this story (DeVorkin 1977). As is well known, the two namesakes of this key tool in understanding the evolution of the stars hit upon the rudiments of the relation independently and serially. In 1905, Ejnar Hertzsprung (1873–1967) was an amateur astronomer publishing in the little-known, Leipzig-based *Journal of Scientific Photography, Photophysics and Photochemistry.* He discerned that certain stars with narrow and weak absorption lines had been placed by Antonia Maury (1866–1952) in her spectroscopic class c among the red stars, but had distinctly small proper motions and parallaxes compared with other red stars. This implied that the "c" stars were much farther away than stars of similar color that also possessed

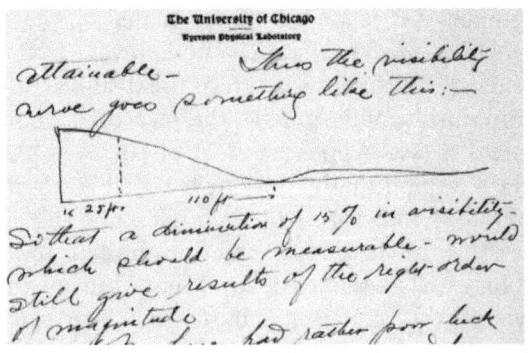

Figure 3.1. In his letter of 1919 Nov 27 to Hale, Michelson shows how the fall-off in visibility at baselines shorter than that required to reach the full null where fringes disappear can be sufficient for a diameter determination. This is standard practice for modern interferometers. (Courtesy of the Caltech Archives George Ellery Hale Collection and the Observatories of the Carnegie Institution for Science Collection at the Huntington Library, San Marino, California).

broad and stronger absorption features. Hertzsprung concluded that red stars come in two species with very different intrinsic brightnesses, one of which would become known as "red giant" stars (Hertzsprung 1905).

In elaborating a literature reference in the endnotes of a 1906 paper whose German title translates as *About the Optical Strength of Black Body Radiation and the Minimum Light Equivalent*, Hertzsprung concludes, in the following translation, that: "*It is of interest to see if Arcturus, whose stellar magnitude, according to Harvard photometry, is 0.24, has an angular diameter of the order of 1/20 arc seconds or not* (Hertzsprung 1906)." This first estimate of a star's angular size based upon radiation laws combined with photometric measurements is, remarkably, imbedded in a footnote of an article in an obscure journal by a chemist with a keen interest in astronomy!

DeVorkin pointed out that the methodology employed by Hertzsprung was first demonstrated by E. C. Pickering a quarter-century earlier but judged unworkable due to the lack of quantitative photometry (Pickering 1880; Wesselink et al. 1972) Pickering would see that this shortcoming was addressed. Interferometry, if carried out with sufficiently large spacings of slits masking the apertures of the largest available telescopes, might be able to resolve this nearby red giant star—a point overlooked at the time by the few who might undertake such work.

On 1913 June 13, Princeton University professor Henry Norris Russell (1877–1957) presented a talk entitled *"Giant" and "Dwarf" Stars* to a London meeting of the Royal Astronomical Society, where he displayed for the first time what would be dubbed the "Russell diagram"—subsequently and belatedly redesignated as the "Hertzsprung–Russell diagram" (Russell 1913). Six months later, he again showed his diagram during a special address to the 1913 December meeting of the Astronomical and Astrophysical Society of America (now known as the AAS) in Atlanta. As a result of the meeting city being "*rather remote or out of the usual track*," his address was heard by only a dozen or so attendees—all men, of course. Fortunately, however, it was published in its 39 page entirety, albeit five years after the meeting—the first published version of the HR diagram (Russell 1918). Much more on the dynamics of this fascinating period in the history of modern stellar astrophysics, including other contributors beyond Hertzsprung and Russell, can be added beyond this abbreviated description (DeVorkin 1977; Hearnshaw 2014; Dick 2013). Previously unpublished photographs of a few of the principle players in this process are shown in Figure 3.2.

All this was transpiring as Hale's great endeavors on Mount Wilson were bringing powerful new telescopes equipped with state-of-the-art spectrographs to bear on stellar astrophysics. The possibility of resolving the disks of the largest of the red giants—and what would later be labeled supergiants—would be a siren call that Hale could not resist.

In Section 2.2, we saw the first attainment of interference fringes during brief stints at the 60 and 100 inch telescopes by Michelson on 1919 September 18. With the feasibility of interferometry thus demonstrated, it was decision time in regard to the stellar-disk requirement of a higher-resolution interferometer than Anderson's. Thus, as 1919 was drawing to a close, the Hale–Michelson correspondence focused mostly on how long the interferometer baseline might be. The tradeoff was length

Figure 3.2. These previously unpublished photographs from the 1932 General Assembly of the International Astronomical Union held during in Cambridge, Massachusetts were taken by the author's mentor Karel Hujer. They show several of the key scientists involved in the revolution in stellar astrophysics that began some twenty years earlier. Shown at left are: MWO Director Walter S. Adams and Henry Norris Russell (with sunlight casting a shadow across his face) during a boat outing; middle: Ejnar Hertzsprung and Raymond S. Dugan, who co-authored with Russell and John Q. Stewart the standard astronomy textbook for a generation; and at right: Arthur S. Eddington and Annie J. Cannon who simplified the Harvard spectral classification according to the strength of the Balmer lines and hence into a temperature sequence. (These copyrighted images are the property of the author.)

(or angular resolution) versus weight. The mass that could be effectively supported atop the telescope required many times that amount below the declination axis to achieve balance.

On 1919 November 11, Michelson pointed out that increased length also requires higher magnification to see the fringes generated at the longest baseline. Hale replied a week later that he had not thought of that but then pivoted to baselines required to "*resolve Capella, Sirius, Antares, and Betelgeux* (sic) *as 110, 120, 150, and 130* feet" [my emphasis] without citing the source of these values. Did Hale mean "inches" rather than "feet"? Apparently not. DeVorkin located a late-1920 letter from Shapley to Russell that blames the exaggerated baselines on an incorrect value for the Sun's absolute magnitude (DeVorkin 1975, p. 8).

Aside from these details, Hale was justifiably worried about embarking on a mission that might not pay off if the estimated diameters were overly optimistic. It was at this point that on 1919 November 7, Michelson sent Hale the sketch we saw above in Figure 3.1 that should have eased Hale's mind over this concern. Although Michelson would subsequently propose observing schemes for measuring V quantitatively as a function of baseline, none were ever successfully pursued.

By 1920 January 3, Hale had rationalized that the risk involved in resolving stellar disks is minimized by "the greatest value" interferometry will have on the field of spectroscopic binary stars as demonstrated by Anderson. After all, the resolution of a spectroscopic binary in which orbits can be determined separately for each stellar component yields those components' masses *and* the distance to the resolved system, the latter leading to the two stars' luminosities (Appendix B.3.4). Those are important constraining data for stellar evolution theory in their own right. Hale also noted that "*it is amazing how sharp and steady the fringes appear under a power of 5000*," giving him more confidence in the workability of a 25 ft beam. Michelson was gratified by this in his reply of January 11 and went on to point out that the fringes for Capella "*should not have entirely vanished*" unless the two stars were of equal brightness, which we now know they very nearly are. He showed Hale how the intensity ratio for a binary star is given by $(V_1 - V_2)/(V_1 + V_2)$ and proceeded to sketch out in detail additional optics that would produce numerical values of V. As an afterthought, Michelson added a vertical note to the margin congratulating Anderson and Pease for their good work.

Momentum continued to build toward fabrication of an actual beam interferometer, and on 1920 January 19, Hale asked Michelson to go ahead and send sketches for a 25 ft beam structure so that a full design could be made and the instrument shop be put to work on its fabrication. Michelson did as requested on January 25 and accompanied the sketches with cautions about the effects of vibration on fringe stability that must be guarded against even at the expense of added weight. He recommended that the initial alignment and mirror position calibration be done by looking at a mercury-arc source a mile away with the telescope and beam pointed horizontally, a non-trivial demand considering the load on the telescope. Michelson also set a limit of 10 arcsec on the deviation of the light beam from a collecting mirror as it moves along the slide. These were serious constraints on the final design, but superb engineering was Michelson's key to success in all he undertook.

Hale wrote on 1920 February 5, once again with nervousness about the implication of Michelson's specifications on the beam's weight. Furthermore, it had come to his attention that because the beam could not be rotated, it could not unambiguously measure double star orientation angles. So, maybe there went his reliance on spectroscopic binaries to justify the risk vis-à-vis diameters.

Their correspondence continued along these lines during the spring with Michelson finding ways to lighten the beam without sacrificing stiffness. But mass calculations again rang the alarm bell and resulted in a shorter beam and hence diminished limiting resolution. On 1920 May 20, Hale told Michelson that Pease was nearing completion of the design "*of the large interferometer, which will probably have a base line of about 20 ft.*" By June 8 the 20 foot[1] design was ready and Hale reported that the required "*channel bars*" would be "*ordered today.*" Eight days later, Michelson replied with the news that the mirrors and their mounts, which he had taken responsibility for at Chicago, would be expressed shipped to Pasadena in a few days. On June 21, Hale announced that the 20 foot base had been riveted together and machining would start soon.

---

[1] Note on abbreviations: We would today routinely refer to the Michelson–Pease interferometer as the "20 ft interferometer." Out of respect for its builders' nomenclature, I will refer to the instrument as the "20 foot interferometer" or just "the 20 foot." Likewise for the 50 foot interferometer as well as for the 60 and 100 inch telescopes.

Thus, with considerable hand wringing, the risk/performance tradeoffs had been made, and a 20 foot interferometer beam was in the works.

Michelson submitted a report of his summer work as a Research Associate in a formal—and uncharacteristically typed rather than hand-written—letter to Hale on 1920 September 1. The report consisted of brief summaries of progress on "The Stellar Interferometer" and "The Velocity of Light" programs. No mention was made of earth tides, which had played third fiddle to the above two efforts in the Hale–Michelson letters in prior years. Most relevant was a brief description of the first experiments with the 20 foot beam mounted on the 100 inch to detect white-light fringes using a double wedge of glass in one beam to provide the requisite small adjustments to path length required to match paths in the two interfering beams. In the other beam, a single plane-parallel plate compensated for the adjustable unit in the other and also allowed beam alignment through a tip/tilt mechanism. These glass plates were near the eyepiece end of the device. A white-light source, presumably the mercury arc mentioned in Michelson's January 25 letter, illuminated the setup from a distance in a manner not described. Pease found fringes at a seven-foot baseline on August 10, and the following day Hale and Michelson joined Pease to see fringes for themselves at eight feet. Michelson ended this part of the report by—oddly enough —concluding that their success was encouraging toward detecting fringes out to a baseline of 100 ft "*or more*"! This optimistic extrapolation could likely resolve the disks of the nearest Sun-like stars. No mention was made in the report of resolving "giant" stars—the *raison d'etre* for this massive apparatus. Michelson was clearly enthusiastic about the future of stellar interferometry.

Whatever it was that Michelson had in mind as he wrote this report, on October 4 Hale brought matters back to the present use of the 20 foot beam by enthusiastically bringing to Michelson's attention an article in the September 20 issue of *Nature* reporting a talk Eddington had given the previous month to the British Association for the Advancement of Science at Cardiff (Eddington 1920). In a non-technical discussion of "The Internal Constitutions of the Stars" Eddington had proclaimed that "*Probably the greatest need of stellar astronomy of the present day, in order to make sure that our theoretical deductions are starting on the right lines, is some means of measuring the apparent angular diameter of stars.*" Eddington stated that it was the surface brightness of a star that mattered in estimating diameters—forget the parallax, which might be, and probably then was in most cases, too small to be measurable. Undoubtedly what excited Hale was what followed: "*Thus the estimation of the angular diameter of any star seems to be a very simple matter. For instance, the star with the greatest apparent diameter is almost certainly Betelgeuse, diameter 0.051"* [arc-second]. *Next to it comes Antares, 0.043". Other examples are Aldebaran 0.022", Arcturus 0.020", Pollux 0.013". Sirius comes rather low down with diameter 0.007".*" Eddington went on to say that a true measurement would show how reliable were these estimates and then cited the ongoing interferometry work at Mount Wilson with successful fringe detection under poor seeing conditions. He concluded by declaring that "*I anticipate that atmospheric disturbance will ultimately set the limit to what can be accomplished. But even if we have to send special expeditions to the top of one of the highest mountains in the world, the attack on this far-reaching problem must not be allowed to languish.*" What better project endorsement could be had from such a preeminent personage as Eddington?

The die was cast. Hale informed Michelson that those diameters *"are so large we are planning to measure Betelgeuse with the 100-inch as soon as possible. Dr. Anderson will make the trial very soon, and will report the results to you."* In fact, the 100 inch observing log shows that November 16 and 17 had been assigned to Pease and Anderson for the "Micholson (sic) experiment." Regrettably, the promised report is not found in the Caltech Hale letters collection.

Eddington was not alone in estimating diameters of stars with the idea that they be targets for interferometric resolution. Among other topics in a long letter of 1920 May 5 to Walter S. Adams (1876–1956), Henry Norris Russell responded to Adams's previous letter—not in the archive—apparently about the enormous luminosities of Betelgeuse and Antares. And, in the same month that Betelgeuse would be resolved, Russell published a list of "The Probable Diameters of the Stars (Russell 1920)." Russell's diameter estimate for Betelgeuse was 40% smaller than Eddington's. It should be noted that the topic under discussion was in Walter Adams' wheelhouse. He and Arnold Kohlschütter (1883–1969), in another landmark MWO achievement, had developed the method of spectroscopic parallax already in 1914 by finding that certain line intensity ratios led to luminosity estimates consistent with those derived from trigonometric parallaxes of nearer stars (Adams & Kohlschutter 1914).

It was high time for a measurement of Betelgeuse's actual size.

3.2 A Historic Night at the 100 inch Telescope

MOUNT WILSON, 1920 December 13—Although Albert Michelson's name is always listed first in references to what Pease and Anderson achieved on this date, Michelson was not present for the observations. He would learn of what transpired in a letter from Hale, to which he replied on 1920 December 21: *"Thanks for the letter with the very satisfactory result of the first measurement of a star diameter. I feel very happy over it and congratulate you and Pease on the result."* At age 68, the Nobel laureate was still in his extended intellectual prime. In addition to getting at the problem to be successfully attacked on this night, Michelson was intensely engaged in planning ambitious experiments. His to-do list is enormously impressive. First came measuring the speed of light with unprecedented accuracy. Next on the list was repeating the measurement of the amplitudes and other properties of Earth tides that he and Henry G. Gale (1874–1942) had started at Yerkes Observatory in 1914 but had only published in 1919 due to the war interruption. Lastly, there was time to be invested in having another go at the Michelson–Morley experiment.

Francis Pease, a month short of his 40th birthday, had played a major role in the design and fabrication of astronomical optics and instrumentation since his 1901 graduation as a mechanical engineer from the Armour Institute of Technology— now the Illinois Institute of Technology (Figure 3.3). His first job was at Yerkes Observatory. There, he had sufficiently proven himself technically under the supervision of G. W. Ritchey as well as demonstrating sufficient skill as an observer at the

Figure 3.3. Francis Gladheim Pease (1881–1938) ca. 1930. (Image courtesy of the Observatories of the Carnegie Institution for Science Collection at the Huntington Library, San Marino, California.)

24 inch telescope, to be among those Yerkes staff Hale spirited off to Southern California in 1904. At Mount Wilson, Pease would become the principle designer in every telescope project that Hale undertook. The telescope Pease would use this night, along with the gangly instrument attached to its top end, was very much the product of his design. He would also be a major contributor in Michelson's forthcoming speed of light experiments. But, it is for the upcoming night's work that Pease is best known, along with his subsequent efforts to extend the resolution reach they were now attempting by a factor of two-and-a-half (Adams 1938).

The two astronomers had likely slept the day away in the observers' dormitory known as the "Monastery" (see Figure 3.4). Following a late Monday-afternoon supper in the Monastery dining room—sunset was about a quarter to five that day—Francis Pease and John Anderson walked uphill along the ridgeline toward the dome of the 100 inch. A long night lay ahead. The observing time had been assigned to Pease, but Anderson brought a second pair of eyes and, as we saw in Section 2.3, considerable experience in making visual interferometry measurements. December is far from being the ideal month for observing on Mount Wilson. Cloudy skies and mediocre to poor seeing too often detract from the potential presented by the long solstitial nights. 1920 December was no exception. The 100 inch telescope observing log is an austere document reporting dates, observer(s) name(s), instrument used, seeing conditions, dome opening and closing times, hours observed, etc. The log shows that Pease was given telescope time starting with the night of December 11. He was preceded by John C. Duncan, a Wellesley College professor who frequently visited Mount Wilson to

Figure 3.4. (Above) The Monastery Night Building, where Pease and Anderson were likely staying, had only been open for a year at the time of their 1920 December observing run to accommodate the additional astronomers using the newly commissioned 100 inch telescope. (Below) They would gather with others in the sitting room before going through the doorway for their early supper prior to walking up to the 100 inch telescope dome. (Author's photos.)

obtain images of galaxies. Duncan lucked out with the weather on only one of this three nights. After Pease came MWO staff member Gustave Stromberg, who was doing spectroscopy. He also fought a losing battle with the clouds. Pease was generously scheduled for seven nights. Of those seven, four would succumb to overcast skies, and the seeing would be rated "poor," "fair," and "good" on the nights of December 12–14 when Anderson came up the mountain to backup Pease's opinion regarding fringe disappearance. Their night assistant was John Kimple who entered "Experiment Micholson" (sic) in the log under the "Kind of Work" column. As we shall see, Kimple would do more than just point the telescope that night.

During their quarter-mile stroll uphill to the telescope dome, Pease and Anderson passed the site where Edward Emerson Barnard (1857–1923) had magnificently photographed the northern Milky Way in 1905. A bit further along was the very spot at which, a few years in the future, Michelson, with Pease's collaboration, would initiate his new speed of light experiment that was already well along in the planning stages. Closer to the top, they would walk by the 60 inch telescope that Harlow Shapley had employed in showing that the Sun was well away from the center of our galaxy—yet another landmark achievement from Hale's telescopes. (Shapley had just participated in the famous "Great Debate" with Heber D. Curtis (1872–1942) over the nature and size of the universe at the National Academy of Sciences the previous April.) After six or seven minutes, they would have completed the steady climb up from the Monastery, passed by the Galley (where they would later partake of night lunch), crossed over the footbridge, and entered the massive 100 inch telescope dome. The dome's shutters had already been opened by Mr Kimple to equalize the interior temperature to that outside. They would soon be at work.

With the 20 foot Michelson–Pease Interferometer mounted atop the 100 inch, as shown schematically in Figure 3.5, Pease and Anderson saw a telescope turned

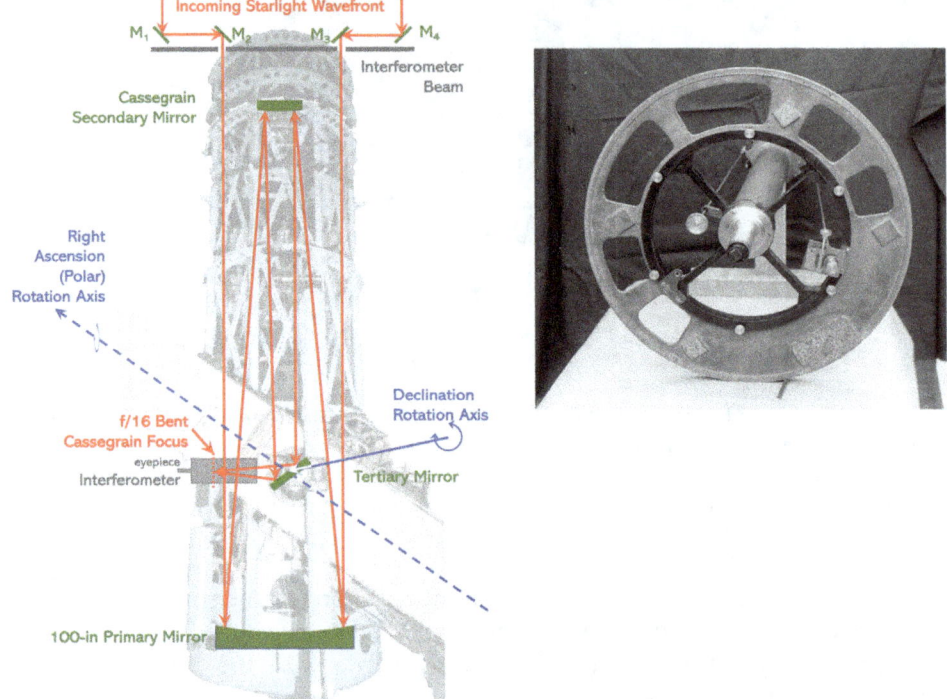

Figure 3.5. (Left) The 20 foot Michelson–Pease Interferometer required physically separated mirrors above the telescope's entrance pupil in order to have a baseline of $\overline{M_1M_4}$ that transcends the 100 inch upper limit to projected separation of slits in the converging beam. The steel beam is shown here as mounted N–S to provide a clear view of the telescope, but it was actually mounted E–W. The dome shutters open to a width of approximately 25 ft to provide adequate clearance. Objects would ideally be observed as they transited the meridian to benefit from the smallest zenith angle and hence least atmospheric dispersion. (Author's photo and diagram.) (Above right) The beam-combining apparatus with the observer's eyepiece lacks the rotation feature of the Anderson Interferometer. It is in the direct prism configuration used to search for fringes. (Photo courtesy of the Observatories of the Carnegie Institution for Science Collection at the Huntington Library, San Marino, California.)

into an odd-looking hybrid device not seen before in the history of astronomy. The light-gathering power of the 100 inch diameter primary mirror was not a feature of the telescope's configuration, as the primary, secondary, and tertiary mirrors all served the purpose of combining the "pencils" of light collected by mirrors M_1 and M_4 whose linear separation determined the system's angular resolution. These optical flats relayed their light beams to mirrors M_2 and M_3 to be sent to the primary mirror. From there, they were directed in two converging beams that would intersect near the Cassegrain focus were they not interrupted by slits and other optics for path length matching and image overlapping incorporated into the combining apparatus shown at the right of Figure 3.5. The full aperture of the telescope was covered with four quadrant frames to which canvas was attached. Four holes admitted the pencils of light from the 20 foot. To the north of the beam, two additional holes were cut to introduce light to the beam combiner from a separate and close slit pair that would have high-contrast fringes to serve as a comparison. For a full description of how the optical system worked, the reader is referred to the 1921 Betelgeuse paper by Michelson & Pease (1921) and Pease's 1931 excellent review article on "Interferometer Methods in Astronomy." Unlike the Anderson Double Star Interferometer shown in Figure 2.7, the focal-end assembly for the Michelson–Pease interferometer has, unfortunately, been lost.

Pease, with Anderson's confirmation, identified the M_1M_4 mirror separation at which the fringes disappeared, thus obtaining a measure of the star's diameter. Initial setup that night involved observations of β Persei (Algol) at a mirror separation of 81 in followed by reobservation of β Persei and γ Orionis (Bellatrix) at 121 in. Both baselines yielded fringes for the check stars. Turning next to Betelgeuse, no fringes were seen even after considerable time was spent making small path length corrections with the wedges in their search. Pointing then to α Canis Minoris (Procyon), fringes were immediately evident. At this point Pease and Anderson were convinced of Betelgeuse's resolution. They then had the baseline reset to 156 in, which would correspond to the visibility maximum in the second fringe lobe for Betelgeuse, which, as did α Canis Minoris, clearly displayed fringes. This was ultimate confirmation of the resolution of Betelgeuse. The remainder of the run was plagued by clouds and bad seeing, but they managed to look at a few more potentially resolvable stars—α Ceti (Menkar), α Tauri (Aldebaran), and β Geminorum (Pollux)—for which they found some indication that they were indeed on the downward slope of the visibility curve just as Michelson had demonstrated to Hale with the sketch in Figure 3.1. The detailed methodology of their observing sequence is described in the publication paper for this breakthrough (Adams & Kohlschutter 1914).

Once down off the mountain and back at the observatory's headquarters on Santa Barbara Street in Pasadena, Pease turned his attention to a careful reduction and analysis of his data. This yielded a value for the angular diameter of Betelgeuse of 0.047 arcsec. No attempt was made to attach an error estimate to it. Using the then available parallax for the star of 0.018 arcsec, Pease found a linear diameter for the star of 2.4×10^8 miles or *"slightly less than the orbit of Mars."*

The modern parallax for the star places Betelgeuse nearly seven times farther away than Pease considered, thus the star's radius would place its photosphere nearly out to Saturn. Subsequent observations of Betelgeuse with the 20 foot will be considered later in this book so as to compare them with the extensive compilation of modern results, some of which were obtained decades later from Mount Wilson.

Fully satisfied with the validity of the final diameter value, Hale lost no time in writing letters to astronomers with whom he had corresponded over the diameter challenge. Of course, Michelson must have been at the top of this list, and he replied on 1920 December 21 with the note quoted from at the beginning of this section. Michelson said that W. S. Adams had told him that Hale was now out of danger and returned to good health—apparently following one of Hale's all too frequent breakdowns—and that he was sorry that Hale would not attend the American Astronomical Society meeting to be held at the Ryerson Lab of the University of Chicago during December 28 through the 30. No doubt there would be plenty of buzz there about the diameter triumph from Mount Wilson. That gathering was conducted jointly with a meeting of the American Association for the Advancement of Science. In the proceedings of the meeting (Fox & Stebbins 1920), the first paper was by Adams and Alfred H. Joy (1882–1973)—Joy was the only Mount Wilson astronomer who attended—with a very relevant title of "Evidence Regarding the Giant and Dwarf Division of Stars Afforded by Recent Mount Wilson Parallaxes." On down the list was "The Diameter of α Orionis by Michelson's Interferometer Methods," by F. G. Pease, who did not attend the meeting.

An accounting of the AAS/AAAS meeting prepared for the Royal Astronomical Society of Canada gives special note that "on *Wednesday* [December 29] *morning the astronomers attended a joint session with the physicists on which occasion Professor Michelson described* his [author emphasis] *measurement of the angular diameter of Betelgeuse...* (Chant 1921)." Presumably it was Pease's paper that Michelson presented.

Buzz there must have been[2] as the *New York Times* reported on 1920 December 30—on page one, no less—with the headline "Giant Star Equal to 27,000,000 Suns Like Ours." The "*Special to the New York Times*" article occupied nearly all of column three on the front page and continued on page two with even more surface area. It was a paean to Michelson and included a fifth sub-headline reading: "By the Famous Physicist Whose Researches Laid the Foundation for Einstein's Theory of Relativity." The paper Michelson read at the meeting was entitled "The Application of Interference Methods to Astronomical Measurements," the original version of which had been published the previous summer and concluded pessimistically that the measurement of a stellar diameter would require a minimum baseline of 10 m (Adams 1938). The timing had been perfect for Michelson to insert the Betelgeuse announcement in a paper that no one

---

[2] A search of Newspapers.com on the word "Betelgeuse" for the interval 1920–1921 yielded 771 hits.

would have taken much note of otherwise. No doubt Michelson gave full acknowledgement of Pease's role in the triumph, but Pease was not mentioned in the *Times'* main article. Oh well, considering Michelson's stature in American science, it is not surprising that he got all the credit for the achievement. But, following that article, the *Times* printed another one, under the byline of George Ellery Hale, with the headline "Pease and Anderson Aided." Patient readers would find an interesting follow-up to the lead article with some solid discussion of the then frontlines of stellar astronomy. Ironically, the hometown *Los Angeles Times* carried a briefer account the same day as the *New York Times* extensive coverage but far less prominently on page 4.

For the first measurement of a stellar angular diameter, Michelson was awarded the Gold Medal of the Royal Astronomical Society. The Medal was presented at the 103rd General Annual Meeting of the RAS on 1923 February 9, a meeting that Michelson did not attend. Instead the presentation was handed to a representative of the American Embassy for transmission to the awardee whose study must have been festooned with such honors. But it is not the award that makes this relevant to us, it is instead the address given by the then RAS President Arthur Eddington—himself having played no small role in motivating Michelson's and Pease's achievement. It is worthwhile to quote extensively from the proceedings in which Eddington's address is given in its entirety (Eddington 1923).

Eddington told his audience that on "*On 1920 July 10, the great 20-foot beam was placed across the telescope. On December 13 success was attained, and the diameter of Betelgeuse was measured. We can guess something of the hopes and disappointments, the difficulties and exasperations of that interval. I think we should all like to know more of the patient struggle which preceded the great achievement. So, I have induced Dr. Pease, who was in charge of the instrument, to lift the veil a little. He has kindly taken great trouble to reply to my request, and what follows is extracted from his letter nearly in his own words.*"

And here follows Pease's wonderful description of the sequence of those events in the second half of 1920. "*On July 10 the beam was placed on the telescope and I attempted to line it up, but no adjustment of mirrors would seem to bring success. The following morning, I found that one of the images was formed by a slot between the mirror and the frame of the beam, so the next day we lined up the mirrors by sighting downward through the system to a lamp at the focus. That night we set on Vega, but could not get any fringes. The mirrors gave distorted images, and it was found difficult to adjust the paths to equality by moving the inner west mirror. Michelson was present this night, and we decided to have the mirrors refigured and to introduce a wedge system and compensator near the eyepiece to adjust for path difference. The next trip was in August. On the 6*th*, after much patient labour, using the new wedge system, I found fringes with the outer mirrors placed 18 feet apart. I left the instrument set for the assistant to observe the fringes. He saw them, shifted the wedges, and endeavoured to reset; but neither of us could find them again though we hunted for hours. These were the first fringes ever seen on any star using the Michelson periscopic system. On the 8*th* Michelson came up, and advised a step by step process which we find he uses with great success. So, with mirrors set 7 feet apart I soon picked up the fringes, and*"

Michelson then saw them for the first time. The next night we placed the mirrors 12 feet apart, and this night Mr. Hale saw them for the first time. On October 16, I spent half a night trying to pick up the fringes, only to find that one of the mirrors had slipped three inches. On the 17ᵗʰ Anderson was present, and whilst we obtained zero fringes from α Andromeda and α Tauri we could not get those from the outer mirrors set 6 feet apart. Fringes did show on Betelgeuse using two openings 7 feet apart. We had hitherto used a very fine motion for moving the wedges, and decided to substitute a quicker motion."

"On December 12, Anderson went up the mountain with me, and his previous experience (on Capella) was of greatest assistance. Using the revised wedge motion, we very easily picked up the zero fringes on Pollux, but did not get fringes from the outer pencils. On the memorable 13ᵗʰ we covered the end of the tube with canvas, except for the necessary openings, and placed a plank across the end of the tube for the assistant to sit upon and adjust the mirrors. We turned on Algol and endeavoured to pick up the fringes, moving the west mirror 1/8 mm at a time and running through with the wedge, but could find nothing. I was working with a 0.34-inch eyepiece and it fell out on the floor; I put in a 1-inch eyepiece—and the fringes were there. The mirrors were then 81 inches apart. With the telescope still on Algol we shifted the mirrors to 121 inches and obtained the zero fringes immediately and the outer fringes shortly. Observations on γ Orionis showed both sets of fringes with but slight adjustment of the wedge and compensator. Turning to Betelgeuse, the beautiful orange and black zero fringes were present; but no trace of the outer fringes showed though we looked again and again."

It was then time to adjourn to the Galley where Pease and Anderson would take their night refreshment of (quoting Eddington) *"cocoa and toast with their heads full of sums."* They were absorbing the irrefutable evidence presented by the persistence of fringes in γ Orionis at the same baseline as the observed full disappearance of Betelgeuse's fringes. The star had been resolved!

Pease went on... *"Presently Anderson speaks up and says, "It's a whale of a thing —let's see, as big as the orbit of Mars." Returning to the telescope, we set on Procyon and found the fringes immediately; the slightest motion of the wedge bringing them in. It was quite certain therefore that the instrument was in adjustment, and that the fringes on Betelgeuse were not present at 10 feet separation."* The 10-ft baseline and an adopted effective wavelength for Betelgeuse of 575 nm led to an angular diameter of 0.047 arcsec, or, in Eddington's analogy, *"about the size of a halfpenny fifty miles away."* What a nice story with which every astronomer who ever worked out the kinks in a new instrument at the telescope can identify.

But, wait a minute... Back up three paragraphs to where it talks about placing *"a plank across the end of the tube for the assistant to sit upon."* That would have been John Kimple up there sitting on a board in the dark at the top of a telescope canted off the zenith and moving during the night with only concrete and steel to catch him should he fall. At the latitude of Mount Wilson, the declination of Betelgeuse tilts the telescope 26.8° off the zenith at the star's transit, and it only gets worse from there. The 20 foot beam would later be equipped with an elaborate system for remotely positioning the flat mirrors, but for now and through the majority of its use, it had to be done by hand with someone perched atop the telescope. No mention of this hair-raising service was made in their *Astrophysical Journal* paper, and I recall hearing at one point that a graduate student was given this potentially deadly job. But, of course, there were then no graduate students at Mount Wilson. That it was Kimple who sat up there in the dark moving mirrors M_1 and M_4 was graciously made clear by Hale himself in the section of his annual report on "Interferometer Measures of Star Diameters" for the 1920 Carnegie Year Book. In a wonderful understatement, Hale says: "*Mention should be made of the patient and helpful assistance given by Mr. John Kimple in the troublesome adjustment of the mirrors* (Hale 1921)."

Because this was a key moment in the history of astronomy, John Elmer Kimple deserves a modest amount of attention.[3] Born in 1878 January in Allerton, Iowa, Kimple was 42 years old when he was up on that plank under circumstances depicted in Figure 3.6. The 1910 U.S. census shows him already at work at MWO as a house carpenter. Single, he boarded at the Mount Wilson hotel along with other unmarried observatory workers. When he registered for the draft in 1918, he had worked his way up at the "Solar Observatory" to being a "machinist," his job title as of the 1920 census when he was still living in one of the hotel cottages. Apparently as a result of the newly opened 100 inch telescope, Kimple was pinch-hitting as a

Figure 3.6. The author's fantasy as to how it might have looked with Kimple on the plank and Pease at the interferometer eyepiece of the 100 inch telescope on 1920 December 13. The telescope was then painted black, making for a gloomy scene even with some modest interior dome lighting turned on. (Author's photo of telescope with Pease inset courtesy of the Observatories of the Carnegie Institution for Science Collection at the Huntington Library, San Marino, California.)

---

[3] Unless otherwise noted, this information is from records accessible through Ancestry.com.

telescope night assistant apart from his machinist job. In the 1924 Carnegie Year Book, Kimple and Sam Jones are acknowledged for their work in "*planning an area of about 125 square feet to an average depth of one thirty-second of an inch*" in the trough holding the mercury for the telescope flotation system. This cured a problem with the drive that occurred when temperature changes would result in the float contacting the side of the trough (Hale 1924). More than a decade later, Kimple was devoting "*much of his time to the work on the velocity of light at the Irvine Ranch, assisting in the mechanical and optical adjustments and in the maintenance of the light-source during the measurements* (Hale 1931)." There is a photograph in the Caltech Archives of Kimple working on the declination drive enclosure of the 200 inch telescope in 1940 when he had been employed at Mount Wilson and then at Mount Palomar observatories for at least 30 years. He died in 1945 August at age 67.

I first encountered Kimple's name during the remodeling of the Kapteyn Cottage in 1995 for use by Georgia State astronomers who would be working on the mountain at the future CHARA interferometric array. A section of a board was found in the little house's attic rafters on which had been written in pencil: "*This house built by M.P. Eckleen, John Kemple* (sic), *painted by C.E. Tice during the year 1910 for Prof. Captyne* (sic)."

All this is a reminder that it takes others along with the astronomers to make for a great observatory.

3.3 Science from the 20 Foot Michelson–Pease Interferometer

After securing of the first directly-measured star diameter, the 20 foot continued to be used to work its way down the list of stars with predictably resolvable diameters. But time is precious, and Pease, who had been left by himself upon the return of Anderson and Merrill to their original subfields, had plenty of his own goals to fulfill. Michelson was intent on getting his new speed of light experiment up and running, and Pease was working on the successor to the 20 foot along with substantial other interests and duties that later in the decade would turn more toward 200 inch telescope planning and design. Nevertheless, Pease intended to continue efforts with the 20 foot with one of the eventual tasks being the motorizing of the optical flats to make the observations easier and more accurate, not to mention eliminating the risk to Kimple or another hapless night assistant of falling from the top of the 100 inch telescope.

In taking a closer look at the circumstances for the Betelgeuse measurement, one can gain a high regard for the care Pease took in homing in on the baseline at which the fringe contrast, i.e., visibility, went to zero. The two visibility curves in the top of Figure 3.7 show in red the curve corresponding to the 0.047 arcsec (47 mas) angular diameter Pease and Michelson settled on along with a modern somewhat larger value of 49.4 mas taken from Table 3.1 determining the blue curve. In the figure's lower plot, we zoom in to the two visibility nulls. If the blue curve is the "correct" one, imagine Pease's challenge in estimating the fringe contrast using "yell net" to communicate with Kimple while he moves the mirrors. At each mirror separation increment, Pease sought to determine whether the visibility had increased or decreased. The search

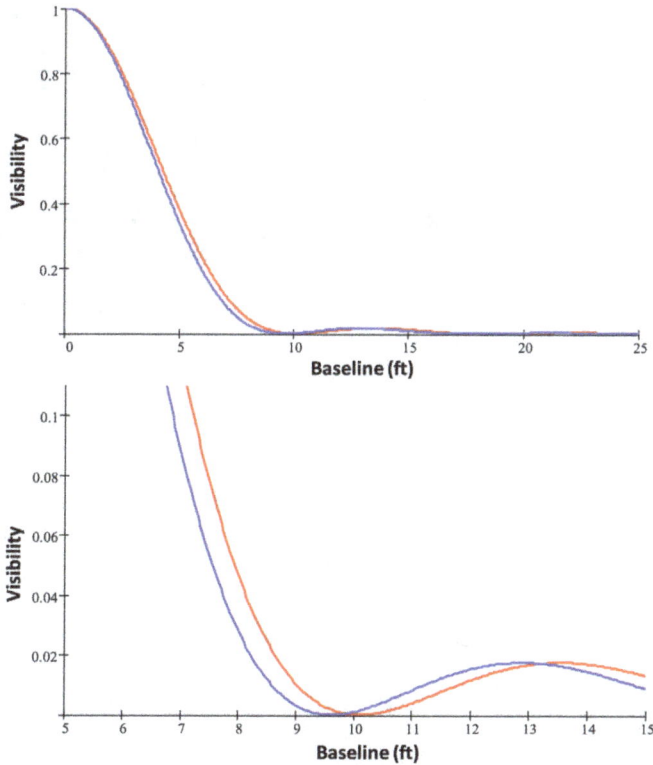

Figure 3.7. (Above) Visibility curves for Betelgeuse that would have been encountered by Pease and Anderson on 1920 Dec 13, with red being the curve for the diameter they determined of 47 mas and the blue corresponding to the modern diameter of 49.4 as listed in Table 3.1. (Below) Zooming in to the vicinity of the first visibility null for each curve, one sees that V is below 2% for baselines longer than about 8.5 ft. (Author's diagram.)

region is quite broad as the contrast drops below 2% beyond baselines of about 8 ft and then slowly increases again toward the next lobe in visibility at ~13 ft. This could lead to the assumption of having arrived at the visibility null when it actually occurs at a somewhat longer baseline. The result would be an over-estimation of the angular diameter. My first impression in looking at this was to marvel at how nearly right they got it! As with most undertakings, patience and perseverance are the key—for the observer as well as the poor fellow up on the plank.

Note that in Table 3.1, the diameters listed are those of disks of constant brightness from the center to the edge or "limb." Real stars diminish in brightness close to their limbs due to the semi-transparency of their outer atmospheres. The uniform-disk diameters Θ_{UD} that we use here are corrected to limb-darkened diameters Θ_{LD} by factors generated from theoretical models of stellar atmospheres and observed properties of the star. For more on this, see Appendix B.4.1.

Betelgeuse was observed on additional occasions into 1922 with some suspicion arising that its diameter was variable. In fact, Pease found the diameter to be as large as 54 mas in the late fall of 1921 compared with 34 mas in 1922. The brightness of the star was known to vary, and real changes in radius could imply that Betelgeuse is a

Table 3.1. Stars Observed in 20 foot Interferometer Diameter Program

Star	Θ_{UD} (mas) 20 foot	Ref. 20 ft	Θ_{UD} (mas) Modern	Ref. Modern	$\Delta\Theta_{UD}$ (mas) 50 foot – modern	Predicted VV (at X ft)
α Ori	47	1	49.4 ± 0.24 (800 nm)	B	−2.4 (4.9%)	—
	Possible increase	4				
	54, 34	5			4.6 (9.3%), −15.4 (31%)	
	47, 34	9			−2.4 (4.9%), −15.4 (31%)	
β Gem	Reduced V	2	7.13 ± 0.08	B	—	1(6); 0.95(10); 0.83(20)
α Tau	Reduced V	2	18.47 ± 0.54	B	—	0(24.1); 0.08(22)
	$V = 0$ at 22 ft est.	4			—	
	20	9			0.3 (1.5%)	
α Cet	Reduced V	2	11.33 ± 0.29	B	—	0.88(10); 0.58(20)
α Boo	23.7	2	19.47 ± 0.27	C	3.7 (19%)	—
	$V = 0$ at 22 ft est.	6			—	
	Unchanged	4			—	
	20	9			0	
α Sco	40	3, 9	39.85 ± 0.56	C	1.3 (3.3%)	Good agreement
	Unchanged	4			—	
β Peg	$V = 0$ >18 ft	3	16.00 ± 0.20	B	4.4 (23%)	0.79(10); 0.33(20)
	21	6, 9				0(29); 0.24(22)
γ And	Est. $V = 0$ at 30–40 ft	6	7.11 ± 0.10	C		0(64); 0.55(35)
α Ari	Est. $V = 0$ at 40–50 ft	6	5.86 ± 0.07	B		0(73); 0.46(45)
α CMa	V near 1 all seps.	6	5.82 ± 0.11	C		0(82); 0.97(10); 0.89(20)
α CMi	"	6	5.32 ± 0.05	C		0(90); 0.98(10); 0.91(20)
α Cyg	"	6	2.26 ± 0.06	C		0(203); 1.00(10),0.98(20)
α Gem	"	6				
α Oph	"	6	1.8 (oblate)	D		0(267); 1.00(10); 0.99(20)
α And	"	6	—			
α Peg	"	6	—			
α Cas	"	6	4.92 ± 0.05	B		0(89); 0.98(10); 0.91(20)
β Leo	"	6	1.25 ± 0.0.09	A		0(380); 1.00(10); 0.99(20)
γ Ori	"	6	0.70 ± 0.04	A		0(675); 1.00(10); 1.00(20)

ε Cyg	"	6	4.34 ± 0.04 (800 nm)	B	—	0(109); 0.96(10); 0.94(20)
α Aql	"	6	3.26 ± 0.05	C	—	0(142); 0.99(10); 0.96(20)
α Leo	V at times near 1	6	1.32 ± 0.06	A	—	0(359); 1.00(10); 0.99(20)
α Vir	"	6	0.85 ± 0.04	A	—	0(151); 0.98(10); 0.97(20)
α Lyr	"	6	3.08 ± 0.07	A	—	—
o Cet	56	7	36.0 ± 1.7	E	Modern meas. @ 2.2 μm. Star has extended shell
	47	9			
α Her	**30**	8,9	36.8 ± 0.6	C	Preliminary result later withdrawn

Notes.

References

20 foot publications:

1. Michelson & Pease (1921).
2. Pease (1921a).
3. Pease (1921b).
4. Pease (1922a).
5. Pease (1922b).
6. Michelson & Pease (1922).
7. Pease (1925c).
8. Adams (1925).
9. Pease (1931).

Modern publications:

A. Hanbury-Brown et al. (1974)—Narrabri Intensity Interferometer $\lambda = 443$ nm.
B. Mozurkewich et al. (1991)—Mark III Interferometer (Mt. Wilson) $\lambda = 450$ nm.
C. Mozurkewich et al. (2003)—Mark III Interferometer (Mt. Wilson) $\lambda = 550$ nm.
D. Zhao et al. (2009)—CHARAI Interferometer (Flagstaff) image—rapid rotator.
E. Ridgway et al. (1992), $\lambda = 2.2$ μm.

pulsating variable star. Pease's paper on these puzzling findings (ref. Hertzsprung 1906 in Table 3.1) included the last 20 foot results for Betelgeuse, so Pease did not resolve the question of this variation. The book was closed for α Orionis, at least for now.

Additional diameter work, the majority of which was qualitative rather than new, hard numbers, had also been undertaken and was published in relatively short papers or even as mere abstracts in conference proceedings. Pease's last attempt at a diameter with the 20 foot interferometer was for the long-period variable star o Ceti—the famous star Mira—which at maximum brightness came into the 20 foot's regime of sensitivity once the original 6 in diameter mirrors were replaced with 12 in optical flats in 1923. Ironically, the star was found to be over-resolved with the 20 foot, and the observations had to be made with the Anderson double-star apparatus.

Mira is a complicated star, and its appearance is likely wavelength dependent, so comparison with the small group of modern diameters is not straightforward. A 2016 result from the European Very Large Telescope Interferometer in the K-band infrared ($\lambda = 2.1$ to 2.3 μm wavelengths) found a diameter of 28.5 ± 1.5 mas (Wittkowski et al. 2016) while an earlier maximum-light K-band observation at the Infrared Michelson Array (IRMA) on Mount Hopkins, Arizona yielded a uniform-disk diameter of 36.1 ± 1.4 mas, which increased to 43.3 ± 1.7 mas when a limb-darkening model was applied (Ridgway et al. 1992).

In his 1931 review article (Pease 1931, p. 90), Pease took an unorthodox way of updating the list of diameters from the 20 foot without explanation other than noting that "*since that time* [i.e., the 1920 December 13 Betelgeuse triumph] *many observations have been made on a number of stars, the list below giving the results of those measures.*" His "list below" consists of seven stars that are folded into Table 3.1 and shown underlined in bold face. The result for α Her was given as preliminary when published in the 1924–1925 Year Book until it appeared in the 1931 review. In a 1938 exhaustive study of the stellar temperature scale (Kuiper 1938), Gerard P. Kuiper (1905–1973) mentioned a private communication from Pease that included a "subsequent unpublished paper" in which the α Her diameter was omitted. It seems that Pease was preparing the final word on his diameters, but its publication was cut short by his death. Thus, we are left with results for six stars lacking any precise epochs although they are presumably averages of multiple determinations. It is likely that Pease's 1931 review was intended as an interim report, but it is all that we have. Conceding that, let's take a look at the cumulative scientific return from the Michelson–Pease Interferometer.

Those results are summarized in the relatively busy Table 3.1. A total of 27 stars were observed by Pease in the diameter program. They are listed in the table in the order in which they were published. The Bayer designation comprises the first column, with the measured diameter or constraint on the value along with the relevant publication in columns two and three. The next two columns give a value for a "modern" diameter measurement and its source paper,[4] from which the difference is given between Pease's measurement and its modern counterpart along

---

[4] Diameter measures in Table 3.1 were obtained from the SAO/NASA Astrophysics Digital Data System online library referencing the VizeaR online data catalog JMMC Stellar Diameters Catalogue—JSDC. Version 2.

with a percentage difference in the sixth column. The final column in Table 3.1 gives baseline and visibility values (as calculated by the author) pertinent to each entry.

Because of the question of the variability of Betelgeuse, we cannot really judge the accuracy of the 20-foot results by comparison with the modern value because the diameter might itself be variable. And, we have already seen how complicated o Ceti (Mira) is. The supergiant α Sco (Antares) is a pulsating, semi-regular variable star that is difficult to compare with modern measurements in spite of its being in close agreement. Pease was ambivalent about α Her, which we will return to in Section 8.1, although measured at a much longer wavelength. In the end, the only fair "then–and–now" comparisons are for the stars α Tau (Aldeberan), α Boo (Arcturus), and β Peg (Scheat), which respectively show discrepancies of +22%, 0%, and +23% from modern values. Except in the case for Arcturus, where Pease got "the right answer," the two large positive residuals likely result from believing he was at the null before actually reaching it. Had Pease had the advantage given by these comparisons, it is probable that he would have understood their origin and developed a methodology for correcting for premature null landings.

The majority of the remaining stars inspected by Pease did not show diminishing visibility with increasing mirror separation, indicative of their being well below the resolution limit for the 20 foot. Pease's conclusions about these stars are supported by their now-known small angular diameters and the corresponding values of V given in column seven of Table 3.1.

Pease's "swan song" with the 20 foot instrument was the resolution of the multiple-star system comprising the star ζ Ursae Majoris—Mizar, the famous second star in the handle of the Big Dipper–with its near neighbor Alcor. Mizar has a visually resolvable companion, Mizar B at 14 arcsec separation, and both Mizar A and B have companions originally discovered through spectroscopically-revealed variable radial velocities. Mizar is thus a quadruple star. If Mizar is a physical system with its naked-eye neighbor Alcor, as their motions suggest, then together they comprise a sextuple star system (Mamajek et al. 2010).

On 1925 April 22, Pease sent Hale a letter that began "*The Algonquin and Iroquois Indian legends portray the seven stars in the handle of the Dipper as the "Hunters" and the four stars in the bowl as the "Bear," the former pursuing the latter. The second of the hunters, Mizar, is the "Chickadee" and he carries a "pot" Alcor along to cook the bear meat in.*"

"*Russell has already written you of his suggestion that I try to measure Mizar with the 20-foot beam and this note is to tell you that the hunter has made his kill and there is lots of meat to cook in the pot.*"

He then transitioned from the charming Native American mythology to an informal description of the rigors of observing that is worth repeating here: "*Astronomically speaking I spent the nights of the 15ᵗʰ to the 19ᵗʰ of April inclusive in successfully measuring the position angle and distance of the historical spectroscopic binary Zeta Ursae Majoris. The weather was not all one wanted; a "sea" was on a storm brewing and it's no jolly job running up with the crane and platform to rotate the whole cage and beam and then spend an hour or two picking up the fringes in the new position since the cage has been scraped in for one position only. But with all this Russell has just done a little figuring and finds that the measures very consistently fall on an orbit having an inclination of between 50° and 60°."* My Mount Wilson colleague Larry Webster believes that the expression *"scraped in"* probably refers to the fact that the top-end cage that supports the Cassegrain secondary mirror was never intended to be rotated. Lifting it with the crane and rotating the cage with the beam attached and re-clamping it to where it had not been "scraped in" was far from straightforward. Nevertheless, Pease had found a way to overcome Hale's concern that binaries required such rotation.

The following month, Pease published the results from those nights with the Chickadee in a brief note (Pease 1925a) followed by a somewhat more detailed report (Pease 1925b). He then set them aside for additional observations, which he only secured two years later. The final word on Mizar contained the cumulative measurements and Russell's second reduction of them to determine elements of the orbit that cannot be determined from radial velocity data (Pease 1927).

The modern orbit for the system, determined interferometrically with exquisite precision, has values for the orbital period 20.5453835 ± 0.000005 days, semimajor axis $a = 9.83 \pm 0.03$ mas, inclination of the orbital plane $i = 60.5° \pm 0.3°$, and nodal longitude $\Omega = 106.0° \pm 0.4°$ (Hummel et al. 1998). Henry Norris Russell, using essentially the same spectroscopically-determined period, found values for a, i, and Ω of 11.5 mas, 60°, and 102°—not bad. (See Appendix B.3.3 for more on orbital elements.)

By the time the Mizar effort was completed, so also was Pease pretty much done with the 20 foot as its successor—the 50 foot interferometer—was going into a shakedown phase that more than required the fraction of his time that he could devote to interferometry. We will see how Pease finally met his match with the 50 foot Pease Interferometer after the following diversion regarding the fate of the 20 foot Michelson–Pease Interferometer.

3.4 The 20 Foot Then and Now

Given its watershed contribution to stellar astrophysics, the Michelson–Pease Interferometer beam—the pièce de résistance of this first era of Mount Wilson interferometry—should be at the top of the list of important 20th Century astronomical instruments. Images of the beam from the years of its use are quite rare. In fact, the only early photograph of the 20 foot close up is the one published in the 1921 Michelson and Pease paper on Betelgeuse—an image, reproduced in Figure 3.8, that appeared in countless textbooks on elementary astronomy and optics over most of the century. Why was it not more photographed? In particular, why is there not another image of it atop the 100 inch? If one was ever taken, it has not come to light in the ensuing century.

Larry Webster, who has worked on Mount Wilson for decades and is the resident expert on the Observatory in all its aspects, found additional photos of the Michelson–Pease interferometer beam taken during the 1920s. They were hiding in plain sight in the backgrounds of photographs of the 100 inch during the period when the interferometer was stored between uses on a rack attached to the dome just below the crane that would fetch and carry it to the top of the telescope. Two examples are shown in Figures 3.9 and 3.10. After its usage (summarized in Table 3.1) tapered off in the 1920s, the beam was taken off the dome floor, wrapped

Figure 3.8. The 20 foot Michelson–Pease Interferometer Beam sits atop the 100 inch telescope in this iconic image. (Courtesy of the Observatories of the Carnegie Institution for Science Collection at the Huntington Library, San Marino, California.)

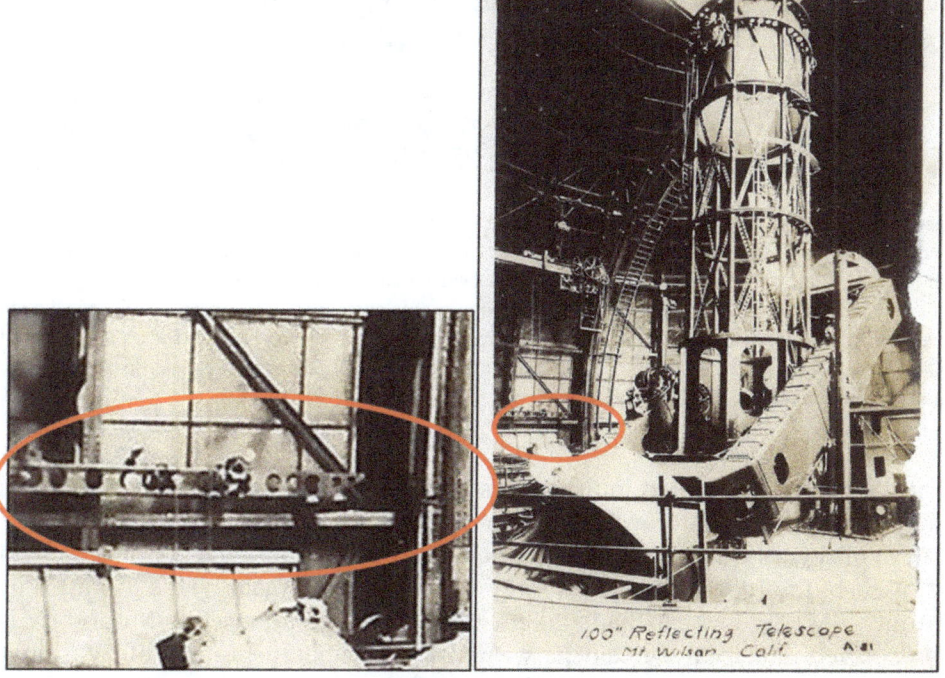

Figure 3.9. This undated vintage postcard shows the 100 inch telescope while the 20 foot was still in service. The interferometer beam can be seen stored on some sort of rack mounted to the bottom of the dome opposite the dome's open shutters. In that era, the telescope and the dome interior were painted black. (Public domain images.)

Figure 3.10. (Above) This fascinating image, taken on 1922 Dec 22, captures the telescope in an unusual orientation, pointing near the horizon just east of north. The interferometer beam is mounted to the dome at right, and an unknown worker sits on the observer's elevated platform with the north Cassegrain focal position just lower and to his front. (Below) A blowup of the interferometer shows that it had been equipped with motor drives for moving the two outer light-collecting mirrors but still has the original mirror cells that would be replaced in 1923. (Image courtesy of the Observatories of the Carnegie Institution for Science Collection at the Huntington Library, San Marino, California.)

in heavy burlap and stored in the rafters on the ground floor of the telescope dome where it would languish for nearly the remainder of the century.

My colleagues Bill Hartkopf and Brian Mason and I had the honor of being the first users of the 100 inch telescope following its refurbishment by the Mount Wilson Institute (MWI), which had taken over the operations of Mount Wilson Observatory under an agreement with the Carnegie Institution of Washington (CIW). A bit of background is in order here. Carnegie had announced the closure of the Observatory in 1984 in order to focus its resources on its dark sky site at Las Campanas, Chile. The non-profit corporation, the Mount Wilson Institute, was formed and assumed management of the Observatory in 1989 January (Vaughan 2005). Robert Jastrow was appointed MWI CEO and MWO Director in 1992 and launched a campaign to modernize the 100 inch and make it available to visiting astronomers with their own itinerant instrumentation while MWI developed an adaptive optics system for the telescope. More on this is forthcoming in Section 7.1.

As we had a speckle camera devoted to high-resolution studies of the orbital motions in binary star systems that we took to telescopes in both the northern and southern hemispheres, the MWI 100 inch telescope opportunity was right up our alley. Thus, in early 1993 December, we arrived to make speckle observations of binary stars with the upgraded 100 inch telescope.

During that first trip to Mount Wilson, someone mentioned to me that Michelson's 20 foot interferometer beam had been wrapped in canvas and stored in the rafters of the 100 inch telescope dome. Brian and I spotted the likely resting place for the beam. To see it close up, we climbed atop a photographic plate-loading

Figure 3.11. (Upper left) CHARA postdoc Brian Mason, now at the U.S. Naval Observatory, with the unwrapped 20 foot on 1993 Dec 1. (Upper Right) A CHARA Control Building image on 1999 May 20 shows the Exhibit Hall annex at right with the pit dug for the 100 inch cage. (Middle Left) Randy Peyton, crew chief for Sea West Enterprises, CHARA's prime contractor, directs the beam's extraction from the dome rafters on 1999 May 20. (Middle right) Visitors in the Exhibit Hall ca. 2015. (Lower left) The 20 foot was on loan to the American Museum of Natural History when CHARA held its annual meeting there in 2007. (Author's photos.)

room over which the beam had been inserted. Recalling that the 20 foot was in use at the time Howard Carter came upon the tomb of Tutankhamun in the Valley of the Kings, I couldn't help feeling a bit like the Egyptologist as we carefully pulled back a portion of the canvas to reveal "beautiful things." Beneath the canvas, which was, of course, covered in decades of dust, were the leavings of countless generations of mice that had nested inside the resting place of the beam. Fortunately, the insulation of the electrical wiring had apparently not been to their taste.

It was clear to us that here was an historic instrument with unanticipated sophistication of design and fabrication that should be appropriately exhibited rather than ignominiously stashed in this dome to serve as cozy rodent residential quarters. That opportunity would come when Mount Wilson was selected as the site for the CHARA Array—a process described in Section 8.2. As construction contracts were being awarded, it became clear that the project could afford a central office/control

Figure 3.12. (Upper left) The M_2, M_3, and M_4 mirrors are seen on the beam with the electric motor added to control the M_1–M_4 separation and the orientation of the M_2 and M_3 mirrors from the observer's position. (Upper right) The observer's hand paddle. (Lower left) The M_2 mirror and its assembly. (Lower right) The M_1 mirror and assembly as it is attached to its drive screw, which is directly coupled to that for M_4. (Author's photos.)

building of modest size. It had been my goal to attach to that building an exhibit hall to describe CHARA facilities and science to the many visitors who come to this historic mountain every year. A gift from Georgia State University alumnus Jack Kelly enabled that wish, and negotiations with Bob Jastrow led to our gaining permission to put Michelson's beam on exhibit. Nils Turner, who had obtained his doctorate in our group and stayed on as a mountain staff member, had the inspired idea of mounting the beam atop the prime focus upper cage, which had been sitting out in the elements for decades, to give the impression that the upper end of the telescope, with the beam attached, was looming up out of the floor on a slight tilt.

And thus, this wonderful relic from the early application of interferometry to stellar astrophysics found a home. Figures 3.11 and 3.12 present photographs to further tell the story.

References

Adams, W. S. 1925, Interferometer Measures of Stars, Carnegie Institution of Washington Year Book No. 24 (Washington, DC: Carnegie Institution of Washington), 116

Adams, W. S. 1938, PASP, 50, 119

Adams, W. S., & Kohlschutter, A. 1914, ApJ, 40, 385

Chant, C. A. 1921, JRASC, 15, 59

DeVorkin, D. H. 1975, JHA, 6, 1

DeVorkin, D. H. 1977, in IAU Symp. 80, The HR Diagram, Memory of Henry Norris Russell Dudley Obs. Rep. No. 13, ed. A. G. Davis Philip & D. DeVorkin (Albany: Dudley Observatory), 61

Dick, S. J. 2013, Discovery and Classification in Astronomy: Controversy and Consensus (Cambridge: Cambridge Univ. Press), 92

Drake, S. 2001, Dialogue Concerning the Two Chief World Systems: Ptolemaic and Copernican (New York: The Modern Library)

Eddington, A. S. 1920, Natur, 106, 14 (also published in Obs **43** 341)

Eddington, A. S. 1923, MNRAS, 83, 309

Fox, P., & Stebbins, J. 1922, PAAS, 4, 199

Hale, G. E. 1921, Interferometer Measures of Star Diameters, Carnegie Institution of Washington Year Book No. 20 (Washington, DC: Carnegie Institution of Washington), 278

Hale, G. E. 1924, Construction Division, Instrument Shop, Carnegie Institution of Washington Year Book No. 22 (Washington, DC: Carnegie Institution of Washington), 213

Hale, G. E. 1931, Construction Division, Instrument Shop, Carnegie Institution of Washington Year Book No. 30 (Washington, DC: Carnegie Institution of Washington), 219

Hanbury-Brown, R., Davis, J., & Allen, L. R. 1974, MNRAS, 167, 121

Hearnshaw, J. B. 2014, The Analysis of Starlight: Two Centuries of Astronomical Spectroscopy (2nd ed; Cambridge: Cambridge Univ. Press), 139

Hertzsprung, E. 1905, Z Wiss Photogr, 3, 442

Hertzsprung, E. 1906, Z Wiss Photogr, 4, 53

Hummel, C. A., Mozurkewich, D., Armstrong, J. T., et al. 1998, AJ, 116, 2536

Kuiper, G. P. 1938, AJ, 88, 429

Mamajek, E. E., Kenworthy, M. A., Matthew, A., Hinz, P. M., & Meyer, M. R. 2010, AJ, 139, 919

Michelson, A. A., & Pease, F. G. 1921, ApJ, 53, 249

Michelson, A. A., & Pease, F. G. 1922, PAAS, 4, 375

Mozurkewich, D., Johnston, K. J., Simon, R. S., et al. 1991, AJ, 101, 2207

Mozurkewich, D., Armstrong, J. T., Hindsley, R. B., et al. 2003, AJ, 126, 2502

Pease, F. G. 1921a, PASP, 33, 171

Pease, F. G. 1921b, PASP, 33, 204

Pease, F. G. 1922a, PASP, 34, 183

Pease, F. G. 1922b, PASP, 34, 346

Pease, F. G. 1925a, PASP, 37, 155

Pease, F. G. 1925b, PNAS, 11, 356

Pease, F. G. 1925c, PAAS, 4, 375

Pease, F. G. 1927, PASP, 39, 313

Pease, F. G. 1931, ErNW, 10, 84

Pickering, E. C. 1880, PAAAS, 16, 1

Ridgway, S. T., Benson, J. A., Dyck, H. M., Townsley, L. K., & Hermann, R. A. 1992, AJ, 104, 2224

Russell, H. N. 1913, Obs, 36, 324

Russell, H. N. 1920, PASP, 32, 307

Russell, H. N. 1918, in Publications of the American Astronomical Society, Vol. 3, ed. P. Fox, & J. Stebbins (Washington, DC: The American Astronomical Society), 22

Vaughan, A. H. 2005, Notes on MWI Founding, unpublished, 5 pp

Wesselink, A. J., Paranya, K., & DeVorkin, K. 1972, A&AS, 7, 257

Wittkowski, M., Chiavassa, A., Freytag, B., et al. 2016, A&A, 587, A12

Zhao, M., Monnier, J. D., Pedretti, E., et al. 2009, ApJ, 701, 209

Seeing the Unseen
Mount Wilson's role in high angular resolution astronomy
Harold A McAlister

Chapter 4

A Bridge Too Far

4.1 The 50 Foot Pease Interferometer

It would have been out of character had Hale not immediately begun planning a follow-up instrument to the successful 20 foot Michelson–Pease Interferometer. The 20 foot had penetrated the parameter space of resolution but was limited to only a handful of stars by its baseline and the insensitivity to fainter stars of its 6 in mirrors. A longer beam with larger mirrors on the 100 inch telescope would reach fainter and smaller stars, but its weight would have been prohibitive. Even if that were not a problem, access to more stars would have put great pressure on the scheduling of the 100 inch resulting in less time for other pressing science programs. An independent interferometer could be made much larger than anything mountable on the 100 inch and would free the telescope from all interferometry, a bonus mentioned explicitly in the 1929–1930 Carnegie Year Book (Adams 1930). All of this argued unquestionably for a stand-alone, dedicated interferometer, and that is what Hale came up with.

In a letter to Francis Pease of 1922 April 5, Hale, pursuant to a previous conversation with Pease, provided *"some notes on the 50-foot interferometer I described to you."* On a single page, Hale outlined the features as he envisioned them for the pier, polar axis, trussed beam to provide the 50 foot baseline, and rails for the mirror carriages on the beam. The beam would be made from standard structural steel components. Hale proposed a 30 inch diameter concave mirror to play the role of the 100 inch primary mirror in the 20 foot arrangement to initiate convergence of the two "pencils" of light toward the combining eyepiece at the observing position. The interferometer would teeter-totter an hour or so either side of the meridian about the polar axis in a plane parallel to the Earth's equator. As such, it could not point to declinations north or south of the celestial equator. To accomplish this necessity, the M_1 and M_4 mirrors must rotate to the required declination of each target star. A schematic diagram of the optical path superimposed on the 50 foot beam is shown in Figure 4.1.

doi:10.1088/2514-3433/abb4dech4

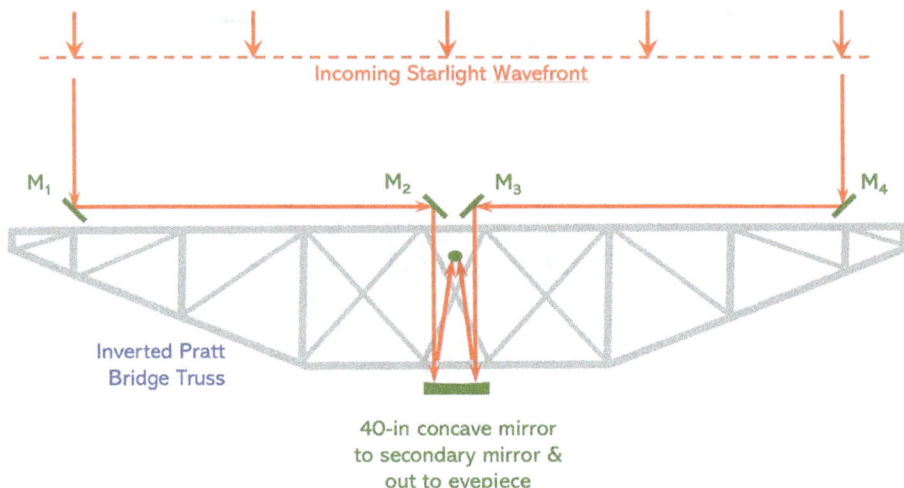

Figure 4.1. The optical path of the 50 foot interferometer is essentially identical to that for the 20 foot working in concert with the 100 inch telescope. Movable mirrors M_1 and M_4 direct starlight to fixed mirrors M_2 and M_3, which then relay the light to the telescope optics for convergence toward the eyepiece. The secondary mirror rotates the beam perpendicular to the plane of this diagram to give a convenient observing position. Note that the primary mirror diameter was larger than what Pease originally suggested it could be if fabricated from an existing, and already purchased, piece of glass. (Author's diagram.)

Hale asked Pease to send him a preliminary design with "*changes as you may wish to make, to show to Dr. Merriam.*" This, to a large extent, summarized a complementary relationship in which Hale would turn a new and perhaps visionary concept over to Pease to fill in the details and make it work. The performance of a new interferometer—or more precisely, an interferometer telescope—was important but equally so was its price tag. Hale knew the cost of this instrument was far from marginal and required some additional funding from the Carnegie Institution. In addition to the cost of the interferometer, there would also be the expense of its housing, which, having to enclose a telescope in excess of 50 ft in length, was no small structure. Hale felt he had to scrimp wherever he could and even asked Pease about the feasibility of covering "*only the sensitive parts and leave the beam exposed.*" That idea did not get a foothold.

Pease jumped on it, and only twelve days later telegraphed Hale at the National Research Council in Washington that prints of his design drawings were in the mail. Figure 4.2 shows one sheet from those plans of a North–South cross-sectional view of the beam in its enclosure. Pease also summarized the budget estimate for shop time, optics fabrication, materials, concrete, transportation, electrical, and housing. The large beam-combining mirror could be made from the back surface of an unused 36 inch flat mirror, yielding a tidy savings. The two dominant costs were the shop time at $4500 and the roll-off shed enclosure at $4000. The total estimated cost was $13,200—about $200,000 today. Hmm, seems like a real bargain—maybe too much of one?

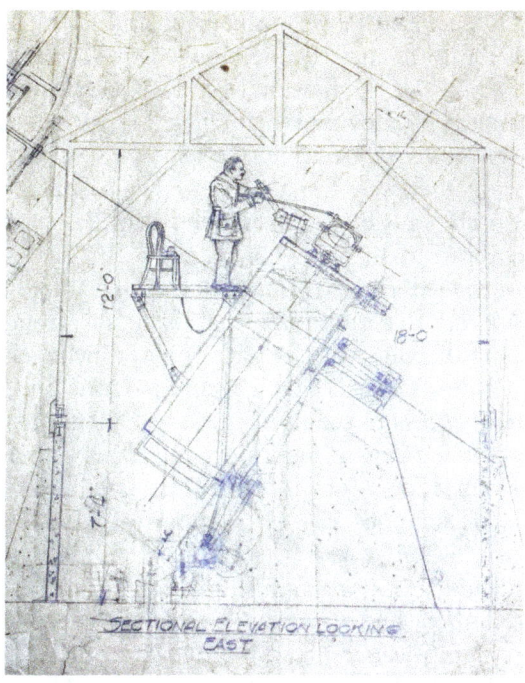

Figure 4.2. This closeup of a portion of Pease's 1922 April 17 "Proposal Drawing," which he mailed to Hale the next day, gives a view looking east through the plane of the beam's long dimension. The observer works from a platform attached to the north face of the beam's central section, a surprisingly unwise choice because of the added load on the drive system and, especially, because of the effect of the observer's moving around would have on the beam's pointing stability. He later put the observer on the south side and not supported by the interferometer itself. (Author's photo of the original drawing.)

Seemingly unaware of the maturity of this planning, Michelson wrote a note to Hale on 1922 April 18 with the preamble that he had *"been thinking over the matter of the larger interferometer and have perhaps not quite understood the scheme you proposed from the sketch."* This was Michelson's polite manner of saying he had a better idea. He continued. *"It seems to me however that something of the order of the mounting of the 100 inch would be quite feasible and not too expensive—as per inclosed* (sic) *sketch."* He went on to describe an approach that had occurred to him of varying the baseline by angular instead of linear displacement of the light-collecting mirrors. Inclining a baseline to the direction to a star, if that is what he meant, to foreshorten the baseline would produce path-length matching challenges, but there is no mention of more details nor does the sketch appear in the Caltech archive. Before changing to another subject, Michelson cheerfully closed out this topic with: *"If this does not develop some unforeseen difficulties perhaps you would like to consider it before going ahead."* One wonders what went through Hale's mind when he read this. Did he discuss Michelson's idea with Pease? Maybe not as they were then on opposite sides of the country.

Hale replied on May 8 without mentioning Michelson's alternative design ideas, and it was clear that the 50 foot interferometer configuration was pretty much nailed down conceptually. He enclosed Pease's blueprint and letter *"giving some details."* Hale was enthusiastic about the instrument's low cost—*"about that of a standard 12-inch equatorial telescope"*—and the fact that the most expensive component was the shop time for which salary money was recurring. He invited Michelson to look it over and send any *"criticisms that may occur to you as soon as possible, as we are checking and completing the design and wish to begin construction."* There is no response to this in the archive. Why the rush, one naturally wonders? Was the faster–better–cheaper snake about to strike?

I had thought that was it until during a telephone interview of Doug Currie pertaining to his work on the mountain that is forthcoming in Section 6.3, he mentioned in passing that he had a binder with copies of Michelson and Pease letters and other materials. When I heard that I figured it was stuff that is now in the Caltech archives. I didn't press to see it when he first mentioned it, but I brought it up again as we were winding down. He kindly mailed it to me.

Turns out Currie's collection is not in Caltech's Hale archive, likely due to Hale not being in the loop. It includes Pease's sketches of Michelson's concept for an alternative to the cantilevered beam the eventual 50 foot. Without this newly appearing correspondence, I had assumed that Pease had considered Michelson's plan before deciding that he had to go with Hale's less-costly design concept. In actuality, it was more than just a budgetary consideration.

Among the newly found Michelson letters is one to Pease dated 1921 January 27—15 months prior to his letter to Hale about his *"thinking over the matter of the larger interferometer"*—in which Michelson wrote: *"Regarding the large interferometer, I would point out that a mounting east–west would have the disadvantage of great sensitiveness to errors of the driving clock, which would not be the case if the mounting were in the meridian. Also it would seem to me that a long axis (to which the interferometer could be clamped for any particular declination) would be better than a short central polar axis."*

After bowing to Pease's greater observational experience, Michelson continued to press his case. *"The chief difficulty I imagine in the future work in this direction will be the disarrangement of adjustment where the distances between the outer mirrors is to be continuously changed, and the enclosed photograph is intended to show how this may be avoided by making this motion about a pivot instead of on a slide."* Before switching to the topic of the velocity of light experiments, he conceded that his approach may be of greater difficulty, i.e. more expensive, than *"for the present style."*

Figure 4.3 shows Pease's sketch of Michelson's concept of having arms swinging out from a polar axis to provide outer mirror separations differently than moving them on a beam. There is a 1921 January 31 note to himself on which Pease wrote the following:

"1. *Excellent for path same* [i.e., pathlength matching]
2. *Bad for declination changes, requiring <u>2</u> rotations as against 1 on straight beam*
3. *Bad requiring counterbalance*
4. *Good requiring only small housing*
    F.g.P."

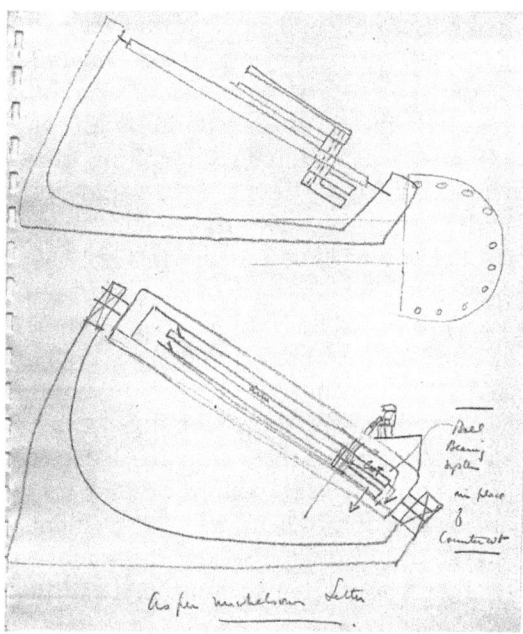

Figure 4.3. (Below) Pease sketched Michelson's concept of swinging mirrors out in opposite directions from the central polar axis using bearings below the observer's position. (Above) Counter weights on the opposite side of the mirror support structure were required for balance. (Courtsey of Doug Currie, Observatories of the Carnegie Institution for Science Collection at the Huntington Library, San Marino, California.)

In the absence of the Currie letters, I had concluded that Pease settled for the cantilevered beam design over Michelson's more robust but costlier concept. I now realize that Pease thought Michelson's concept to be inferior. It seems though that Pease never expressed that opinion to the august Michelson as the subsequent attempt by the physicist to change Hale's mind would imply.

One developing scheduling factor that would further lock in Hale's interferometer design came from a completely unrelated direction. The path of the upcoming 1923 September 10 total solar eclipse would sweep down the Southern California coast and into Mexico. Mount Wilson astronomers wanted to take advantage of this nearby eclipse and planned a number of experiments that all required pointing at the Sun. One of those would be an attempt to measure the displacement of starlight around the Sun's limb arising from Einstein's General Relativity Theory. Eddington had already triumphantly succeeded in measuring the effect in the 1919 May 29 eclipse, but more accurate values would further refine the confirmation for a more accurate check on GR. As a result of this and other experiments, a means of attaching the assemblage of instruments on a common mount for pointing at the Sun would be very useful. It occurred to someone—Pease?—that if they had the central section of the 50 foot beam assembly in hand, it would very nicely serve that purpose while also kickstarting the new interferometer project. That did it. The 50 foot, or at least the central rectangular box section of its beam, would likely get underway on the basis of the Observatory's full-throttle response to an eclipse in its back yard.

In the fall of 1922, Hale was in England recuperating from one of his periods of illness while also anticipating a trip to Egypt, a place and subject of deep interest to him (see Figure 4.4) and where Howard Carter would come upon Tutankhamun's tomb on November 22. Pease sent Hale a letter on 1922 October 11 in care of the Royal Astronomical Society at Burlington House, London, to bring the boss up to date on various of Pease's projects. The 50 foot was first on the list, and Hale learned that the final drawings of the beam were being done and would soon be sent to the machine and optical shops *"as a major order"* to ensure that the center section of the interferometer beam was available for the upcoming eclipse. Pease also mentioned an important additional feature indicative of the hardcore engineer inside him. He planned to use a single, common drive to point the two outer mirrors at declinations away from the celestial equator in order to help preserve path length and diminish pointing discrepancies that would otherwise inevitably make the observer's job an unpleasant one. One wonders how often such insights over Pease's years at Mount

Figure 4.4. Hale had a keen and long-held interest in Egypt and decorated the interior and exterior of his personal solar laboratory on Holladay Road, just south of the Caltech campus, with imagery associated with the Sun god Ra. In this photo, UCLA solar physicist Roger Ulrich and Mount Wilson Institute director Robert Jastrow stand at the entrance to the Hale Laboratory on 1999 April 13, with the radiating hands of the solar disk Aten descending over the entrance. Another of Hale's and Michelson's faculty colleagues at the University of Chicago was the great Egyptologist James Henry Breasted (1865–1935). Hale and Breasted corresponded (at the uncharacteristic "George" and "Jim" level) frequently and at length until Breasted's untimely death. (Author's photo.)

Wilson measurably improved an instrument's workability unbeknown to the concept's originator. Pease updated that report on 1922 October 27 with news that the beam would go to the shop at the beginning of December.

Pease's design was quite advanced by then, although it would continue to evolve in the interim up to its ultimate construction. The first relatively detailed description of the plan for the 50 foot was already in the literature by late 1922. The 1922 October 7 issue of *Nature* appeared in which Hale had written a two-page article under the title "A Fifty-foot Interferometer Telescope." The piece had originally been presented at the September 11 meeting of the British Association for the Advancement of Science and was read in Hale's absence by Oxford astronomy professor Herbert H. Turner (1861–1930). Turner showed photographs—presumably as lantern slides—of a model of the interferometer Pease had produced as a funding enticement to the Carnegie management. Figure 4.5 displays a photo of that model along with an image of the eventual interferometer in the flesh.

Walter Adams, who was effectively the observatory director by then, reported to Hale on 1923 January 10 that the interferometer model, along with a model of the entire mountain, and a 4 in diffraction grating had *"attracted much attention"* in

Figure 4.5. (Upper) As would be seen from the north, Pease's 1922 model of the 50 foot interferometer shows a gossamer structure for the beam, the central rectangular portion of which would be used at the 1923 Southern California total solar eclipse. (Lower) The actual interferometer is seen from the south in this photograph on 1930 March 7, probably taken by Ferdinand Ellerman. Pease is the taller of the two men standing on the observer's platform. (Images courtesy of the Observatories of the Carnegie Institution for Science Collection at the Huntington Library, San Marino, California.)

Washington. However, the Institution was pressed for funds so there would be no new resources for its departments—they must continue to pull off the 50 foot with existing dollars. The central section of the interferometer beam had become essential to the eclipse effort, which would be carried out at Point Loma, in San Diego. An Adams-to-Hale letter of 1923 March 8 informed the observatory's founder that: *"Preparations for Eclipse work are progressing in great shape, almost everyone knows what he will do and practically all details for such parts of the interferometer that are to be used at the eclipse are in the shop. Sam Jones and Johnny Kimple have the central portion of the Bridge well under way. It is going to keep us hustling though to be in ship-shape by September."* The use of the word "bridge" will below become clear to the reader.

And, hustle they did. On the first day of August, Adams told Hale he wished that *"you could see the extraordinary structure back of the office building today, bristling with tubes like a German Big Bertha."* The critical beam section was complete, and the various spectrographs and direct cameras were mounted on it in the trial configuration as shown in Figure 4.6. They would be ready for the eclipse through the all-hands-on-deck efforts of most of 1923.

Sadly, the weather gods would not smile favorably on the astronomers, and clouds covered the region on 1923 September 10. All the resources in personnel and funding expended for eclipse work came to nothing—except the center part of the interferometer "bridge" was in existence. Now that the shop was free of all the work given it for the solar effort, momentum toward the 50 foot could continue—except that it didn't.

Figure 4.6. The 50 foot central beam section is nearly obscured by the assemblage of instruments mounted to it in this trial assembly at the Observatory headquarters in Pasadena. (Image courtesy of the Observatories of the Carnegie Institution for Science Collection at the Huntington Library, San Marino, California.)

The 1923–1924 Carnegie Year Book indicated that *"comparatively little work has been done on the 50 foot interferometer in the instrument shop, since the mechanical part is now in advance of the optical work required for completion of the telescope."* But, there was also nothing in the optical shop for the interferometer.

Here is a brief summary from a succession of Year Books on progress toward completion:

1924/1925—A 40 inch mirror blank for beam convergence was received, and the four flat mirrors to feed it were already in stock.

1925/1926—Construction of footings and foundations for the interferometer was delayed for lack of mountain water to mix concrete. The 40 in mirror was being parabolized and interferometer operations were anticipated in Summer 1927.

Pease and Anderson also investigated a technique using elliptically polarized light from an interferometer that had been put forward in 1912 by the Russian physicist Sergei Ivanovich Pokrovsky (1872–1939) (Pokrovsky 1912).[1] Three years later, Pokrovsky published a practical means of designing a stellar interferometer to employ his concept (Pokrovsky 1915). His work is remarkable in that there was little stirring in the field in the first two decades of the 20th century until Michelson accepted a Mount Wilson research associateship. Unfortunately, Pokrovsky's method had a fatal flaw. It went unnoticed at Mount Wilson until brought to Pease's attention by a visiting astronomer. Pease promptly built an interferometer utilizing calcite crystals to induce the prescribed polarization but found that the optical effect did not materialize. In looking over the concept's theory, Anderson found *"the existence of serious theoretical difficulties* (Adams 1926)."* The flaw was also noticed and described by none other than Eddington (1926).

1926/1927—Observations with the 20 foot Michelson–Pease interferometer were discontinued in anticipation of the 50 foot.

1927/1928—Optical figuring of the 40 in mirror was completed.

1928/1929—In spite of progress on many aspects of the 50 foot, the telescope was not yet operational, but the *"difficult work"* of aligning the tracks mounted to the top surface of the bridge beam was underway. This would not be the last use of the word "difficult" in describing this radically new facility.

1929/1930—Six years had passed since the core section of the beam had been completed for the eclipse, and at long last *"Preliminary tests of the 50-foot interferometer were obtained during the spring months and seem to leave little doubt of the successful performance of this unique and powerful instrument. Fringes have been observed without difficulty at nearly the full separation of the outer mirrors, and the results obtained are in good accord with those to be expected from an extension of the values measured with the 20-foot interferometer on the 100-inch telescope."* Pease had seen fringes at mirror separations of 25 and 34 ft. There is no report as to the separations at which he did not see fringes, but there was *"no reason to believe that*

---

[1] Biographical information regarding S. I. Pokrovsky was kindly provided to the author by Andrei Tokovinin of the Cerro Tololo Inter-American Observatory who wrote of him in his 1988 book *Stellar Interferometers.*

Figure 4.7. At the 50 foot Pease Interferometer on 1930 March 3. (Upper) A cropped closeup of the lower photograph from Figure 4.5 clearly shows Pease with his left hand resting on the beam truss close to the eyepiece and push-button control panel. The west light-collecting 15 in flat mirror is seen at upper left while the two downward directing relay mirrors M_2 and M_3 are in the housing to Pease's left, just above the control panel. (Lower left) Seated on the observer's platform on the south side of the instrument, Pease is imitating fringe observing with his right hand on the focusing knob for the eyepiece. The control panel is an easy reach from there. The M_4 mirror is behind and to Pease's right. (Lower right) Looking at ground level from the west, one sees the telescope primary mirror cell just above the technician, the telescope's drive sector and motor to his left, the instrument's inclination parallel to Earth's rotation axis, and the stairway leading to the observing platform. (Images courtesy of the Observatories of the Carnegie Institution for Science Collection at the Huntington Library, San Marino, California.)

they will not be well seen up to the full separation of 50 feet." Figure 4.7 shows several photographs of the Pease Interferometer taken soon after its completion.

Measurements had been made under moderate seeing of the diameters of Betelgeuse and Arcturus. Both stars were within the reach of the 20 foot with which the new measurements agreed. Some flexure effects were seen (and fixed) as the beam pointed away from the meridian.

1930/1931—"*The difficult work of adjusting and using for measurement the 50-foot interferometer has been carried on by Pease from time to time during the year, but the demands of other investigations, especially that on the velocity of light, have interrupted the regular use of the instrument.*" There's that "difficult" again. Pease had detected fringes out to 44 ft. The first actual diameter measurements from the 50 foot was reported as about 0.040 arcsec for Betelgeuse and 0.016 arcsec for β Andromedae, the latter being a new result.

1931/1932—Although Pease had made a few observations with his interferometer, he was tied up with Michelson's in-vacuum speed of light experiment then being conducted at a site well away from Mount Wilson. Some bracing to the 40 in mirror support had been added.

1932/1933—Michelson's speed of light work had been completed, so Pease could direct more effort toward measurements with the interferometer, which had had some additional minor bracing installed to improve rigidity. New results were obtained for: α Ceti (Menkar), fully resolved at 40–42 ft corresponding to a diameter 0.0115 arcsec; α Scorpii (Antares), 15–17 ft, 0.029 arcsec; and, α Boötis (Arcturus), 25 ft, 0.019 arcsec.

1933/1934—Although Pease invested some time at the 50 foot, no new results were reported.

1934/1935—Diameters were measured for three stars: γ Aquilae, 55 ft, 0.0084 arcsec; ε Pegasi, 55 ft, 0.0084 arcsec; and, α Scorpii (Antares), 11.5 ft, 0.041 arcsec.

1935/1936—"*Pease has given considerable time to observations with the 50-foot interferometer, but the extremely exacting requirements of this instrument in the way of stellar definition have prevented many measurements of the diameters of stars.*" This likely only partially conveys the frustration Pease was having in his attempts to extract usable results from his interferometer. He was able to see fringes disappear for Betelgeuse at 13 ft giving a diameter of 0.035 arcsec, an unusually small size compared with all previous results for the supergiant.

1936/1937—This year—the last that any results would ever be described from the 50 foot—produced the following diameter values: β Andromedae, 44 ft, 0.0108 arcsec; α Ceti, 51 ft, 0.0094 arcsec, Betelgeuse, 14 ft, 0.034 arcsec; and, Antares, 17 ft, 0.028 arcsec.

It is telling that the only quantitative diameter measurements from the 50 foot appeared in Carnegie Institution Year Books rather than in the science literature. The one exception is a 92 word abstract of a paper Pease presented at the 1930 June Eugene, Oregon meeting of the American Astronomical Society (Pease 1930). The abstract described the interferometer as being "*in the form of a cantilever 'bridge' of structural steel....*" The only science alluded to was that: "*Fringes have been seen at a mirror separation of 34 feet on Regulus with the seeing 1 to 2.*" Aside from additional details of the form of the instrument, that is all that was ever published in a scientific journal regarding the 50 foot interferometer.

This reticence can only reflect Pease's low confidence in the numbers. If so, he was correct in judging that those diameters lacked the level of accuracy that would be useful for astrophysical purposes. Table 4.1 contains a summary akin to that in

Table 4.1. Stars Observed in the 50 foot Interferometer Diameter Program

Star	Θ_{UD}(mas) 50 foot	Baseline (ft)	Year Book	Θ_{UD}(mas) 20 foot[a]	Θ_{UD} (mas) Modern[b]	$\Delta\Theta_{UD}$(mas) 50 foot—Modern
α Ori	—	—	'29–30	47,54,34	49.4 ± 0.2	—
	~40	—	'30–31			−9.4 (19%)
	35	13	'35–36			−14.4 (29%)
	34	14	'36–37			−15.4 (31%)
β And	16	—	'30–31	—	12.8 ± 0.1	+3.2 (25%)
	10.8	44	'36–37			−2.0 (16%)
α Cet	11.5	40–42	'32–33	Found V	12.3 ± 0.3	−0.8 (6.5%)
	9.4	51	'36–37	reducing with baseline		−2.9 (24%)
α Boo	19	25	'32–33	23.7	20.0 ± 0.2	−1.0 (5%)
α Sco	29	15–17	'32–33	40	38.7 ± 0.4	−9.7 (25%)
	41	11.5	'34–35			+2.3 (5.9%)
	28	17	'36–37			−10.7 (28%)
γ Aql	8.4	~55	'34–35	—	6.8 ± 0.1	+1.6 (24%)
ε Peg	8.4	~55	'34–35	—	7.0 ± 0.3	+1.4 (20%)

Notes.
[a] See references for 20 foot results in Table 3.1.
[b] Mozurkewich et al. (2003)—Mark III Interferometer @ 800 nm (Mt. Wilson).

Table 3.1 for the 20 foot. There are thirteen 50 foot diameter measurements for seven stars, three of which were also observed with the 20 foot. As seen in the last column of Table 4.1, the 50 foot data, when compared with modern values, are all too frequently discrepant by double digits in percentage. There is no doubt that Pease had indeed resolved these stars, but his results are more qualitative than quantitative.

For the non-scientist audience, Pease wrote two very nice articles. The first—"The Fifty-foot Interferometer Telescope of the Mount Wilson Observatory"—was featured in the 1925 May issue of *The Armour Engineer*, a quarterly publication of the College of Engineering of the Armour Institute of Technology, Pease's *alma mater* as a member of the class of 1901. Pease wrote for engineers with lots of dimensions and specifications and even individual part numbers, but the date was early and so there were no scientific results.

Five years later, Pease authored "The New 50-foot Stellar Interferometer" for the 1930 October *Scientific American*. That month's magazine was chock full of astronomy ranging from a front-cover Russell W. Porter rendering of the future 200 inch telescope, which looks little like the actual 200 inch, to a contribution on amateur telescope making by Clyde Tombaugh, made famous earlier that same year by his Pluto discovery. Pease's well-illustrated account again focused on the historical build-up to the present status of interferometry and dwelt primarily on a description of the 50 foot, which was noted as weighing nine tons. No new science was reported.

That's pretty much it for the limited legacy of Pease's 50 foot interferometer. Why was it not the great success that Hale had predicted it would be in his 1929–1930 annual report? We will attempt to answer this in the next section.

4.2 A Postmortem of the 50 foot

The difficulty of making these measurements visually with a baseline as long as that of the Pease Interferometer cannot be overstated. We do not know the full observing strategy Pease used to search for fringe disappearance with the 50 foot. How did he choose the starting M_1M_4 baseline from which to begin his "scan"? What step size in baseline did he use, recalling that at each change in baseline he had to find fringes yet again due to inevitable, instrumentally-induced, path-length changes? What is the threshold in fringe contrast below which the eye can see low-contrast fringes superimposed on a quavering Airy pattern? What kind of record did he keep to remind him where he last saw fringes? Once through fringe minimum, could he see the climb up to the second lobe? If so, how much wider in baseline did he go on the other side of fringe disappearance to try to detect the rise in contrast toward the next lobe in visibility? Would he then go back to the other side of minimum to reverse his direction to home in on the vanishing fringes more precisely? All this had to be done in the presence of atmospheric seeing that would cause the "beautiful orange and black fringes," exquisitely colored by Betelgeuse's intrinsic redness, to wash in and out. In asking these questions, the one thing that is clear is that the tedium and frustration encountered at that eyepiece hour after hour—while sitting out in the open night air—were palpable and exhausting.

To get an idea of what Pease was up against, we can take a look at two results from Table 4.1 as displayed in Figure 4.8, where only the bottom 5% in visibility is shown. These stars have similar errors of about 25% in the measured diameters, where "errors" are defined here as the difference between Pease's values and the modern measurements. In the case of β Andromedae, Pease found a diameter of 0.016 arcsec, corresponding to a baseline of 29.7 ft at which he had decided that the fringe visibility had gone to zero. The modern diameter value of 0.0128 arcsec puts that baseline at 37.1 ft, a full 7.5 ft farther away. That's a lot of path to scan through in search of the elusive fringe minimum. Where Pease decided he had found the minimum, the fringe contrast actually should have been up around 5%, which is a pretty subtle value of modulation. The situation only worsens for γ Aquilae for which Pease decided fringes had vanished at 55 ft—a foot past the separation obtainable for the M_1 to M_4 mirrors—whereas the actual contrast minimum is at nearly 70 ft. Pease had run out of baseline, but had that not been an obstacle, he would have had to go another 15 feet to the point of disappearance and then search on the other side to make sure visibility was coming back again. This is nearly twice the search distance compared with that for β And.

Both of these are suggestive of Pease visually losing the fringe across the seeing-distorted Airy pattern somewhere in the vicinity of a visibility of 0.05. This assumes that he started fringe searches from small baselines and watched for disappearance

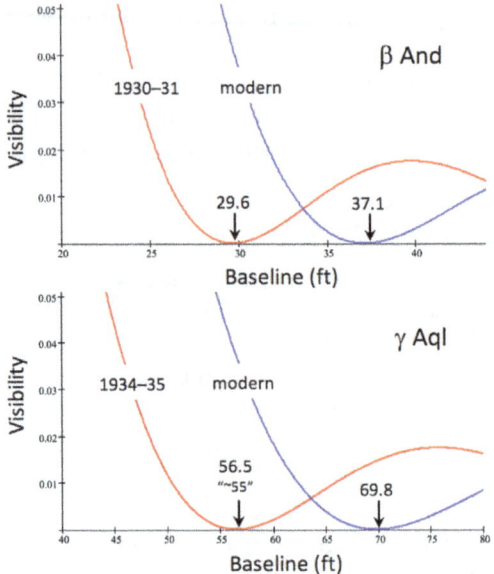

Figure 4.8. Theoretical visibility curves for two stars whose diameters were measured with the 50 foot interferometer. The blue curves correspond to the measured diameters while the red curves are for modern diameter values. The baselines at which fringes vanish are indicated by downward-pointing arrows. See text for more discussion. (Author's diagram.)

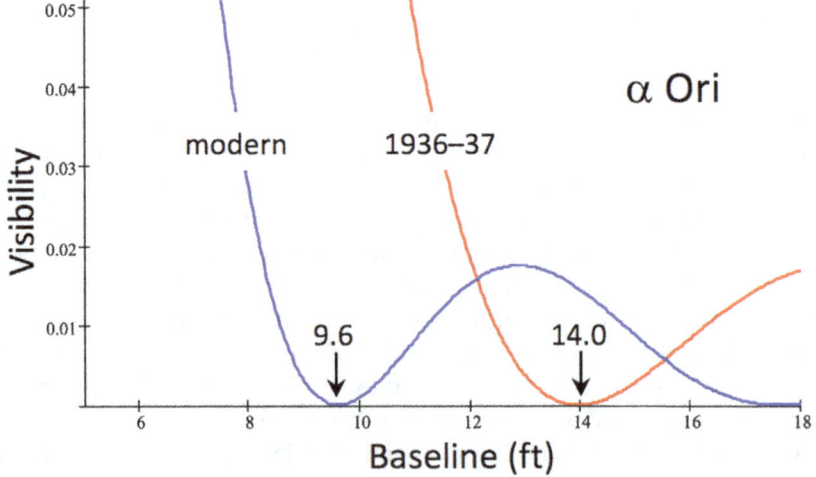

Figure 4.9. Visibility curve for the 1936–1937 diameter measure of 0.034 arcsec for Betelgeuse (red) compared with that for the larger 0.049 arcsec modern value (blue). (Author's diagram.)

as he incremented his mirrors outward. If he were unable to distinguish fringes below 0.05, then he would assume they had vanished.

Things get complicated, however, when considering Pease's suggestion that the angular diameter of Betelgeuse might be variable. Figure 4.9 presents visibility curves for the 1936–1937 50 foot diameters that were small in comparison with our modern reference result. Pease still saw fringes well beyond the minimum he would have encountered had the star then had the modern diameter used as a reference here. It is easy to find reasons why fringes might appear to have vanished prematurely in one's search but not so easy to conjure up fringes that should not be there. Was Betelgeuse really smaller in 1936–1937 as well as many years earlier when observed with the 20 foot? Perhaps so.

In a short note published in Spring 1922 (Pease 1922a), Pease reported without assigning a corresponding angular diameter that *"Betelgeuse this season has shown no fringes beyond 8¼ ft."* He concluded that perhaps the star's diameter had increased in comparison with previous years, *"possibly related to the variability in brightness of the star."* He also offered up the possibility of this apparent increase arising from poor seeing. Later that same year and with new data available, he elaborated further that the disappearance at 8½ ft corresponded to a diameter of 0.054 arcsec and clarified that those data were obtained in September and November of 1921 (Pease 1922b). Those observations had been supplemented with observations on 1922 October 14 and 15 when fringes went away at 14 feet from which the corresponding diameter was 0.034 arcsec. The seeing on the 14th had been "excellent," but deteriorated to "poor" the next night. He also recalled that 1920 December 13 observations pointing to a diameter of 0.047 arcsec and had assigned a 10% level of uncertainty to these values.

Before considering that Betelgeuse might actually vary in size, Pease again cited seeing effects as a possible cause of an apparent but unreal variation. He then went on to show how his diameter measures followed the phase of the variations in the star's radial velocity as well as its brightness as can be seen from Figure 4.10. Radial velocity and brightness do not correlate here as in the case of Cepheid variables for which velocity maximum coincides with brightness minimum, and vice versa. Pease was well aware of the decrease in fringe contrast caused by atmospheric effects and, in a stand-alone effort, was attempting to characterize visibility *"during all sorts of seeing in the endeavor to form a visibility–seeing scale so that all measures might be reduce to seeing 10 (Pease 1923)."* This effort was apparently never brought to a useable conclusion.

Thanks to the American Association of Variable Star Observers (AAVSO), there is a readily accessible database from which the more than 35,000 mostly visual estimates of Betelgeuse's apparent magnitude can be extracted. Those data are plotted in Figure 4.11 and show that the star has varied by more than 2.0 magnitudes—a factor of six—in brightness in the century since Michelson and Pease first observed it. Overlaid on the brightness record are the epochs of the 20 and 50 foot diameter measurements as well as the "modern" value selected as a reference point for our immediate purposes. A suggestive but inconclusive aspect that stands out in Figure 4.11 is the location of the 1936–1937 diameter measurements where the brightness upper envelope is taking a downward dive lasting well into the 1940s.

There are today dozens of measurements as well as images of Betelgeuse obtained using several ground- and space-based techniques over a wide range of wavelengths. Betelgeuse is enormous in size and complexity, and the apparent diameter of its

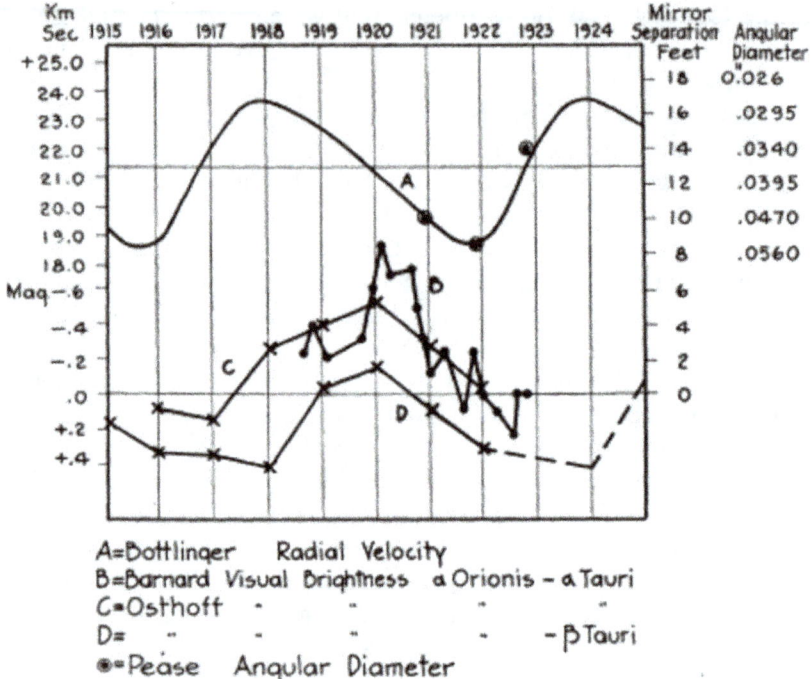

Figure 4.10. Pease's plot from 1922 showing the correlation of Betelgeuse's measured diameters with radial velocity and brightness. (From Pease, 1922b. © IOP Publishing. Reproduced with permission. All rights reserved.)

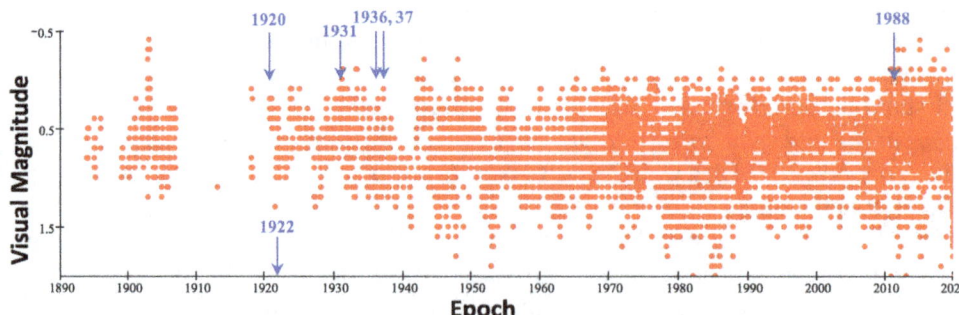

Figure 4.11. The 1920–1937 epochs of the Mount Wilson interferometer measures of Betelgeuse's diameter along with the 1988 measurement by Mozurkewich et al. (reference B in Table 3.1) are flagged by arrows in this plot of the brightness of Betelgeuse as recorded in 35,000 visual magnitude estimates from observations in the AAVSO international database (www.aavso.org).

photosphere is wavelength dependent. Much is now known about the star's variation in size (Townes et al. 2009) as well as its asymmetry with the presence of bright spot (Haubois et al. 2009). Other than to suggest that Pease may have indeed seen variation, further discussion of this is deferred to later in this book in the discussions of two subsequent Mount Wilson interferometers.

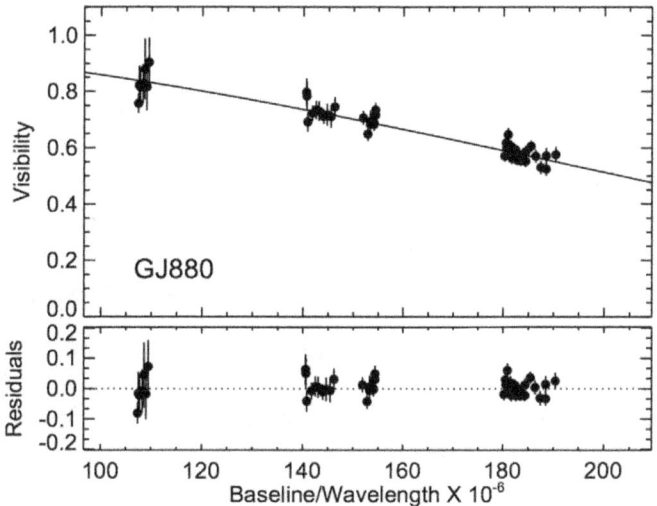

Figure 4.12. The fit to CHARA Array visibility measurements of the nearby M1.5 V star GJ 880 by Tabetha Boyajian found the limb-darkened diameter to be from 0.744 ± 0.004 mas. (Boyajian et al. 2012 © IOP Publishing. Reproduced with permission. All rights reserved.)

We no longer measure star diameters on the basis of locating the visibility null. We now do as Michelson suggested to Hale in his already-mentioned letter of 1919 November 27 and measure visibility at numerous baselines that descend toward the null without reaching it. A numerical fit to the baseline versus visibility curve then reveals the diameter and its formal uncertainty. An example of this approach from the CHARA Array, described in Section 8.2, is shown in Figure 4.12 for the nearby M dwarf star GJ 880. Those data, obtained by Tabetha Boyajian over the period 2009–2011 revealed a diameter, corrected for limb darkening, of 0.744 ± 0.004 mas (Boyajian et al. 2012). The abscissa of the data plotted in Figure 4.12 is in units of B/λ with the star being observed in the K'- and H-band infrared regions ($\lambda = 2.12$ μm and 1.64 μm). In this plot, $B/\lambda = 200 \times 10^{-6}$ corresponds to $B = 328$ m. The longest baseline of the CHARA Array is 331 m, whereas the null for GJ 880 would occur at an unavailable baseline of 553 m. We can measure sub-milli-arcsecond diameters with 0.5% accuracy because we have the luxury of sensitive and linear photon detectors to accurately sample $V(B)$ as opposed to visually locating the baseline for which $V(B) = 0$. Pease would have loved this kind of observing!

In wondering why Pease's 50 foot interferometer failed—if it truly was a failure—to produce useful (i.e., precise) diameters, a number of contributing factors arise. In his historical notes on interferometry (Lawson 2006), Peter Lawson identified several problem areas:

1. The apertures were large enough to produce speckle images, which would result in multiple displaced and moving fringes, making judgment about fringe contrast challenging and wash out fringes as they approach the null.
2. Longer baselines amplify fringe motion and instrumental vibrations.

3. The non-identical reflection sequences produce polarization mismatches that decrease fringe visibility.

All of the above undoubtedly bedeviled Pease. Let's take a closer look at other factors that Pease had to deal with.

For an accomplished engineer who showed all the signs of a perfectionist, why could Pease not bring this instrument to perfection? He knew how difficult interferometry is—a few microns of displacement through flexure and/or vibration will obliterate fringes. Could one really expect that what amounted to a non-vertical, upside-down Pratt bridge truss would perform to interferometric specifications under the load of the rails with their moving mirrors while tilting the whole shebang on either side of the celestial meridian to catch a target star in transit?! As someone who has participated in designing and building a large and complicated optical interferometer, I can think of a few reasons why Pease's 50 foot interferometer failed, starting with the cause most obvious and most frequently-cited.

1. **Inadequate technology**—Whenever I give a talk on interferometry, I like to show the images in Figure 4.13 that put the state of technology from Pease's time in perspective with that for my lucky generation. Those photographs speak loudly to the advantages we have today that were the stuff of science fiction to Pease, starting with better delivery vehicles. We are certainly no smarter than he was—not to mention Michelson. Rather, we are the beneficiaries of, for example: fast computers for analysis, data reduction, and closed-loop, real-time device and subsystem control to compensate for seeing and maintain equal path lengths; laser interferometers for ultra-precise metrology that can feed back to those controls; high-performance materials, alloys, and coatings for optics and structural fabrication; and, as we've just seen, solid-state detector arrays with practically 100% quantum efficiencies that provide access to wavelengths outside the visible spectrum. This is only a

Figure 4.13. A comparison of means of delivery available for installation of the Pease Interferometer ca. 1925 with that for the CHARA Array telescope enclosures in 1999 sets the technology context for related events occurring three-quarters of a century apart. (Image at left courtesy of the Observatories of the Carnegie Institution for Science Collection at the Huntington Library, San Marino, California; Image at right by the author.)

partial list to which must be added nearly a hundred years of experience to which Pease greatly contributed. Much has been learned over those decades about how not to build an interferometer.

Figure 4.14 presents a greatly simplified engineering perspective on Pease's design of the interferometer beam. The top sketch in the figure shows the classic Pratt Bridge Truss, patented in 1844, which was one of several truss schemes widely employed for road bridges for nearly a century (e.g., Barker & Puckett 2013). The middle sketch shows that the primary difference between a highway bridge and Pease's "bridge" is that the latter is inverted from the former and supported in the center rather than on the ends. This results in the tapering ends being cantilevers, which can sag if not properly supported. At the bottom of the figure is an application of the basic statics method of joints considering zero force members showing how the cantilever reduces to a two-member structure if one assumes that the two support points are rigid, i.e., that the central box section is itself absolutely rigid. The point here is that this underlying structure is easily analyzed, at least to first-order performance.

This is, of course, an overly simplified depiction of Pease's far more elaborate arrangement with two of these sections joined by additional

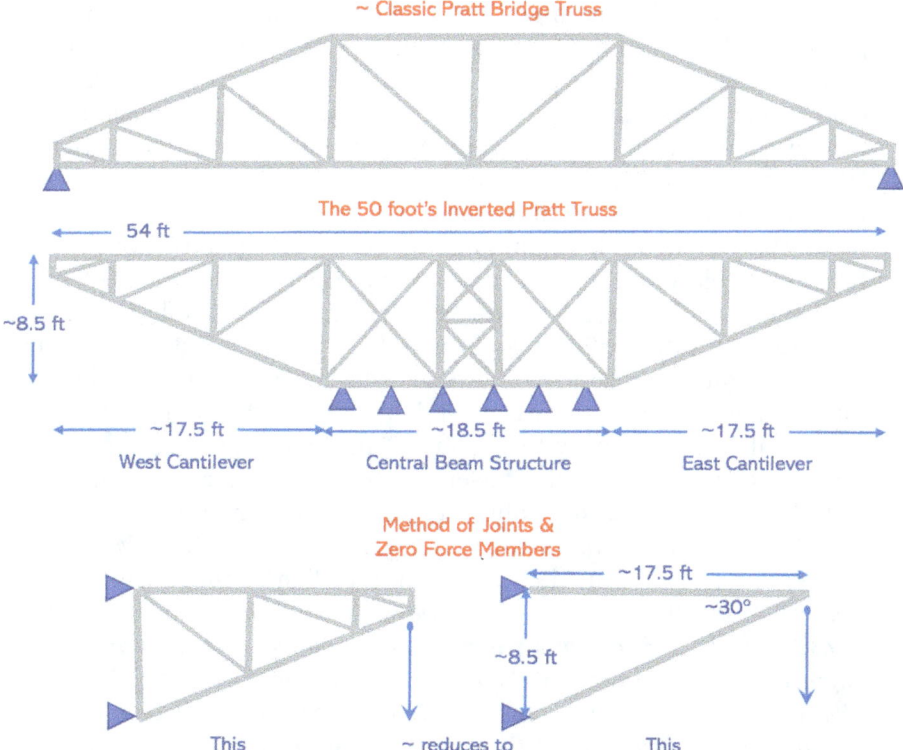

Figure 4.14. A very simple comparison and analysis of the Pease Interferometer "bridge" with a commonly used highway design of the day. See text for more discussion. (Author's diagram.)

support structures on their top and bottom surfaces and rails running along the top surface. Pease was an excellent engineer and knew very well what he was dealing with both from the engineering performance of a modified Pratt truss as well as the micron-level requirements interferometry insists upon.

Inadequate technology is usually blamed as the reason the 50 foot did not produce science. Granted, modern technologies have certainly enabled interferometers with baselines 50 times that of the 20 foot, but was the state of technology that Pease had at hand really so limited at to prohibit the successful performance of an interferometer only 2.5 times the size of one that actually worked? I have heard it said that the 50 foot performed satisfactorily out to the baseline obtainable from its predecessor 20 foot. But, that's not really the case. Five of the seven stars in Table 4.1 are beyond the reach of the 20 foot, and they are of comparable quality to those accessible to the smaller interferometer. Was technology really the problem here? Granted, Pease's measurements were imprecise, but they were not wildly errant.

2. **Inadequate budget?**—All projects managers have faced the reality of being encouraged to undertake a highly promising new venture as long as it doesn't cost any money. You just have to redirect resources already available to you to get the job done. It was no different for Hale who knew that circumstances were such that while he might get splendid encouragement from Carnegie president Merriam, no new money appropriate to the scope of the under-taking would be forthcoming. So, Hale chose the inevitable route of redirecting salary funding for designers, opticians, and machinists away from other efforts in order to produce the components for an interferometer. What else could he do? Well, he could have sought private funding for which he had a powerful gift in his earlier days. But, Hale's frequent breakdowns were taking a huge toll on him, and he was intensely drawn to the 200 inch telescope project and its enormous funding needs by the mid-1920s. He had burned his candle at both ends for decades, and, on 1923 July 1, Hale became Honorary Director with his long-time deputy Walter Adams stepping into the observatory directorship (Adams 1924). Hale's interest in a larger interferometer would not wane, and he wanted the 50 foot to exceed in both quality and quantity the output of its 20 foot predecessor and figured that Pease could pull it all off even under severe budget restraint.

In his authoritative first volume contribution to the *Centennial History of the Carnegie Institution of Washington* entitled *The Mount Wilson Observatory*, renowned MWO astronomer Allan Sandage wrote this about Pease: *"In the 1930s, it was often said in the halls of the office building that Pease had saved Hale's and Adams' bacon countless times in instrument and engineering excellence. Without him, most agreed, the Mount Wilson Observatory would never have enjoyed the degree of success it ultimately attained. This much is clear: Pease is one of the underrated—and now virtually unknown—giants who built Mount Wilson* (Sandage 2004)." It isn't surprising that Hale likely felt Pease would yet again save the bacon.

3. **Too many spinning plates?**—Pease had far too much work to do. He was a one-man advanced engineering design department with virtually all new instruments coming through him for review and guidance if not for detailed design. Additionally, he was continuing a program of lunar and planetary photography with the 100 inch telescope in support of the Carnegie Institution's Committee on the Surface Features of the Moon that had yielded magnificent imagery of the lunar surface since 1919 (Adams 1938a, 1925). He also regularly photographed planets, nebulae, and even collaborated with Hubble and Humason on the "extragalactic nebulae." We have seen that Pease was right-hand man to Albert Michelson in Michelson's highly successful velocity of light experiments from Mount Wilson as well as the follow-up Irvine Ranch measurement in vacuum. (When Michelson died in Pasadena on 1931 May 10, newspaper accounts all over the country told of Michelson calling Pease to his bedside to write down Michelson's introduction to the eventual paper on their final speed of light result; see Michelson 1931.) Pease also was a key player in the Michelson–Morley experiment repetition at Mount Wilson in the late 1920s (Michelson et al. 1929). To top all this off, Pease had become deeply involved at the half-time level with 200 inch telescope planning by the time the 50 foot was itself becoming a major effort (Adams 1938b). If only Pease had had a postdoc and a couple of graduate students.

Ironically, help did arrive in 1938 June. Raymond H. Wilson, Jr. (1911–1989) (Taibi 2016) came out to Pasadena to work as a volunteer assistant at the 50 foot during the summer of 1938, but he was on his own by then as Pease had died unexpectedly a few months earlier (Adams 1938b). As his dissertation research, Wilson had built an interferometer for binary star observations and used it productively for a number of years at the 18 inch refractor of the Flower Observatory of the University of Pennsylvania (see Wilson 1936; Billings 1959). Other than a one-sentence statement regarding Wilson, a search of the Year Books through 1950 finds no further mention of the 50 foot interferometer. In a talk to Philadelphia's Rittenhouse Astronomical Society, Wilson, unlike Pease, commented on the challenge of using the 50 foot (Wilson 1950). Toward the end of that 1938 summer, Wilson was getting the hang of it before being distracted by the need to calculate the orbit of a new Jovian satellite discovered by Mount Wilson astronomer Seth B. Nicholson (1891–1963). Wilson apparently set aside the 50 foot work and spent the remainder of his time on the mountain attending to the Jupiter-moon problem. Twelve years later, he told the Philadelphia amateur astronomers of his brief attempt to use Pease's interferometer that *"The large dimensions of this instrument made its operation so clumsy and adjustments so time-consuming that it was generally impractical. This is probably the reason that no results from it have been published since."*

This is the only published remark on this perspective, and there is likely considerable truth to it. But, Pease was not there to lift Wilson well up the steep learning curve for observing with this "clumsy" instrument, and it is not clear if Wilson ever actually even saw fringes or ever came close to locating a visibility null. It is highly likely that an indefatigable observer like Pease would have discovered the optimal way to get the most out of the 50 foot. In his 1931 review article, Pease waxed optimistically about his unique instrument for which he had made "a few preliminary measures" of diameters that had already been done with the 20 foot. Those initial results were consistent with the earlier diameters, and Pease had a list of 25 other stars for which the null in visibility should be within his 50 ft baseline span (Pease 1931).

4. **Engineer or astronomer?**—While Pease trained as an engineer and not as an astronomer, he was unquestionably an accomplished and devoted astronomical observer. From 1919 September 11, when nightly scheduling commenced at the 100 inch, until the end of 1921, Pease observed there on one-sixth of the assigned nights. As there were ten or so routine users of the telescope, Pease had more than his fair share of access to the world's largest telescope.

Insight into Pease's observing propensity appears in the eulogy by his colleague Gustave Stromberg, who, we have seen in previous chapters, had an interferometric bent himself (Stromberg 1938). Stromberg described Pease's roles in designing and using the equipment for both of Michelson's speed of light experiments from Mount Wilson during 1924–1928 and near Santa Ana, California during 1930–1934 as well as for their 1929 Michelson–Morley repeat. Stromberg said of Pease: *"The observations themselves involved the highest precision known to physical technique and required vast patience and perseverance on the part of the observer."* No doubt Pease brought a very large dose of patience and perseverance to his nights on the 50 foot interferometer. One can only imagine the frustration and disappointment he endured night after night searching for the minimum visibility as he varied the M_1 to M_4 mirror spacing. He knew that the uncertainty in locating subtle visibility minima would not lead to stellar diameters suitable for publication. Pease invested less and less effort trying to make it work, temporarily setting it aside around 1930 in favor of the far more promising velocity of light work. That major effort had the value added of collaborating with Michelson, who had himself pretty much turned away from astronomical interferometry. And, yet, Pease returned to the interferometer in 1932.

I believe that this last causative factor has no solid basis and may be dominated by projection, but putting one's self in Pease's shoes is hard to resist. I can well imagine walking by the closed 50 foot structure, located prominently at the center of the observatory grounds, and having a sinking sensation after having spent many frustrating nights trying to make it produce stable and measurable fringes. But, then, I am not a Pease.

The bottom line—Had Pease designed the beam today, he would have run a point-deflection analysis of the truss on his MacBook to verify and refine the design. Those calculations would show where he could place steel members to decrease flexure and dampen vibration, particularly in the cantilevered ends of the bridge. He might have found it doable, or he might have concluded it to be a sow's ear that should be dropped from consideration. Or, he might have decided that the design he had come up with was adequate to the financial resources given him.

With an appropriate budget, Pease might have produced a far more robust something or other from scratch with no resemblance to an inverted bridge and more like a scaled-up version of the 20 foot and 100 inch telescope combination. But a very heavy structure from custom steel components costs far more than a simple truss assembled from standard steel members. There was no budget for such a project. Sure, if the 50 foot still existed, modern technology could provide layers of active and adaptive controls to enable and stabilize fringes. And, modern detectors would replace the human eye. The price for that, however, would be many times greater than the inflation-adjusted amount Hale allocated Pease.

While all of these were factors contributing to a kind of misery index for Pease, in the long run I conclude that the 50 foot interferometer performed in the sense of providing observable fringes, but, consistent with Raymond Wilson's 1950 assertion about 50 foot operation, demanded more from the observer than even a person of such grit as Pease could muster for extended periods. Recall that no one stepped up to the plate to inherit the stellar diameter program after Pease died, nor did any outsiders besides Wilson seek time on the 50 foot. How interesting it would have been to sit around the dining room table in the Monastery listening to Francis Pease talk about his interferometer prior to trudging up the ridgeline for a night of Herculean effort to nail down a null in visibility.

The 50 foot Pease Interferometer worked—even at its longest baselines, but visual measurements in the presence of seeing, flexure, vibration, pointing and tracking errors, and other unknown imprecisions stressed the observer's endurance to an extreme while only yielding results of mediocre accuracy. Nevertheless, had the pulls on his time been from fewer directions, I believe that Pease—and likely no other Mount Wilson astronomer—would have refined the instrument and the observing and calibrating processes to ultimately make a success of his interferometer telescope. But, that was not to be.

4.3 The Curtain Lowers

On 1938 February 7, Francis Pease died in Pasadena's St. Luke's Hospital a week after intestinal surgery. He was only 57 years old. His passing made newspapers all over the world including the front page of *The Los Angeles Times.* Two weeks later, Mount Wilson Observatory suffered another severe loss with the death of its founder George Ellery Hale on 1938 February 21. Pease's role in realizing Hale's vision cannot be overstated, particularly in regard to developing giant new telescopes in general and the Palomar 200 inch in particular. Pease was an engineer to his core with more than ample layers of observing competency making him also a fine astronomer. He was to a certain extent a dreamer like his boss. In the Figure 4.15 drawing from 1926, Pease envisioned a 300 in (25 ft) telescope. Across the top of the behemoth in the side view at right can be seen—drum roll—no less than a 70 ft

Figure 4.15. Pease's drawing of a 300 inch telescope topped off with a 70 ft interferometer beam. (Author's photograph of an original drawing.)

Figure 4.16. (Above) A sad end for the 50 foot. In this 1985 May photograph, a pile of scrap metal awaiting removal was all that was left of the Pease Interferometer. The instrument's roll-off roof structure was converted into a building to house office and shop space for the observatory superintendent. (Below) Retrieved as relics from this pile by the author were this bolt and rivet. (Author's photos.)

interferometer! When this concept design was made, Mount Palomar was eight years away from being selected as the site for the 200 inch.

MWO director Walter Adams noted in his memorial to Pease that his "*death came as a great shock to his associates at the Observatory and to his many friends throughout the community both because of the rapid development of his sickness and his normal condition of rugged health and vigor* (Adams 1938a)." Prior to his illness, Pease must have had every expectation of seeing the 200 inch in action. At its dedication on 1948 June 3, he would have been 67 years old.

With Pease's passing, his 50 foot interferometer would never again be opened, and half a century later would be salvaged for scrap and its enclosure converted to space for an office and carpentry shop (see Figure 4.16). This was the end of the line for interferometry carried out by Mount Wilson Observatory astronomers.

The field's future on this mountain would be in the hands of other institutions whose scientists were drawn to Southern California by this developed and easily accessible site's reputed good seeing conditions—considered in detail in Appendix A. As the Los Angeles basin grew beyond imagination, the resulting light pollution over the observatory had no effect on the bright objects that future interferometers there, both prototypes and facilities-class, would target. These instruments would make, and are still making, important strides in advancing the field and maintaining Mount Wilson's forefront role in high-resolution astronomy into its second century. Hale and Pease would have been proud.

References

Adams, W. S. 1924, Carnegie Institution of Washington Year Book No. 22 (Washington, DC: Carnegie Institution of Washington), 181

Adams, W. S. 1925, Carnegie Institution of Washington Year Book No. 24 (Washington, DC: Carnegie Institution of Washington), 14

Adams, W. S. 1926, Carnegie Institution of Washington Year Book No. 25 (Washington, DC: Carnegie Institution of Washington), 129

Adams, W. S. 1930, Carnegie Institution of Washington Year Book No. 29 (Washington, DC: Carnegie Institution of Washington), 166

Adams, W. S. 1938a, PASP, 50, 119

Adams, W. S. 1938b, Carnegie Institution of Washington Year Book No. 37 (Washington, DC: Carnegie Institution of Washington), 191

Barker, R. M., & Puckett, J. A. 2013, Design of Highway Bridges (New York: Wiley), 544 pp

Billings, C. M. 1959, History of the Rittenhouse Astronomical Society (Philadelphia, PA: Rittenhouse Astronomical Society), 59

Boyajian, T., von Braun, K., van Belle, G., et al. 2012, ApJ, 757, 112

Eddington, A. S. 1926, MNRAS, 87, 34

Haubois, X., Perrin, G., Lacour, S., et al. 2009, A&A, 508, 923

Lawson, P. 2006, Michelson Summer Workshop, https://nexsci.caltech.edu/workshop/2006/talks/Lawson.pdf

Michelson, A. A., Pease, F. G., & Pearson, F. 1929, JOSA, 18, 181

Michelson, D. 1931, Democrat and Chronical May 10 (Rochester, NY), 1

Mozurkewich, D, Armstrong, J. T., Hindsley, R. B., et al. 2003, AJ, 126, 2502

Pease, F. G. 1922a, PASP, 34, 183

Pease, F. G. 1922b, PASP, 34, 346

Pease, F. G. 1923, PA, 31, 654

Pease, F. G. 1930, PASP, 42, 253

Pease, F. G. 1931, ErNW, 10, 95

Pokrovsky, S. 1912, ApJ, 36, 156

Pokrovsky, S. 1915, ApJ, 41, 147

Sandage, A. 2004, Centennial History of the Carnegie Institution of Washington, Vol I. The Mount Wilson Observatory (Cambridge: Cambridge Univ. Press), 101

Stromberg, G. 1938, PA, 46, 357

Taibi, R. 2016, Charles Olivier and the Rise of Meteor Science (Berlin: Springer), 479

Townes, C. H., Wishnow, E. H., Hale, D. D. S., & Walp, B. 2009, ApJ, 697, L127

Wilson, R. H. 1936, PASP, 48, 195

Wilson, R. H. 1950, PA, 58, 334

Chapter 5

The Interregnum

5.1 A Summation

We emerge from the last chapter depressed at the early loss of Francis Pease as well as at what seems like only modest return from the whole Mount Wilson interferometry investment starting in 1918 with Hale's recruitment of Michelson as a Research Associate and ending in early 1938 with Pease's death. Just what had been bagged scientifically during these two decades? In terms of stellar diameters to feed into the developing astrophysical theory of stellar structure and evolution, there were size measurements of mixed levels of accuracy for seven stars from the 20 foot and a handful of qualitative diameters from the 50 foot. From Anderson's interferometer came much higher quality measurements of binary stars and a fine orbit of the Capella system including the masses, distance, and luminosities of its nearly twin giant components. There was also the resolution of the Mizar system and determination of its geometrical orbital elements. That's pretty much it, but we must remember that there were no directly-measured stellar diameters prior to this time.

On the cup half full side, great things came from that effort. First, of course, was the very fact that the photospheres of stars can indeed be resolved and stellar diameters measured—a goal that went back three generations to Fizeau. Even more importantly was the vindication of the fledgling astrophysical theory of the structure of stars through Betelgeuse's resolution. The measured size of this supergiant star was in reasonable agreement with what the savants had predicted it to be. This was the so-called "scientific method" at work, and it captured the public's imagination when hundreds of newspapers in the United States and around the world spread the news of astronomers measuring the size of this stupefyingly large star. Hale & Company were lucky to have the likes of Eddington and Russell to guide them to likely candidates for resolution, and all parties were lucky to have a Pease to pull it off. It was the perfect confluence of the right people and access to observational tools sufficient to the task. There was no other observatory with a telescope capable of supporting an interferometer like the 20 foot. Hale's philosophy that the best instruments in the hands of the best people could accomplish great things surely paid off on that 1920 December 13 night. Who knows if all this would have happened as early as it did were it not for George Ellery Hale.

Equally key to the reconciliation of theory and observation were the advances in the practice of interferometry created by Michelson, Anderson, and Pease. They extended the learning curve they inherited from the Fizeau–Stephan–Schwarzschild–Michelson lineage by adding new mathematical tools, interferometers carefully designed to optimize visual measurements, and a set of stars with roughly determined diameters that cried out for improvement.

Eddington continued to be a strong promoter of interferometry, and in a 1923 non-technical review of the technique for *Nature* he was particularly appreciative of the mass results for Capella (Eddington 1923). He recalled Michelson's tantalizing supposition that interferometry could conceivably determine the *"distribution of light over the disc,"* i.e., produce an actual image of a star. Soon after Betelgeuse's resolution, Harold Spencer Jones (1890–1960) published a fine overview in *Nature* of the development of interferometry culminating with its application on Mount Wilson (Spencer Jones 1921). This was a prelude to his subsequent detailed theoretical treatment of the field (Spencer Jones 1922). Jones, who would become Astronomer Royal in 1933, developed the mathematical basis for eight scenarios for slit geometries and the measurement of star diameters and double-star parameters. His papers remain elegant overviews of astronomical interferometry. Regrettably, Spencer Jones did not pursue the field further.

Another 1921 article appreciative of the accomplishments of interferometry (Roe 1921) was written by Syracuse University mathematics professor Edward D. Roe, Jr, who founded the Pi Mu Epsilon mathematics honor society (Harwood 2020). Like Eddington, Roe was particularly appreciative of the Mount Wilson Capella orbit and was the first to point out that the discovery plane for interferometry would surely complement the working areas for visual, photometric, and spectroscopic applications and would uncover otherwise undiscoverable double star systems. He presciently closed his "Conservative Appreciation of Expectations from Michelson's Interferometer" with the recognition that like *"all products of the finite human mind, while it now works farther than its predecessors, it too will have it limits of applicability, and will in time be surpassed by a more far-reaching instrument."*

After reading newspaper accounts regarding Betelgeuse, Thomas Jefferson Jackson See (1866–1962) wrote a letter to *Nature* in which he described the enormous size and luminosity along with an extremely low density implied by a parallax of 0.023 arcsec (See 1921). See used these constructs in the context of then current theory of stellar evolution to conclude that the star was in *"an early stage of development, which confirms Lockyer's views first put forth about 1886."* Aside from the now very different view of Betelgeuse's evolutionary status, See misstated the parallax, which was actually 0.024 arcsec.

This is a trivial error, of course, but it inspires one to ask just how far away is Betelgeuse? The only trigonometric parallax then available goes back to an 1887 determination made with the Yale heliometer by William L. Elkin (1855–1922).

He obtained a value of $\pi_{\text{trig}} = +0.024 \pm 0.024$ arcsec, where probable error is used to express the uncertainty (Elkin 1887). This is clearly a measure of distance possessing little usefulness, but it was the best to be had in 1920 December. More would come. In 1921 January, Frank Schlesinger (1871–1943) contributed Yale's second parallax of the star, this time from photographic plates that yielded $\pi_{\text{trig}} = +0.013 \pm 0.007$ arcsec (Schlesinger 1921). This nearly doubled estimates of the distance and physical size of Betelgeuse. Then, in 1921 February, Oliver J. Lee (1881–1964) expedited the reduction of a photographic plate series initiated in 1916 to arrive at $\pi_{\text{trig}} = +0.017 \pm 0.007$ arcsec (Lee 1921). In press at the time, and known to Lee, was a large collection of Mount Wilson spectroscopic parallaxes (Adams et al. 1921) among which was a value of $\pi_{\text{spec}} = +0.012 \pm 0.010$. Lee reduced the trigonometric parallaxes from relative to absolute and included the Adams et al. spectroscopic parallax to an overall mean of $\pi = +0.0154$ arcsec corresponding to a distance of 65 parsecs.

A century later, we have a distance to Betelgeuse that incorporates the very high-precision parallax from the European Space Agency's 1989–1993 HIPPARCOS mission.[1] That particular result acknowledged systematic effects arising from the fact that the angular size of the star, which is not circularly symmetric, is significantly larger than the parallax itself. Using radio astrometric data in addition to the space astrometry, G. M. Harper et al. (2017) have derived a parallax of 0.00451 ± 0.00080 arcsec, thereby increasing the distance by a factor of 3.4 in comparison with Lee's 1921 result. It was an even bigger "*whale of a thing*" than what Anderson commented to Pease over their 1920 December 23 night lunch of toast with their heads "*full of sums.*" This puts Betelgeuse at a distance of 222 +48/−34 parsecs; with a mean angular diameter of 0.044 arcsec, the star's diameter would be 9.8 au. With this radius of 2100 times the Sun's, the star would occupy our solar system out nearly to the orbit of Saturn.

5.2 Early Followers: Maggini, Finsen, Wilson, & Jeffers

Anderson and Merrill found that interferometers worked very well for measuring double star angular separations and orientations with considerable accuracy. This inspired new binary star interferometry programs around the world, and interfero-metric methods today provide the primary means of resolving the components of binaries and measuring their relative motions on the sky. Table 5.1 is a summary of this activity. By contrast, it would be decades before anyone would again go after stellar diameters interferometrically.

As early as 1922, Mentore Maggini (1890–1941) developed an Anderson-like interferometer that he used at Catania on Sicily for six or so years before returning to other matters (Maggini 1922, 1925, 1929). His research interests had long included various solar system objects, novae, and variable stars. He was a frequent observer of Mars and a skeptic of Schiaparelli's canals. A 1934 paper, written well after his venture into interferometry, had the intriguing title (roughly translated from

---

[1] See https://cosmos.esa.int/web/hipparcos/home.

Table 5.1. Some 20th Century Double Star Visual Interferometrists

Observer	Location/Telescope	Years Active	No of Pubs	No. of Meas.
M. Maggini	Catania Obs, Sicily/13 in	1922–1929	5	223
W. S. Finsen	Union (later Republic) Obs, Johannesburg/26 in	1933–1969	65	13,000 inc. 73 discoveries
R. H. Wilson, Jr.	Flower Obs, Philadelphia/18 in	1934–1954	31	394
H. Jeffers	Lick Obs, San Jose/36 in	1939–1941	5	206

the original Italian) "The Psychological Foundation of Visual Investigation." It attempts to understand the discrepancies of what various observers report from visual observations—e.g., canals or no canals—as psychological rather than physiological in origin (Maggini 1934). Such a discussion might also pertain to visual interferometry, as we shall see shortly.

Matters then languished for a decade until Raymond Wilson, who spent that frustrating 1938 summer on Mount Wilson trying to coax the 50 foot interferometer into performing, began his work at the University of Pennsylvania's Flower Observatory. Unlike Maggini, Wilson continued double star measurements for 20 years. Another entrant into this game was Hamilton Jeffers (1893–1976)—brother of the poet John Robinson Jeffers (1887–1962)—a meticulous and skilled observer and cataloguer of double stars at the Lick Observatory, who undertook a brief venture into interferometry at Lick's 36 inch refractor. Most notable among this group was William S. Finsen (1905–1979) of the Union Observatory in Johannesburg, South Africa. Finsen began experimenting in interferometry about the same time and switched back and forth between Anderson-type interferometry and classical micrometry prior to a complete conversion to the former around 1950.

A close colleague of Finsen at Johannesburg was Willem H. van den Bos (1896–1974), an indefatigable observer and discoverer of double stars who holds the individual record for the number of double-star observations—73,000 measurements and 3100 new discoveries—over his half century at the telescope (Argyle et al. 2019). There is no close second in these rankings. Van den Bos was also a prolific calculator of orbits from his and others' data and never hesitated to critique the observations he folded into his calculations. Thus, he addressed the work products of those who continued Anderson's style of double-star interferometer on more than one occasion, with Maggini receiving the harshest criticism. For example, in 1927 van den Bos pointed out that for the 15 Eridani system Maggini found a position angle and angular separation of the two stars of 180.7° and 0.43 arcsec when, the actual orientation was 240° and 0.2 arcsec, with the discrepancy attributed to atmospheric dispersion (Van den Bos 1927), a topic already raised in 1921 by another great double-star observer George Van Biesbroeck (1880–1974) (Van Biesbroeck 1927).

Van den Bos returned to interferometric scrutiny in 1951, by which time he had had enough and directed his more than adequately severe appraisal and comparison at the results of Maggini, Wilson, and Jeffers (Van den Bos 1951). He found *"no erroneous results"* in Jeffers' work, but was not so kind to Maggini and Wilson: *"In view of the numerous spurious and erroneous results contained in these lists, even for objects which should not give any trouble, I find it rather difficult to accept Wilson's discoveries and measures of pairs with distances of the order of 0.1 [arcsecond] or less at their face value. As in Maggini's case, rather more caution in the interpretation of the phenomena observed, or believed to be observed, would seem desirable."* He also recommended an aperture of at least 20 inches for any telescope employed in interferometry.

To be sure, van den Bos was not alone in his judgment of Maggini's interferometry work. Already in 1933, Willem J. Luyten (1899–1994), who would become famous for his own work as well as for his stark criticism of the work of others, pointed out that Maggini's orbit of the components of the star 13 Ceti did not obey Kepler's law of equal areas, indicative of a fundamental flaw in the observations (Luyten 1933).

So, let's take a look at Maggini's measurements of 13 Ceti (Maggini 1928). Figure 5.1 shows those data plotted on the modern orbit of Mason & Hartkopf (2005) and accompanied by speckle interferometric measurements made by me and

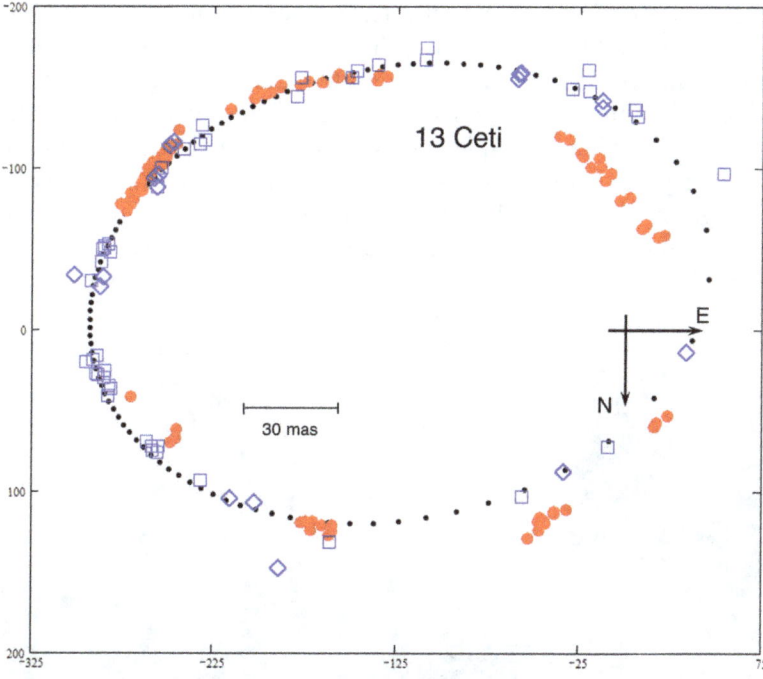

Figure 5.1. The visual interferometric measurements by M. Maggini (red fill circles) of the components of 13 Ceti are here plotted on the orbit of Mason & Hartkopf (2005). For comparison, speckle measurements by the author and his CHARA colleagues are drawn as blue squares along with speckle results from Elliott Horch as blue diamonds. Aside from a few outliers, the speckle observations fit the orbit well. Maggini's results, which Luyten had pointed out are non-Keplerian, show unusual systematic effects that may have been caused by some sort of drift in calibration. (Author's diagram.)

my CHARA colleagues as well as by Elliot Horch of Southern Connecticut State University (Mason & Hartkopf 2006). The speckle observations fit the orbit, which has an angular semimajor axis of 0.25 arcsec, quite well and with moderate scatter. Maggini's results—the red filled circles—are inexplicably peculiar, showing some consistency among themselves but not fitting this orbit—or any orbit, really—at all. It appears likely that Maggini's instrument's calibration was not stable.

Table 5.1 makes clear the dominant role that W. S. Finsen played among this small group who followed in the footsteps of the Mount Wilson observers. Much more could be said of this work and about these gentlemen, and the reader is referred to the extensive descriptions of their efforts given by Daniel Bonneau in his 2019 book on optical interferometry's first century (Bonneau 2019).

Instead, let's dwell a bit more on Finsen's career. His debut in double-star work was a determination of the orbit of α Centauri in 1926 to which he added two of his own measurements to the long list of observations going back to the system's first measurement in 1752 (Finsen 1926). Two years later, he reported his initial list of 262 μm measurements of 115 doubles (Finsen 1928). By 1933, Finsen had devised an interferometer (Finsen 1933) along the lines of Anderson's instrument on the 100 inch telescope. Finsen used this device as well as a classical micrometer in Johannesburg, South Africa at the Union Observatory's 26½ in refractor until he later fabricated a remarkably compact interferometer built into an eyepiece that he designed and fabricated himself. As seen in Figure 5.2, his interferometer would fit into one hand (Finsen 1951). With it, he initiated an observing program in 1954 that continued into his retirement at which point Finsen had made some 6000 measurements of close visual binaries while discovering 73 new systems among more than 8000 stars he inspected for duplicity (Finsen 1971). This enormously-productive, long-term program was inspired directly as a result of the Mount Wilson efforts of Anderson, Merrill, and Pease. Finsen's body of work far exceeds the cumulative efforts of all other visual interferometrists. Along the way, Finsen wrote two very

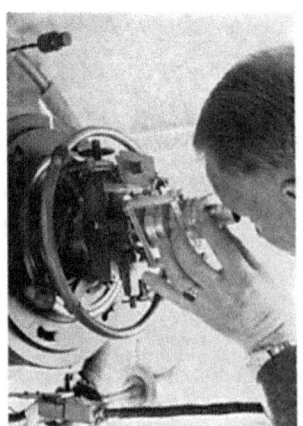 

Figure 5.2. Left: W. S. Finsen with his eyepiece interferometer at the 26 ½ inch telescope of the Union/ Republic Observatory. Right: The interferometer and its accessory eyepieces. (Finsen 1964 © IOP Publishing. Reproduced with permission. All rights reserved.)

fine descriptions of how to carry out visual interferometry of double stars (Finsen 1951a, 1964).

I had the honor of corresponding with Bill Finsen in his retirement during the early years of my binary star speckle interferometry work, a subject we will get to later in this book. In a letter to me dated 1976 September 9, Finsen described how *"the skilled, experienced visual observer is able to seize on instantaneous manifestations of the image, an ability which seems to be inherent rather than acquired and which is often described by the phrase 'double-star eye,' but which is as much mental as physiological... For instance, I recall many occasions when van den Bos and I discussed a phenomenon that puzzled us more than a little: the fact that we had no difficulty in measuring very close <u>bright</u> pairs in even the poorest seeing when instead of just one image we would see perhaps dozens in rapid motion but each recognizable and measurable as a very close double star and resolved <u>as with the full aperture of the telescope</u>. We found this last fact difficult to reconcile with atmospheric cells considerably smaller than the telescope aperture. But now of course I know: like Monsieur Jourdain who spoke in prose without knowing it, we exploited the speckle technique without knowing it."*

In a 1971 article entitled "Twenty Years of Double-Star Interferometry and its Lessons (Finsen 1971)," Finsen followed his statement of the advantages of the technique with a discussion of its disadvantages that is worth repeating here: *"But, what of the disadvantages? Unfortunately, it is impossible to gloss over the fact that they are serious enough to make one think twice before putting it to use."* What follows relates specifically to the nature of the criticism van den Bos leveled at Maggini's work.

"The greatest disadvantage of the interferometer is certainly the fact that one does not see duplicity directly but must deduce it from the delicate and elusive fringe manifestations. One must be on continual guard against misinterpretation, and observing with the interferometer is therefore both slow and tiring. Three or four complete measures in an hour is a good output, but often a measure may take an hour or more to complete. In carrying out a survey a star may be dismissed as single after only a minute or two, but, if variation of fringe visibility is suspected it may again be an hour or more before one is able to make up one's mind. The interferometer is therefore much more prodigal of valuable telescope time than is the micrometer."

"It follows from this that casual or occasional use of the interferometer, without the benefit of considerable experience, is fraught with danger. In general, one would be well-advised to leave the interferometer alone altogether unless one is prepared to become a specialist in its use."

Francis Pease would have said "Amen" to this.

5.3 An Unknown Genius

While there were mixed results of Anderson-style double-star interferometry from a very modest list of practitioners, what about the Michelson–Pease periscopic interferometer scheme? The fact that no one other than R. H. Wilson would attempt to use the 50 foot interferometer telescope on Mount Wilson answers the

Figure 5.3. Edward Hutchinson Synge (1890–1957). (Courtesy of Living Edition Publishing.)

question to first order. But a trail would eventually be blazed from the moribund 50 foot to today's optical/infrared, long-baseline interferometers that have produced hundreds of papers on a wonderful variety of stellar astrophysics topics. It appears, though, that the earliest potential trailblazer attracted no attention whatever.

If you scan down the tables of contents for *The London, Edinburgh, and Dublin Philosophical Magazine and Journal of Science* published during the period 1928–1932, you will find nine entries by one E. H. Synge.[2] You'll probably pass him by as his is a name you've likely never heard of. But, do pause to read the titles of his articles. Among them, you will find detailed concepts that later materialized as near-field scanning microscopy, LiDAR, and multiple-mirror telescopes such as the Multiple Mirror Telescope (Beckers et al. 1981) and the Large Binocular Telescope. Only decades after their appearance were their origins finally attributed to Edward Hutchinson Synge (1890–1957), shown in Figure 5.3 (Donegan et al. 2012).

Among Synge's overlooked contributions to *Philosophical Magazine* is "A Modification of Michelson's Beam Interferometer (Synge 1930)." In just over six

---

[2] You will not find them in the SAO/NASA Astrophysics Data System digital library. Searching on "Synge, E. H." draws a blank in spite of relevant key words in his astronomically-related papers in *PMag*. I stumbled on Synge through an ADS search on the title word "interferometer" during the 1930s, which, among many other papers, led me to W. A. Calder's 1931 publication citing Synge.

stunningly remarkable pages, Synge—pronounced "sing"—lays out a plan for an interferometer consisting of two coelostats, each a rotating mirror that reflects astronomical light onto a flat mirror that in turn feeds its beam through evacuated light pipes to a central beam-combination telescope. He proposes placing this interferometer in a north–south orientation on a hillside so as to be parallel to Earth's rotation axis while noting that the equator would be an ideal location. He considers what we would today call optical path-length compensation and suggests, for the case of a very long-baseline system, placing the beam-combination tele-scope on rails to eliminate the need for a long path length matching system (see Appendix B.5.1). Synge treats atmospheric effects and devises a scheme of varying wavefront retardation by continuously adjusting the pressure inside a rubber tube so as to change the air density in a closed portion of the light path. In essence, Synge laid out the subsystems one now finds in modern long-baseline optical interfero-metric (LBOI) arrays.

Finishing Synge's *Philosophical Magazine* article elicits one of those rare "*Wows*" from the reader followed by the question "*Why I have I never heard of this guy?*" One contributing factor is found in the 2012 article "Unknown Genius (Weaire et al. 2012)." In a nutshell, Synge was born into an extraordinary Dublin family. His uncle was the prominent playwright John Millington Synge (1871–1909) whose early death left his nephew Hutchie, as E. H. was called within the family, independently wealthy. Hutchie then dropped out of Trinity College Dublin and retired to his inheritance. His brother John Lighton Synge (1897–1995) was a prominent physicist and mathematician with whom Synge never shared his written contributions to science and technology, choosing instead to correspond with Albert Einstein who encouraged him to publish his ideas. Synge worked at home with no laboratory aside from the one in his head—his withdrawal perhaps a consequence of Asperger's syndrome. Within a few years of his 1928–1932 productive period, he was involuntarily confined to a Dublin mental institution wherein a stroke took him in 1957.

Although unknown to those who today practice in the LBOI field, their instru-ments and facilities were envisioned by this eccentric physicist during a burst of creative genius even while Michelson and Pease were still alive. Tragically, the link in the chain back to E. H. Synge was never forged, and his anticipation of today's star-imaging interferometers is not even a footnote in the field's history.

Synge's interferometry work did not go entirely unnoticed at the time, however. In what must be one of the few citations of Synge's 1930 paper—I have found no others—Harvard graduate student William A. Calder (1906–1998) wrote in 1931 on his "Results of Experiments with an Interferometer Used with the Fifteen-Inch Refractor" (Calder 1931)—the very telescope on which Michelson had requested observing time 40 years earlier to secure interferometric measurements of the diameters of the Jovian satellites. Calder's first paragraph pretty much tells the whole story: "*In view of the theoretical versatility of interference methods in astronomy, as well as their great simplicity and beauty, it is surprising that their adoption has apparently been confined to some half dozen workers. Recent experience at Harvard may possibly point to an explanation of the paucity of results. It also raises doubt as to the worth of various ingenious proposals for increasing enormously the dimensions and effectiveness of stellar interferometers.*" He next cites Synge's paper as an example of such a proposal.

The litany of problems Calder encountered at the 15 in refractor commences with an account of Maggini's work in Italy for which Calder—apparently having overlooked van den Bos's 1927 pertinent remarks—labeled Maggini as "*at present, the outstanding figure in this field.*" And so, Calder had embarked on "*a repetition of some of Maggini's experiments.*" An aperture mask was devised for the telescope that employed slits of adjustable separation and variable position angle. Calder then selected Jupiter's moons as his initial targets, expecting them "*to be relatively easy to observe.*" But they were not at all easy because of poor seeing. He experimented with various magnifications to benefit the location of fringe disappearance, finding that the best seeing allowed some eight such locations per hour, while poor seeing might lead to never concluding just where the fringes disappeared. In no instance was he able to climb out the other side of a disappearance to find the second visibility maximum.

Over a five-month period ending in 1931 May, Calder was able to obtain useful diameter measures for three of the four Galilean moons. Satellite IV (Callisto) was always too faint. Referring back to Table 1.1 we can add Calder's measurements to that list by comparison with subsequent spacecraft measures. The mean "residual" to the modern diameter determinations for Calder's measures is $+0.130 \pm 0.132$ arcsec, which places them well short of the quality of the diameters found by Struve, Englemann, Michelson, and Barnard. Even so, they were real measurements.

Experiments with double stars and with a method to determine the effective wavelength were entirely fruitless, and Calder presciently lamented that "*these results tempt one to doubt the refinements of Maggini.*" The one positive was that atmospheric dispersion did not contribute to these difficulties, which all apparently accrue to seeing conditions. Calder's lone venture into interferometry concludes "*that the statements frequently found to the effect that, in contrast to what might be expected, the interferometer does not require excellent seeing conditions, are unduly optimistic. Atmospheric conditions appear to be the controlling factor, and seriously restrict the possibilities of interference methods.*" While not widely cited, Calder's critique of interferometry made it into the important reference work *Handbuch der Astrophysik* for 1933 (Lundmark 1933) and perhaps extinguished some sparks of interest that might have otherwise advanced the field.

I met Bill Calder on a few occasions during his years at Agnes Scott College in Atlanta, where he served on the faculty from 1947–1971 and was a much beloved astronomy teacher. But, I was then unaware of Calder's early involvement in interferometry. I am sure that he would have subsequently changed his opinion of the worth of "Hutchie" Synge's ideas.

5.4 Robert Hanbury Brown's Intensity Interferometer

It was inevitable that, among the scientists and engineers who created radio astronomy following their WWII efforts to advance communications and develop radar, some would apply these technologies at much shorter wavelengths. Preeminent in this group was Robert Hanbury Brown (1916–2002) (Radhakrishnan 2002), the grandson of an early British radio pioneer. For three years during the War, Hanbury Brown came to the U.S. to work on radar development at the Naval Research Laboratory. He then participated in efforts to convert the new technologies into peacetime applications until deciding to return to England to study for a PhD in radio astronomy at Manchester. In 1956, in collaboration with Richard Q. Twiss (1920–2005) (Tango 2006), Hanbury Brown, shown in Figure 5.4, discovered that, as a result of the odd quantum mechanical particle/wave nature of light, there is a temporal correlation between the fluctuations in intensity of separate but mutually coherent wavefront samples—the Hanbury Brown and Twiss Effect (Hanbury Brown & Twiss 1956). Measurement of the correlations in turn provides a measure of the source geometry. Since the intensities from two or more separate apertures can be measured independently with no need to combine the wavefronts optically, the baseline(s) between the apertures can be much larger than might be practical for Michelson-style interferometers. Later in 1956, Hanbury Brown demonstrated the effect, which became known as *intensity interferometry*, by measuring the angular diameter of Sirius with two detectors 6 m apart (Hanbury Brown 1956). Although the measurement was not particularly accurate, it relit elsewhere the interferometry flame that had been extinguished on Mount Wilson with Pease's death.

Hanbury Brown emigrated to Australia in 1962 for an academic position at the University of Sydney. An intensity interferometer was constructed at Narrabri,

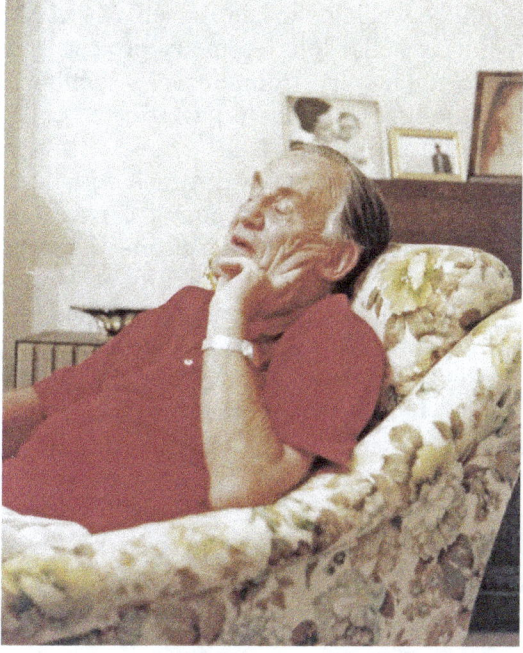

Figure 5.4. Robert Hanbury Brown caught in mid-blink in a relaxing after-dinner moment in the home of John and Madeline Davis in Sydney while reminiscing about Narrabri. (Author's photo from 1997 January.)

New South Wales. Among its significant achievements was the measurement of the diameters of 32 hot, bright stars with useful accuracy (Hanbury Brown et al. 1974). Although limited to very bright stars, the Narrabri Intensity Interferometer launched one of the world's major programs in astronomical interferometry that continues to this day in the form of the Sydney University Stellar Interferometer (Davis et al. 1999). As a result of the properties of the then available detectors, it was decided by John Davis, who succeeded Hanbury Brown, not to build a second-generation intensity interferometer. Thus, SUSI, which possesses the longest base-line of any optical interferometer, adopted Michelson's technique of amplitude interferometry that is the current mainstay of the field. That said, detector advances have led to renewed interest in intensity interferometry, and a modern successor to the Narrabri Intensity Interferometer seems inevitable. In addition to producing productive interferometers, the Sydney University group spawned a number of talented interferometrists who have gone on to enrich other groups' efforts to design and build facilities for high-resolution studies of stars.

5.5 Willet I. Beavers' Second-generation Michelson–Pease Interferometer

Were there other potential interferometry flames to be lit besides that of Hanbury Brown? Yes, indeed. The first of these new pioneers after Hanbury Brown was Willet I. Beavers. In a letter dated 1962 August 6, Lowell Observatory director John S. Hall wrote 28 yr-old Willet Beavers, an astronomy graduate student at Indiana University, that he would be welcome to use Lowell's 24 in Morgan telescope later in the year to implement his stellar interferometry project. Hall's invitation, which had been in response to an inquiry from Beaver's faculty advisor Marshall H. Wrubel (1924–1968), is the first of some 36 pages in the collected correspondence between Hall and Beavers from 1962 to 1967.[3]

From these letters and subsequent conversations with Dr Beavers, I have put together the story of Beavers' courageous venture—carried out more-or-less on his own—to create a modern version of the Michelson–Pease 20 foot interferometer. His specific goal was to be able to quantify visibility at various mirror separations and to fit the visibility vs. baseline curve numerically to determine stellar diameters. Not totally naïve as to the obstacles inherent in interferometry, he would happily settle for measuring a single point along the descending curve. Thus, Beavers was intending to attain what Michelson had envisioned from the earliest days of his guidance to Anderson and Pease on Mount Wilson. Although not involving Mount Wilson, this is a previously untold episode in interferometry's history that is overdue for telling.

Unlike most graduate students, this was a project conceived by the student and not by the faculty advisor. Wrubel, who had studied under Chandrasekhar at

---

[3] These letters were kindly provided to me by Lowell Observatory Historian Kevin Schindler and Archivist Lauren Amundson. They were also much appreciated by Willet Beavers after I was able to locate him in his retirement from a career in astronomy and electro-optics.

Chicago, was a well-known stellar interiors theorist (and Julliard-trained pianist) with little experience in observational astronomy and none in interferometry.[4] Nor was any other Indiana astronomy professor knowledgeable in the area. In order to determine if Beavers' project was worthwhile and if he was adequately prepared in optics generally and interferometry in particular, they consulted with the eminent optical physicist Emil Wolf (1922–2018) of the University of Rochester. Wolf graciously agreed to meet with Beavers in Rochester to talk over the problem with him and let Wrubel know if this young man was up to the task.

So off to Rochester Beavers went and stayed a couple of days. Much of his time was spent talking optics, and especially coherence and interferometry, with the short-statured and perpetually cheerful Professor Wolf while they wandered over the Rochester campus. Their path seemed a random walk with Wolf often and suddenly taking shortcuts through buildings, including on one absent-minded occasion a woman's dormitory. Their conversation was nonstop and more than sufficient for Wolf to reach the requested judgment on his young visitor. Beavers returned to Bloomington with a nod of approval from his vetting by a world-class expert in optics for whom Beavers still holds the highest respect.

Thereafter, Beavers was let loose to pursue his self-designed dissertation project of building a next-generation periscopic interferometer beam, developing an electronic fringe-detection system, and measuring stellar diameters. This was no small order, and, to top it off, Beavers was married and had two small children. To support his family beyond what a grad student's stipend provided, he would soon take up employment as an instructor of physics at the University of Missouri while continuing his dissertation research.

Beavers still has a thick folder containing copies of papers from the Mount Wilson interferometrists along with others from that era. Those collectively comprised his preparatory interferometry textbook. As a start-up experiment with a small interferometer, Beavers cobbled together from war-surplus items around IU's astronomy department a 3-ft interferometer and mounted it on the 36 in reflector at Goethe Link Observatory. He never saw fringes with that configuration. Undeterred, he next assembled a 12 ft instrument, to be equipped with 6 in wavefront collecting mirrors, from two sections of 3 in diameter aluminum tubing held rigidly parallel with spacers. He utilized the department's machine shop, a limited facility that was better equipped than it was staffed. Nevertheless, with the help of a janitor/machinist, he was able to build the interferometer that he would transport to Flagstaff along with his family. He looked forward to astronomical seeing conditions on the high plateau of northern Arizona that are far more stable than what the Indiana heartland's night skies had to offer.

He originally targeted the fall of 1962 to kick off the Lowell observing effort, but the IU astronomy machine shop was preoccupied with repairs to the Goethe Link reflector, resulting in a delay. Hall was pleased to learn in a 1962 September 17 letter that Beavers would eventually bring all that was needed for his work—"*the 12-ft*

---

[4] Lee R K Biographical note to the Marshall H. Wrubel papers, Indiana University Archives, http://webapp1. dlib.indiana.edu/findingaids/welcome.do.

interferometer, compensating wedge system, photometer, amplifiers, power supplies, recorder, and a reasonable amount of spare parts and tools"—as the Lowell shop was already stretched to its limits. However, Hall's offer to provide assistance in mounting the interferometer "*would indeed be appreciated.*" The Beavers family arrived in Flagstaff on 1962 November 12. Fifty-eight years later, I asked him to put his recollections of this winter-long observing run at Lowell down in words.

"*We had no idea what to expect, so we just checked out of our Indiana apartment and stored our possessions. Once in Flagstaff, we stayed in an apartment upstairs in the Observatory's main building. I spent each clear night at the telescope and then around dawn I would walk down the hill to the apartment and get into my low bed several seconds before either or both of the children would run to the bedside and jump on my head. Valerie, aged two, and one year-old Bill did pretty well about staying at home where most of the first two months there was snow. Valerie enjoyed opening the top of the Dutch door to the kitchen while the wind whistled past her ears and she uttered the words her mother had said most mornings 'Nice Day' only she drew it out a bit, with snow hitting her in the face.*" (A taste of a Flagstaff winter is presented in Figure 5.5.)

"*It seemed like most days had some cloud activity, and I often began to hope they might give me a night free. Yet as we finished dinner around sunset I could usually see the beautiful clearing sky telling me to get warm coats on and head up the hill. It didn't take long to open things up and get the observing room cooled off. The Morgan telescope and its attached building complex of dark room, instrument development room, and living quarters with a kitchen was more than one could expect. The observing room was perfect for working with an interferometer. The roofs flap back on both sides, and a panel can hinge down on the south, exposing the telescope to the entire sky.*"

"*I spent most of the days just getting used to the telescope while mounting the rather bare interferometer through the upper portion of the frame. It did fit.* The interferometer support beam *consisted of two main 3-inch diameter, 12-foot long aluminum tubes* (as shown in Figure 5.6). *Every part of the system was hand controlled. The*

Figure 5.5. Looking out of a window of the Lowell Observatory main-building apartment, the author and his wife had this late-afternoon view of the San Francisco Peaks on 2007 Dec 13. (Photo courtesy of Susan J. McAlister.)

Figure 5.6. Beavers' interferometer mounted on Lowell Observatory's 24 inch Morgan telescope. (Photo courtesy of Willet Beavers.)

prisms (that served as mirrors 2 and 3 in the Michelson–Pease arrangement) *had identical reflecting lower sides at right angles to each other at the bottom of the central prism. From there they were directing starlight from the two outer flats through the center section of the telescope between the interferometer tubes and reflecting beams downward to be focused by the telescope primary mirror. Both of those narrowing two beams then passed through the 'stovepipe' vertical light baffle mounted in the cassegrain hole of the primary.*"

"*At the top of the stovepipe, I placed a metal disc that held some optical glass components for matching the path lengths of the two interferometry beams of light. One was a plain clear rectangular boxlike piece. The other consisted of two identical flat-surfaced wedges that could introduce a variable thickness to add or subtract optical path while looking for fringes.*" A schematic diagram of the interferometer integrated into the telescope is shown in Figure 5.7.

The 24 inch Morgan Telescope that would be the foundation of Beavers' experiments was manufactured in 1955 by Tinsley Laboratories, then in Berkeley, California. It was donated to Lowell Observatory in 1960 by Ben O. Morgan of Odessa, Texas (Wilson 1956). The *f*/4 primary mirror fed both cassegrain and Newtonian foci, and the telescope featured a massive mount fully up to supporting the interferometer. Beavers had the fine idea of mounting the components of his interferometer so at to run through the middle of the telescope rather than atop the secondary mirror support structure. This reduced the necessary counter-weight for balancing and minimized flexure of the secondary support struts.

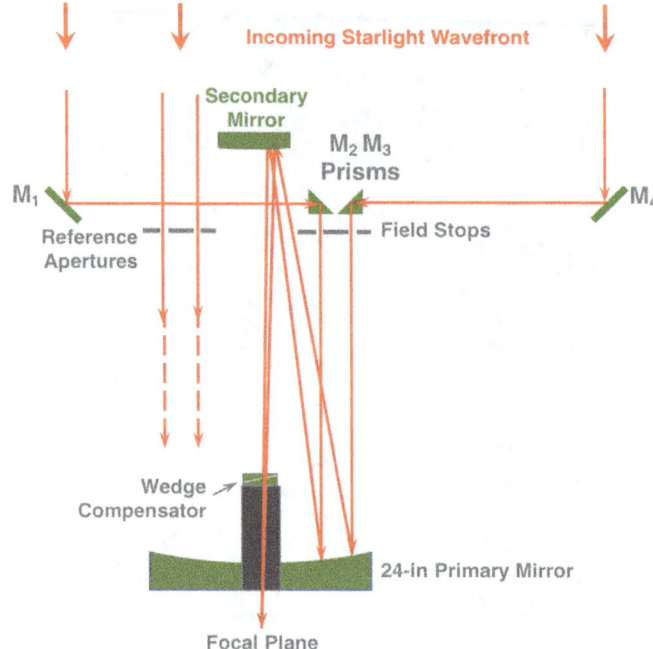

Figure 5.7. Schematic diagram of the interferometer light paths for the science and reference beams integrated into the Morgan telescope. (Author's adaptation from unpublished PhD dissertation courtesy of Willet Beavers.)

Figure 5.8. The opposing wedge mechanism later added to Beavers' interferometer introduced a variable path length in one beam of the interferometer. A fixed glass block ensured that the other beam passed through a nearly equal thickness of glass. (Photo courtesy of Willet Beavers.)

Notice that in the diagram of Figure 5.7, the M_2 and M_3 mirrors that inject the light collected from the interferometer beam's outer mirrors are arranged so as not to interfere with the cassegrain optical path of the telescope. Symmetric to the high-resolution converging beams from M_1 and M_2 are a set of reference apertures that act as slits for a low-resolution and hence high-visibility comparison. The wedge compensator described above by Beavers is shown in Figure 5.8. It allowed the path in one beam to be precisely adjusted compared with the other until fringes are seen. The fixed segment of glass introduces dispersion into the non-adjustable beam equivalent to that in the adjustable side.

Unsurprisingly, observing was a tedious process not fully revealed in Pease's descriptions of his own observing experience. But, at least Pease had an assistant to manipulate mirrors on the beam whereas Beavers worked alone. Beavers would make countless trips up and down a ladder to reposition one mirror and then the other and then back to the eyepiece to look for fringes. Much of his Northern Arizona observing took place in bitterly cold winter conditions, but Beavers, who was born in 1933 in Billings, Montana and grew up in Joplin, Missouri, was accustomed to the cold and had no complaints about the rigors of long winter nights working under the open sky. Besides, he had a family to raise and was anxious to advance his dissertation research and get on with life.

Beavers reported on the results of his first observing experience at the spring 1963 meeting of the American Astronomical Society hosted in Tucson by the Kitt Peak National Observatory and the University of Arizona (Beavers 1963). The first of the two paragraphs in the abstract of his talk crisply describes his research goal: *"The present availability of sensitive photoelectric cells and fast films not available to the early workers using the 20- and 50-ft interferometers has suggested the possibility of making quantitative measurements of the fringe strength. Such measurements would reduce the mirror separation needed to measure any given angle by at least a factor of 4, and would make it possible to measure the limb darkening of a number of stars."*

The second paragraph succinctly summarizes his accomplishments to date. *"A 12-ft interferometer has been built and used on the 24-inch Morgan telescope at Lowell Observatory. Although the fringe motion is too rapid for the present dc photoelectric system, it has been possible to obtain numerous photographs of the fringes."* This brief abstract is one of only two entries in the scientific literature describing Beavers' work. The second has even fewer words. A photograph of the fringes he obtained for Sirius from the high-resolution and reference sides of Beavers' interferometer are shown in Figure 5.9. These may be the first such images of stellar interferometer fringes ever recorded.

The next observing run was set for late 1964 January for which Beavers would come out from Columbia, Missouri, where he had resettled as a physics instructor at the University of Missouri the previous summer. The Missouri Physics department had a small but efficient shop with very cooperative machinists who made a number of improvements in the interferometer, including upgrading the wedge apparatus that made the interferometer easier to work with. Beavers had assembled a set of neutral density filters in an attempt to calibrate the photographs using known steps in transmission through one side of the short-baseline reference aperture pair that mimic quantifiable steps in fringe visibility. This would allow him to attach reliable numbers to the visibilities he would sample at various baselines. Experience soon showed him that, due to night-to-night seeing variations, calibration data must be taken each night. He attempted this approach on Sirius and told Wrubel that his *"data yields a value for Sirius of 2.5 solar diameters."* Using the logic that all degrading effects tend to diminish visibility, the not fully-calibrated data would lead one to deduce a larger than actual diameter, hence his result for Sirius should be regarded as an upper limit. Beavers was right about that as the diameter of Sirius is closer to 1.7 solar diameters. He went on to explain that *"Until I can work on fainter*

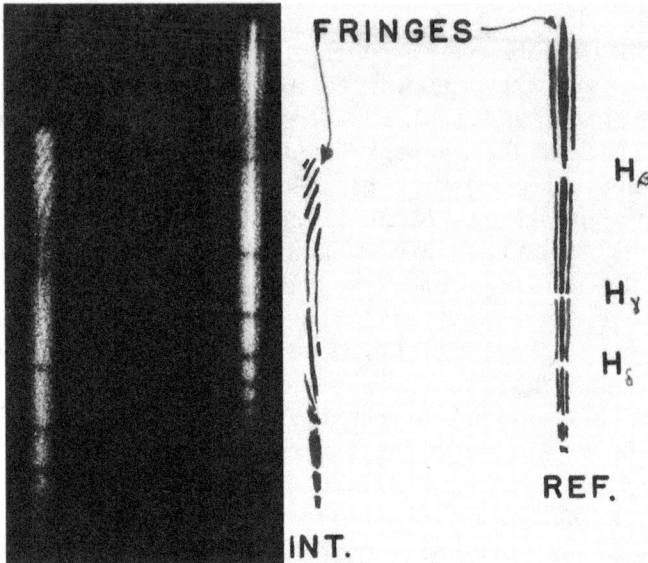

Figure 5.9. (Left) Interference fringes from the star Sirius at left were obtained during Beavers' 1962–1963 winter stay in Flagstaff. A low-dispersion spectrograph dispersed light to allow for spectral sensitivity in resolution. The curvature of the fringes resulted from dispersion in the wedge glass. (Right) His hand-drawn key indicates the fringes from the interferometer set at a 12 ft baseline along with the reference beams' fringes. As would be expected, the fringes fade toward shorter wavelengths where the degrading effects of atmospheric seeing grow with decreasing wavelength. These may be the first photos of stellar interferometric fringes ever recorded. (Courtesy of Willet Beavers.)

stars such losses cannot be evaluated. Of course, Pease had no means of checking this point. Neither did Hanbury Brown and Twiss attempt such evaluation in their first experiment. I look forward to hearing the results of their most recent work." Thus, he planned to bring along a new photoelectric scanning system that incorporated a fixed picket-fence mask whose square-wave apertures were the same frequency as the fringes produced by a specific mirror spacing on the interferometer beam. An octagon with mirrors mounted on its eight surfaces rotated at 100 rpm to scan the fringe pattern across the picket fence. A photomultiplier tube then recorded the modulation of the light passing through the mask. The output of the tube was amplified and low-pass filtered en route to display on an oscilloscope where photographs were used to record selected scans. Each scan took 0.02 s and was sufficient to freeze a fringe pattern that was subjected to atmospheric scintillation noise. The intent of this demonstration experiment was to increase signal strength and, hopefully, improve the signal-to-noise ratio in the fringe detection at one interferometer mirror spacing. The scanning photometer is shown in Figure 5.10.

On 1964 April 7, Beavers described the above progress from his winter observing experience in Flagstaff in writing to Marshall Wrubel as well as to John Hall. He subsequently discussed his research at the Midwest Astronomers Meeting held at Yerkes Observatory on 1964 April 25. John Hall's son Richard spoke there as well and told his father that Willet's paper on stellar interferometry was the highlight of

Figure 5.10. The scanning photoelectric photometer introduced during Willet Beavers' second observing session at Lowell Observatory in 1964. (Photo courtesy of Willet Beavers.)

the meeting.[5] In the 1964 annual report for Lowell Observatory under the heading *Measurements of Stellar Diameter*, Hall wrote: "*Willet Beavers, working as a guest investigator on the 24-in Morgan telescope, succeeded in photographing interference fringes of bright stars which had been detected with a photoelectric photometer and presented on an oscilloscope* (Hall 1964)."

Beavers closed his 1964 April 7 progress report with a plan to replace the present photomultiplier with a more sensitive and stable tube and bemoaned the fact that an increasingly complex instrument that had to be hauled from Missouri to northern Arizona was subject to such frustrating single-point failure devices. During his first observing run with the photometer, his photomultiplier failed. Fortunately, Lowell astronomer Peter Boyce, who was working on spectroscopic applications of interferometry, had a spare that he loaned Beavers.[6] Beavers planned also to experiment with combinations of electronically amplifying only the relevant frequencies in the fringe pattern with the idea of filtering atmospherically induced frequencies as well as using faster scans and an improved recording system. John Hall was impressed with progress and plans for improvements. The Lowell director told Beavers that in a recent letter to

---

[5] Beavers 1964 May 1 letter to Hall and Hall's reply of 1964 May 12.
[6] Letter from Beavers to Hall 1964 November 30.

Marshall Wrubel he had "*mentioned the impressive progress which you have made during the last few months.*" One could only deduce that Beavers' success in interferometry was near at hand.

On his fourth and final observing run at Lowell, Beavers observed Arcturus and Vega during August 20–22 at a mirror separation of 6 ft at which Vega would still be a point source whereas Arcturus would be partially resolved. He recorded scans by photographing them on an oscilloscope screen and developed a methodology for extracting visibility information from a comparison of the scans from the two stars. Figure 5.11 shows a sample of those scan photographs in which fringes were certainly captured.

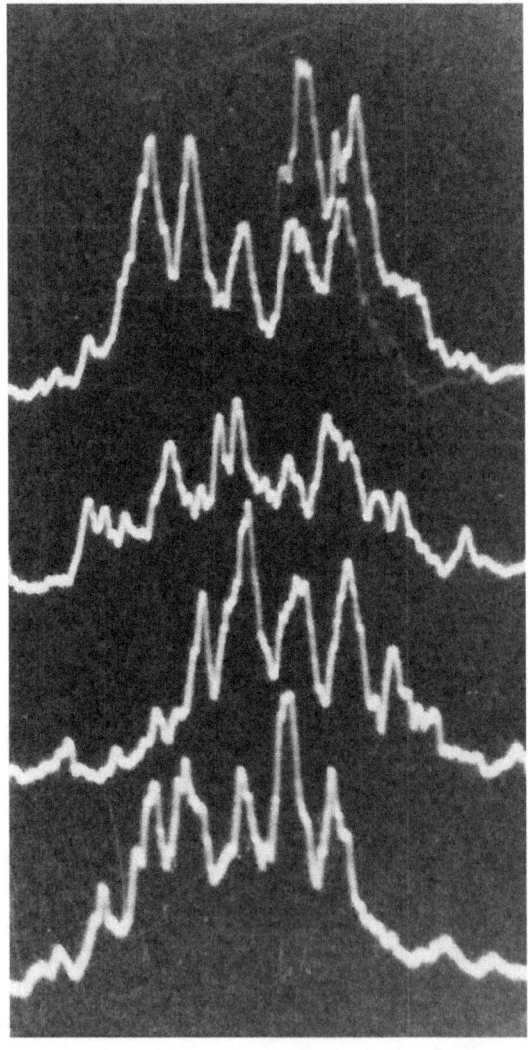

Figure 5.11. A sample of scans from the square-wave photoelectric scanner Beavers developed for recording fringe scans in order to calibrate data from the partially resolved Arcturus with that from the unresolved Vega. (Photo courtesy of Willet Beavers.)

In reducing data such as these, Beavers had to contend with atmospheric scintillation for which a full understanding was still to be developed. Nevertheless, he concluded that losses arising from astronomical seeing were likely correlated in the close reference beam whereas they were uncorrelated in the 6 ft mirror separation of the science beam. This leads to complications in the calibration of interferometric data, and, in the end, Beavers concluded that no quantitatively useful measures of stellar parameters resulted in spite of an angular radius for Arcturus from one data set indicative of a linear diameter 28 times solar. This was consistent with the Mount Wilson results obtained some 40 years earlier. Beavers very conservatively concluded that the inability to satisfactorily account for atmospheric effects that changed from night-to-night and with telescope pointing angle made the calibration process unreliable.

By the time he successfully defended his doctoral dissertation on 1966 January 3, Willet Beavers was already an assistant professor at Iowa State. Working virtually alone and starting from scratch, he had built only the third Michelson–Pease type stellar interferometer and succeeded in obtaining fringes with it. He devised a technique for producing electronic scans of fringes for which he created a calibration procedure, albeit with considerable reservation about its efficacy.

He decided that stellar interferometry was too dicey a field for a lone investigator to entirely stake his tenured future on. Besides, he would need a local telescope analogous to Lowell Observatory's Morgan telescope for instrument development and for carrying out a long-term observing program. Combining NSF and ISU Alumni Fund grants together for land and telescope purchases led to the construction of the Erwin W. Fick Observatory, home to a new 24 in telescope in a slide-off roof enclosure. Instrument developments dominated the next few years after Beavers' last Lowell trip for the interferometer. Beavers was quickly attracted to Roger Griffin's then new radial velocity spectrometer (Beavers & Eitter 1977) and, with the assistance of Joe Eitter, a 12 ft long vacuum photoelectric radial velocity spectrometer was fabricated and installed at the 24 in telescope's coudé focus. It would gather 16,000 stellar radial velocity measurements in its first two years of use.

But, he couldn't quite shake the interferometry urge altogether. Using an experimental setup with an artificial star, Beavers and electrical engineering graduate student W. D. Swift further developed the picket-fence fringe scanning photometer from his dissertation (Beavers & Swift 1968). They developed a practical method for extracting a calibrated measure of fringe visibility from an ensemble of scans. Using a heating source at different locations in their apparatus, the effects of seeing and scintillation could be introduced. They found that as seeing deteriorated, so did the fringe contrast. On the other hand, scintillation decreased measurement accuracy through the addition of noise that was highly correlated between the two interfering apertures, a situation that would not normally hold for an actual instrument.

The imperatives had turned to the velocity and lunar occultation programs, which required hundreds of nights of observing by both Beavers and Eitter. As a result, interferometry fell by the wayside as they pursued these highly productive programs.

Beavers never published the results of his doctoral research. As a result, his work went largely unnoticed except for inclusion in a handful of review articles, one of which acknowledged the photographic record of fringes and the development of the photoelectric fringe scanner (Code 1976). In another, it appears as the first in a list of "a number of authors" who have addressed how modern technology might overcome the limitations encountered in the Mount Wilson Interferometry (Davis 1976). When I mentioned Willet Beavers in a subsequent conversation with Doug Currie, whose own work will be discussed in Chapter 6, he mentioned that when he embarked on interferometry he obtained a copy of Willet's dissertation as it was a rare modern resource with actual observational experience to connect with the much earlier Mount Wilson work.

Beavers left Ames after 21 years for a position at the MIT Lincoln Laboratory. Today, in his Florida retirement (Figure 5.12), he is once again thinking about and experimenting with stellar interferometry. If anyone can pull this off, Willet can.

Willet Beavers came very close to successfully reintroducing the Michelson–Pease interferometer into the field equipped with the means for accurately sampling fringe visibility at multiple points descending toward the null. Who knows? With an appropriate detector, which would come in the near-term future, the 50 foot Pease interferometer might have been taken out of retirement and recommissioned with capacities for resolution, sensitivity, and accuracy beyond Pease's wildest imaginings. Instead, circumstances led Beavers to an entirely rational decision to change his research directions to techniques complementary to interferometry that would measure stellar motions and diameters and the vector separations of binary stars at angular scales that would elude Michelson interferometry for many years to come (see Beavers et al. 1982; Eitter & Beavers 1977; Beavers & Salzer 1983).

Figure 5.12. Willet I. Beavers, 2020 March. (Author's photo.)

5.6 Evgeni Stepanovich Kulagin Resolves Capella

Another isolated interferometry flareup was made at the Pulkovo Observatory in St. Petersburg, Russia (then Leningrad, USSR) by E. S. Kulagin. His work overlapped in time with Beavers' but they were unaware of each other. Kulagin's venture into interferometry—a departure from his normal spectroscopic work to which he would soon return—was enabled through the instrument's design by Vladimir P. Linnick (1889–1984). I will be brief here as this interesting episode is not directly relevant to this book's theme and has recently been well documented by Bonneau (2019). In essence, Kulagin used the fixed 6 m baseline Pulkovo interferometer to measure orbital motion in the Capella system. As with Anderson and Merrill, his work was done visually. He sampled the baseline at projected separations afforded by Earth rotation, carefully watching for the disappearances of the fringes for several hours on either side of the meridian. Kulagin collected 12 observations of the binary from early 1968 December into 1969 mid-February and reported his results in a brief paper in 1970 (Kulagin 1970).

If we examine Kulagin's Capella results as we did Anderson's and Merrill's measurements in Section 2.3, we find mean residuals from the modern orbit (Torres et al. 2015) of $\overline{\Delta\theta} = 0.2° \pm 1.7°$ and $\overline{\Delta\rho} = -2.7 \pm 1.5$ mas. Referring back to Table 2.1 and glancing at Figure 5.13, we see that Kulagin's results exhibit the same behavior as did the Mount Wilson observations. Once again the mean of the residuals in position angle shows no systematic effect, but that in angular separation is systematically small. We attributed this to an uncertainly in the effective wavelength of the Mount Wilson observations, and that is likely to be the case for Kulagin's visual observations as well. Kulagin plotted his measurements against Merrill's orbit, and, unsurprisingly, found good agreement.

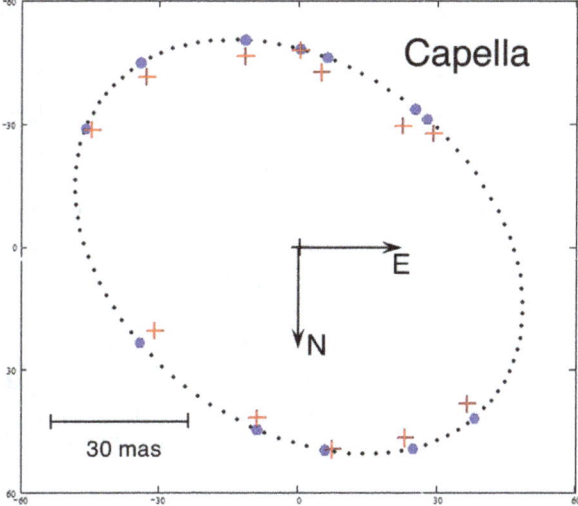

Figure 5.13. The 1968–1969 Measurements of Capella by E. S. Kulagin plotted against the orbit of Torres et al. (2015) show the same systematic trend in angular separation as did the Mount Wilson observations from the early 1920s. (Author's diagram.)

5.7 Looking to the Future—The Woods Hole Summer Study

In the summer of 1916, George Ellery Hale's efforts to establish an organization under the auspices of the National Academy of Sciences (NAS) to buttress the Nation's security came to fruition. As a result of his chairing the committee for this purpose and becoming personally engaged in lobbying President Wilson—a challenge as Wilson was anxious to maintain U.S. neutrality in the War—Hale was naturally selected as the new National Research Council's first Chairman. To this day, the NRC acts to coordinate scientific, engineering, and educational activities for national and international benefit. Much of the NRC's work is done through study committees that target important national issues and develop relevant goals and strategies toward their solution.

In another summer, some forty years after its founding, the NRC convened a panel at the NAS–NRC Woods Hole Summer Studies Center on Cape Cod. The topic of "Synthetic-Aperture Optics" was considered during 1967 August 7—September 1 by 13 "full-time participants" and 43 "visitors." The study, which was sponsored by the Air Force Systems Command, was directed by Stanford's Joseph W. Goodman. The panel's charge was *"to consider the feasibility of increasing optical resolution by what may be termed synthetic-aperture optics. This term is defined here in the very broad sense to include any technique for achieving, with one or more small apertures, the resolution normally associated with a single large aperture."* The work product subsequently appeared in two volumes.[7] The first amounts to an overview of interferometry followed by sketches of what would become known as adaptive optics, imaging with partially-filled apertures, imaging by laser illumination of objects, super-diffraction limited imaging, and other a posteriori imaging methods. The 440 page Volume II consists of 34 appendices contributed by the participants and forming the real meat of the Summer Study.

I go into considerable detail in these paragraphs because this specialized enclave on Cape Cod in the summer of 1967 was a milestone for optical interferometry. For the first time,[8] national goals would be set for the field because of its national-defense relevance, and specific recommendations would be made to foster interferometry's advancement toward next-generation deployable instruments. While astronomy was not the primary intended beneficiary, it would inevitably undergo huge boosts in capability, especially after new technologies developed for military applications would be declassified. Charge-coupled devices (CCDs) are the obvious example. Hale really knew what he was doing when he invested so much time in establishing the NRC!

The specific recommendations for interferometry may seem modest from today's perspective, but the summer study was picking up where Michelson and Pease left off, i.e., with visual detection of fringe nulls. The only interim effort to modernize interferometric observational technique was that of Willet Beavers. Thus, the first priority was to develop electronic fringe-amplitude detection. That effort "should not exclude" techniques for measuring phase, which along with amplitude is needed for imaging. The second recommendation called for exploration of "Jennison three-

---

[7] Summer Study Participants 1967 Synthetic-Aperture Optics *Woods Hole Summer Study August 1967* NAS–NRC Pub, 2 Vols.

[8] In the first of the NAS–NRC decadal reviews of astronomy and astrophysics—the so-called "Whitford Report" of 1964—interferometry was only addressed in the radio astronomical context.

Figure 5.14. Richard H. Miller in 1966. (Courtesy of the Special Collections Research Center, University of Chicago Library.)

element interferometry techniques." This refers to the measurement of what is more commonly called "closure phase," an elegantly simple tool (see Appendix B.5.5) that extracts the phase information from atmospherically- and instrumentally-induced phase errors. Closure phase was developed by Roger C. Jennison (1922–2006), who worked at the Jodrell Bank radio observatory under Hanbury Brown (Jennison 1958). As we shall see, the method would be utilized for imaging of close binaries from Mount Wilson decades after the Woods Hole gathering. Lastly, the development of a delay line was recommended for future work.

Among the Summer Study's participants were two of particular relevance to this book. The first is Richard H. Miller (1926–2020), shown in Figure 5.14 at a blackboard on the University of Chicago campus in March 1966. I had met and corresponded with him years ago and intended to talk with him about his early work in interferometry. I was very saddened to learn that he passed away just a month prior to my starting this section about him.

Dick Miller's primary specialty was computational astrophysics with a particular interest in modeling galaxies and large-scale structure of the universe. His expertise in information processing enticed him to other sidelines, one of which was optical interferometry. In an article published in *Science* (Miller 1966) the year prior to the Summer Study, Miller described how advances in technology since Michelson and Pease led to new opportunities for interferometric measurements of stellar diameters. He pointed out that the field had not been addressed in the "Whitford Report"[9]—the first of what have become known as the NAS–NRC "decadal reviews" of astronomy and astrophysics. This 1967 Summer Study would correct that oversight, and Miller would play a major role as one of the invited full-time participants.

---

[9] Various Panelists 1964 Ground-Based Astronomy—A Ten-Year Program NAS–NRC Pub.

In that *Science* article is Miller's concept of a 1 km baseline interferometer that he would further develop for the Summer Study in his Appendix V on a "Description of a Large Stellar Interferometer." His basic concepts for a two-element, north–south oriented interferometer are shown in Figures 5.15–5.17. Miller refined this design

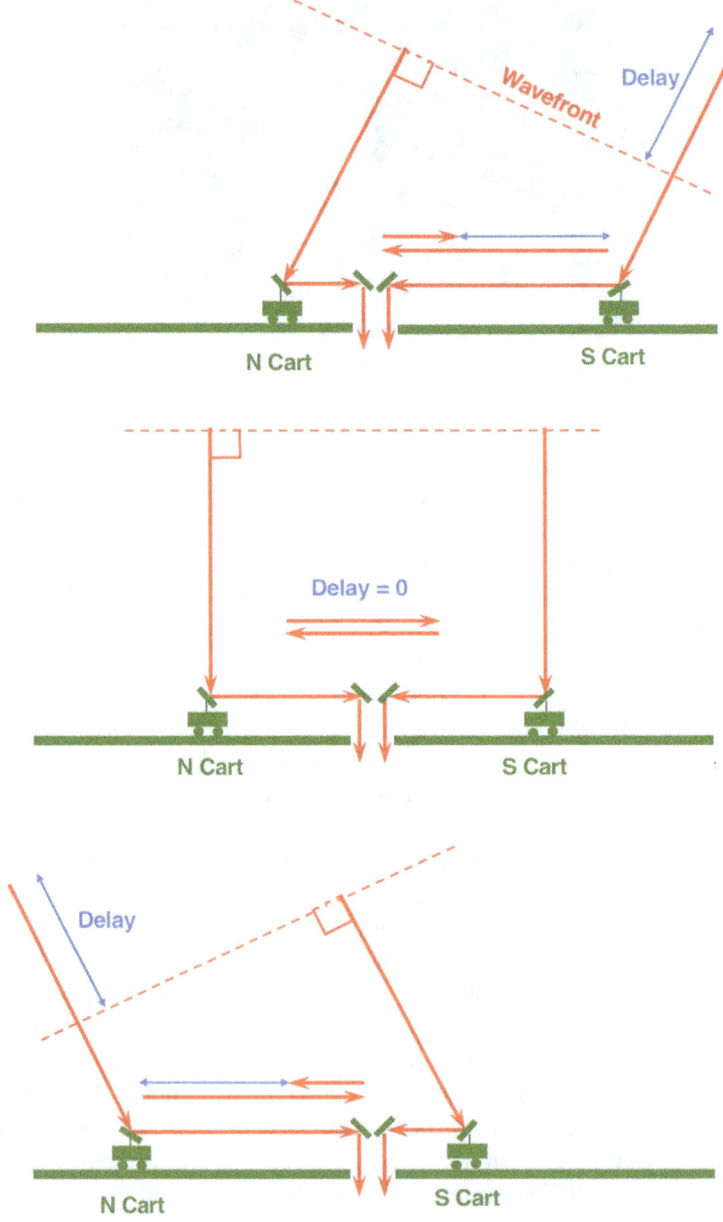

Figure 5.15. Miller envisioned two north–south oriented "trolley" carts with light-collecting mirrors directing beams toward center turning mirrors that inject them into a stationary beam combiner. The carts are positioned with respect to combiner so as to compensate for the declination dependent optical "delay." (Author's diagram.)

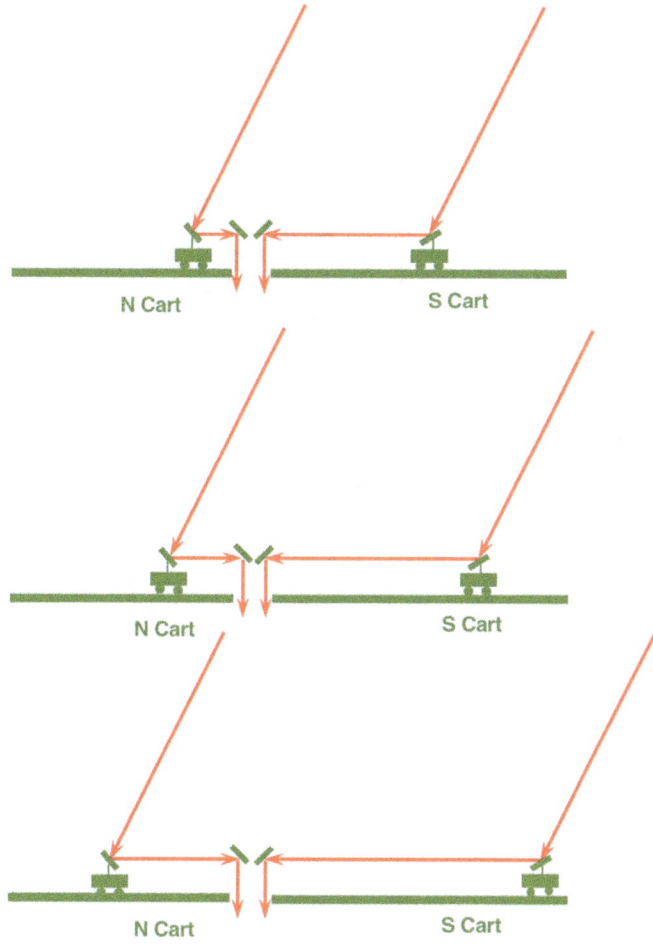

Figure 5.16. The cart spacing can be increased or decreased to sample different baselines while preserving delay compensation as shown above. (Author's diagram.)

during the spring of 1970 and the first quarter of 1971 as a visiting scientist at Kitt Peak National Observatory in Tucson. His final iteration on the interferometer (Miller 1971) was published as an engineering technical report, a medium he used for additional reports on fringe detection (Miller 1970a) and seeing modeling (Miller 1970b). Regrettably, these early publications are not easily obtainable.

Following the concept of Figure 5.15, Miller's design employed two 36 inch diameter flat mirrors to collect light and send it to the central beam combiner through a series of evacuated pipe sections to mitigate ground-level seeing effects. It would later be shown that such a system would be subject to losses in visibility arising from the polarization mismatch in the beams coming off the two flats, a shortcoming that can be fixed by a multi-element beam collector rather than a single mirror (Traub 1988). Figure 5.18 shows a layout for a facility with surrounding embankments to protect the telescope trolleys from local wind effects. The

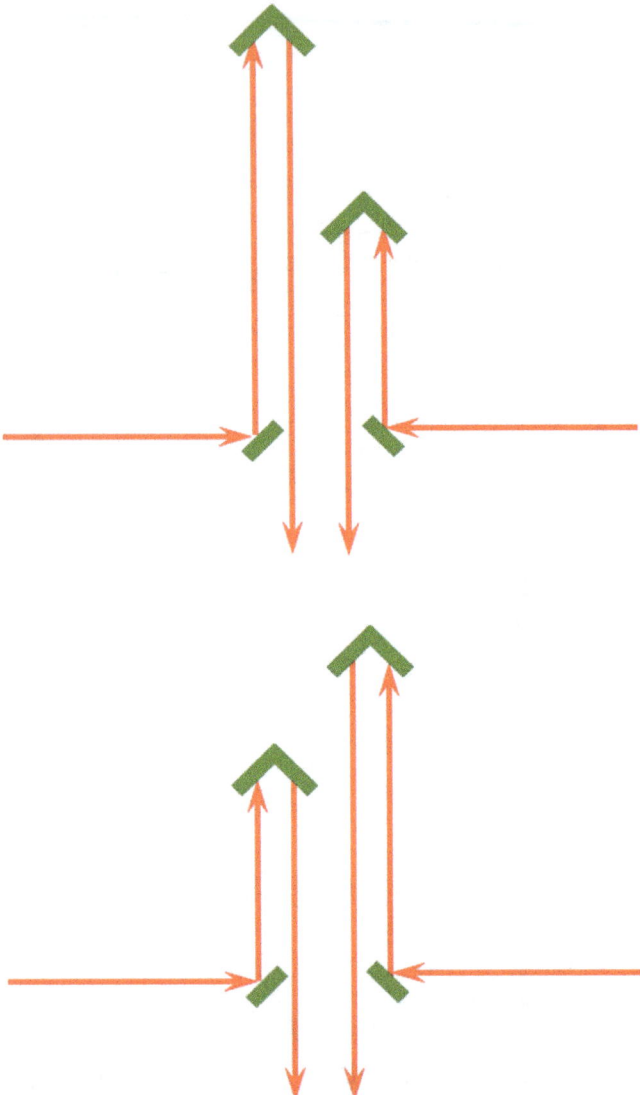

Figure 5.17. Miller's optical "trombone" scheme provides final correction to the first-order delay compensation shown in Figures 5.15 and 5.16. (Author's diagram.)

interferometer included such features as atmospheric dispersion compensation and laser metrology for delay measurement. A system error budget was calculated for each of the instrument's components. The final product was a highly-refined concept for a modern interferometer ambitious in terms of its long baseline.

The scope of this analysis conveyed a high degree of feasibility. Miller's status as a visiting astronomer left the national observatory without an in-house proponent of interferometry at a time when many initiatives were being advocated by the constituents of a wide variety of pressing science goals. Most astronomers considered

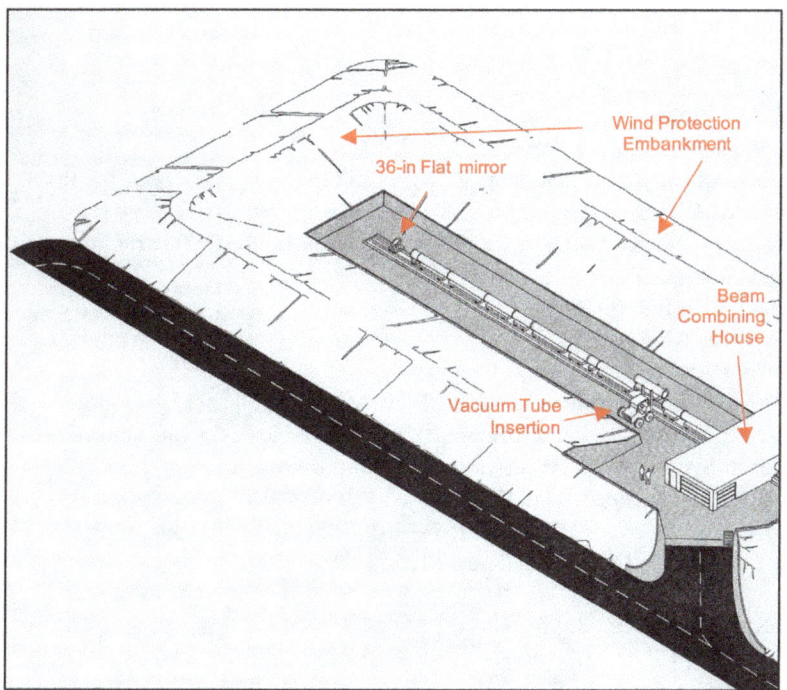

Figure 5.18. One half of Miller's concept for an optical interferometer is shown in this 1971 Kitt Peak National Observatory engineering drawing. Note the tractor removing a section of the vacuum tubing so that the 36 inch flat can be relocated to create a shorter baseline. (Courtesy NOIRLab/NSF/AURA.)

interferometry to be an arcane, niche technique capable of making important but narrow scientific contributions to stellar astrophysics. His concept was never pursued by the national observatory. Nevertheless, many of the features he developed have been incorporated into the current generation of operational interferometers, including the CHARA Array on Mount Wilson. Dick Miller's contribution to the field are therefore enduring.

Douglas G. Currie, of the University of Maryland, was the second participant at Woods Hole relevant to the Mount Wilson interferometry story. Whereas Dick Miller laid out the components of long-baseline interferometers on paper, another generation would lay them out on the grounds of Hale's mountain. Doug Currie would himself bring to reality the interferometer concept he presented at the NRC Summer Study and mount it on the venerable 60 inch telescope. We'll save the full story for the next chapter to place it in the timeline of Mount Wilson's awakening to interferometry.

References

Adams, W. S., Joy, A. H., Stromberg, G., & Burwell, C. G. 1921, ApJ, 53, 13

Argyle, B., Swan, M., & James, A. 2019, An Anthology of Visual Double Stars (Cambridge: Cambridge Univ. Press), 28

Beavers, W. I. 1963, AJ, 68, 273

Beavers, W. I., Cadmus, R. R., & Eitter, J. J. 1982, AJ, 87, 818

Beavers, W. I., & Eitter, J. J. 1977, PASP, 89, 733

Beavers, W. I., & Salzer, J. J. 1983, PASP, 95, 79

Beavers, W. I., & Swift, W. D. 1968, ApOpt, 7, 1975

Beckers, J. M., Ulich, B. L., Shannon, R. R., et al. 1981, in Telescopes for the 1980s, ed. G. Burbidge, & A. Hewitt (Palo Alto, CA: Annual Reviews), 63

Bonneau, D. 2019, Mieux Voir Les Étoiles Ier Siècle de L'interférométrie Optiq (France: EDP Sciences), 209 pp

Calder, W. A. 1931, BHarO, 885, 8

Code, A. D. 1976, ARA&A, 11, 264

Davis, J. 1976, PASAu, 3, 26

Davis, J., Tango, W. J., Booth, A. J., et al. 1999, MNRAS, 303, 773

Donegan, J. F., Weaire, D., & Florides, P. 2012, Hutchie: The Life and Works of Edward Hutchinson Synge (1890–1957) (Austria: Living Edition), 155 pp

Eddington, A. S. 1923, Natur, 111, 572

Eitter, J. J., & Beavers, W. I. 1977, ApJS, 34, 493

Elkin, W. L. 1887, Yale Univ. Obs. Trans., 1, 255

Finsen, W. S. 1926, CiUO, 68, 343

Finsen, W. S. 1928, CiUO, 78, 35

Finsen, W. S. 1933, CiUO, 90, 379

Finsen, W. S. 1951a, Obs, 71, 42

Finsen, W. S. 1951b, PA, 59, 399

Finsen, W. S. 1964, AJ, 69, 319

Finsen, W. S. 1971, Ap&SS, 11, 13

Hall, J. S. 1964, AJ, 69, 686

Hanbury Brown, R. 1956, Natur, 178, 1046

Hanbury Brown, R., Davis, J., & Allen, L. R. 1974, MNRAS, 167, 121

Hanbury Brown, R., & Twiss, R. 1956, Natur, 177, 27

Harper, G. M., Brown, A., Guinan, E. F., et al. 2017, AJ, 154, 11

Harwood, M. N. 2020, Biography of Professor Edward Drake Roe, Jr., Founder of Pi Mu Epsilon, http://pme-math.org/2515-2

Jennison, R. C. 1958, MNRAS, 119, 3

Kulagin, E. S. 1970, SvA, 14, 445 (English translation of original from 1970, AZh, 47, 557)

Lee, O. J. 1921, AJ, 33, 146

Lundmark, K. 1933, in Handbuch der Astrophysik—Das Sternsystem: Erster Teil, ed. H. D. Curtis, B. Lindblad, & K. Lundmark (Berlin: Springer), 587

Luyten, W. J. 1933, ApJ, 78, 225

Maggini, M. 1922, Contributi del R. Osservatorio Astrofisico de Catania, no. 4

Maggini, M. 1925, MmSAI, 3, 231

Maggini, M. 1928, AN, 233, 97

Maggini, M. 1929, AN, 235, 135

Maggini, M. 1934, MmSAI, 8, 167

Mason, B. D., & Hartkopf, W. I. 2005, IAUDS, 156, 1

Mason, B. D., & Hartkopf, W. I. 2006, in Proc. IAU Symp. 207, ed. W. I. Hartkopf, E. F. Guinan, & P. Harmanec (Cambridge: Cambridge Univ. Press)

Miller, R. H. 1966, Sci, 153, 581

Miller, R. H. 1970a, A Fringe Detector for Use with Michelson Stellar Interferometers, AURA Engineering Technical Report No. 29

Miller, R. H. 1970b, A Phenomenological Representation for Seeing, AURA Engineering Technical Report No. 31

Miller, R. H. 1971, A 100-meter Michelson Stellar Interferometer, AURA Engineering Technical Report No. 40

Radhakrishnan, V. 2002, PhT, 55, 75

Roe, E. D. 1921, PA, 29, 546

Schlesinger, F. 1921, AJ, 33, 130

See, T. J. J. 1921, Natur, 106, 663

Spencer Jones, H. 1921, Natur, 107, 685

Spencer Jones, H. 1922, MNRAS, 82, 513

Synge, E. H. 1930, London Edinburgh Dublin Philos Mag J Sci, 10, 291

Tango, W. 2006, A&G, 47, 438

Torres, G., Claret, A., Pavlovski, K., & Dotter, A. 2015, ApJ, 807, 26

Traub, W. A. 1988, in NOAO-ESO Conf. High-Resolution Imaging by Interferometry, Part II, ed. F. Merkle (Garching: ESO), 1029

Van Biesbroeck, W. H. 1927, ApJ, 54, 78

Van den Bos, W. H. 1927, BAN, 4, 103

Van den Bos, W. H. 1951, JO, 34, 85

Weaire, D., Donegan, J. F., & Florides, P. S. 2012, PhyW, 25, 26

Wilson, A. G. 1956, AJ, 61, 328

Chapter 6

The Great Reawakening of the 1970s

6.1 Antoine Labeyrie Lights a New Torch

The year 1970 was a watershed for astronomical interferometry. The 27 year old astronomer Antoine É. H. Labeyrie published a paper based on his doctoral dissertation at l'Observatoire de Meudon presenting his invention of a new and elegantly simple technique for thwarting the blurring effects of atmospheric turbulence (Labeyrie 1970; Figures 6.1 & 6.2). Labeyrie's Speckle Interferometry is one of those why-didn't-I-think-of-that ideas, the inklings of which had been under the noses of astronomers for generations. We have already seen in Section 5.2 how double-star observers like Finsen and van den Bos had spoken about visually seeing "images within the image" in the form of brief manifestations of multiple occurrences of the binary populating the seeing disk's core. Labeyrie's insight resulted from a convergence of understanding the underlying cellular structure of turbulence-fed "seeing" along with simple Fourier techniques for analyzing the analogous phenomenon of laser speckle. He credited the French optical designer and engineer Jean Texereau (1919–2014) with the first accurate description of the speckle phenomenon in the context of large-aperture telescopes (Texereau 1963). But, it was Labeyrie's inspired jump toward diffraction-limited imaging that created a brand-new observing methodology in astronomy. That gift continues to give to the present day. Measuring orbital motions in binary star systems has been the hallmark of speckle interferometry with thousands of measurements appearing in the literature since Labeyrie's first demonstration of his technique.

Radial velocity studies of binaries can often characterize orbits about the center of mass for both components of a binary in a plane perpendicular to the plane of the sky. Techniques that provide angular resolution can specify the relative orbit in the plane of the sky. If these complementary methods are applicable to a specific system, it is then possible to garner all that is needed to solve Kepler's third law for the masses of the component stars. The primary pathway toward determining stellar masses is via binary star orbit determination. The catch was that observational

Figure 6.1. Antoine Labeyrie relaxes supine in the snow on Mount Palomar while his colleague Laurent Koechlin looks on after observing at the 200 inch telescope in 1977 January. (Author's photo.)

Figure 6.2. Deane M. Peterson (SUNY Stony Brook) and the author were the speakers at the 2000 April 28 symposium in honor of Antoine Labeyrie's award of the 2000 Benjamin Franklin Medal in Engineering for his invention of speckle interferometry. Deane and I flank Antoine in this photo taken after the ceremony at the Franklin Institute in Philadelphia. (Photograph courtesy of Susan J. McAlister.)

selection effects typically led to visual binaries with velocity variations too small to be measured spectroscopically. Now comes along Labeyrie's new method that could achieve diffraction-limited resolutions of 30–50 mas with the largest telescopes. New opportunities suddenly arose for reaching into the realm of the spectroscopic binaries and making a substantial increase in the collection of accurately determined stellar masses (McAlister 1976).

Speckle interferometry of binary stars would be carried out by visiting astronomers at Mount Wilson's 100 inch telescope as early as 1985 and continue periodically after that. We will focus a bit later in this chronological compilation on that application of this powerful technique. Labeyrie's broader impact has been well acknowledged elsewhere, including in a number of the reference sources recommended in Table B.1, so we will not describe his ideas and achievements in detail. This in no way diminishes Antoine Labeyrie's remarkable contribution to astronomy, which incidentally launched my own career. A semi-technical overview of binary star speckle interferometry is given in Appendix B.7.

That first speckle paper was soon followed by a series containing measurements of diameters of stars that had not been measurable since Pease's time and orbital motions in spectroscopic binary stars. This seminal work led to considerable effort in developing speckle imaging techniques that could be applied to a wide variety of astronomical objects with complex extended structures such as spotted stars. The challenge was to reliably reconstruct the phase information lost in the original speckle interferometry process as shown schematically in Figure B.29. In his 1982 review article on "Astronomical Speckle Imaging," Richard H. T. Bates (1929–1990) commented that "*While speckle imaging shows promise, none of its manifestations seems to be as robust as speckle interferometry* (Bates 1982)." Another important early review of the field is that of J. C. Dainty (1984).

In the interim since those reviews, reliable methods have been developed for imaging binary stars, which, as a pair of δ-functions typically with unequal amplitudes, are the simplest of imaging problems. This progress has been particularly enhanced by detector advances such as electron-multiplying CCDs (EMCCDs; Scott 2018). Speckle interferometry/imaging has now taken over the angular measurement of visual binaries, with the obsolete adjective "visual" essentially meaning something like *angularly resolved on the plane of the sky*. Of course, astronomy is notorious for hanging onto obsolete terms. The bottom line is that Labeyrie's insights led to a technique that replaced its predecessor—the visual micrometer—on the basis of high angular resolution, higher accuracy, and higher observing efficiency. What more could one ask for? I emphasize this as a prelude to taking a closer look in Section 7.1 at the binary star speckle interferometry that has been carried out on Mount Wilson.

Soon after launching speckle interferometry, Labeyrie expanded his attention to include interferometry from telescope pairs while also undertaking additional speckle observations using a digital speckle camera (Blazit 1977). He had put together a very talented group of other young French astronomers to tackle all the interferometry he had in mind. By the summer of 1974, he had already built and recorded fringes from the bright star Vega using two 25 cm aperture telescopes with a 12 m baseline (Labeyrie 1975). This facility at the Nice Observatory was a potentially expandable prototype, but Labeyrie was already thinking well past that instrument to an array of larger telescopes of a radically different design. Accomplishing this goal required more real estate than was available at Nice, and only two years after the Vega fringes, Labeyrie had relocated the Nice interferometer to the 1300 m altitude Calern Plateau in southern France. The maximum baseline had been increased to 20 m (Blazit et al. 1977). One throwback of the first observations from Calern is that they were done visually rather than electronically. With this approach, Labeyrie and his colleagues were able to estimate the diameters of the A and B components of Capella doing exactly what Pease had done by locating the baseline at which fringes disappeared. The fringes arising from the individual star diameters were modulated by the fringes from the binary, but, regrettably, their results report does not include any images of

Figure 6.3. One of the pair of boules telescopes of the GI2T. (Author's photo.)

those fringes. Their estimated diameters for the Capella stars are $\Theta_{UD,A} = 5.2 \pm 1.0$ and $\Theta_{UD,B} = 4.0 \pm 2.0$ mas.

The successor to Labeyrie's "small" interferometer, which would be called the Grand Interferomètre avec deux Tèlescopes or GI2T, took a decade to develop (Figure 6.3). Its design is, to say the least, novel. To point the 1.5 m aperture telescopes around the sky, Labeyrie housed the primary mirror in a large sphere made of ferroconcrete with a cylinder of the same material supporting the secondary mirror. The sphere rests in a ring with actuators that point it around the sky. Labeyrie calls this design his "boules" telescopes (Labeyrie et al. 1986).

And, of course, Labeyrie has conceived of space-based interferometers, but we must stop here. Antoine Labeyrie deserves an entire book, which will no doubt come at some point in the future. For now, though, this particular book must get back on its Mount Wilson track.

6.2 William C. Wickes' Perfection of the Anderson Interferometer

After a 35 year hiatus, interferometry returned to Mount Wilson just after Thanksgiving 1972 when Princeton postdoc Bill Wickes brought out an instrument he had developed under the supervision of Robert H. Dicke (1916–1997).[1] Dicke

---

[1] Unless other noted, the details provided in this section result from conversations and correspondence with Bill Wickes.

had a wide range of interests that he attacked with brilliance and depth, ranging from lock-in amplifiers (he founded Princeton Applied Research) to gravitation theory (e.g., the Brans–Dicke scalar-tensor theory) and cosmology. His instrumentation efforts had typically been at microwave frequencies, but the cosmological challenge of measuring the primordial helium abundance led him to consider a means of measuring the mass of a very old (and very rare) stellar population II binary star system—μ Cassiopeiae. In combination with spectroscopy, one could then determine the mass of helium in the star, which was predicted to near zero by the Brans–Dicke theory's rapid Big Bang rather than the ~25% predicted in the Einsteinian scenario. This required determining the orbital elements of the system by resolving the companion, which is more than 100 times fainter than the primary star. The challenge was not so much in terms of resolution but rather in the very large dynamic range in sensitivity required of the instrument. Toward this end, Dicke thought some sort of scanning double-star interferometer would do the trick. He went looking for a graduate student, and Bill Wickes stepped up to the plate in 1969.

Wickes was born into a physics and technology heritage—his father, William H. Wickes, was a mechanical engineer with an undergraduate degree in physics who was part of the Rockwell International team that developed the environmental control system for the Apollo command module. His mother, Nancy R. Castles, was an Australian war bride who immigrated to the U.S. with her Air Force veteran husband and Bill's 6 month old sister in 1945 September. Bill was born the following November.

After growing up in the Los Angeles area, Wickes entered UCLA to major in physics. As with many of us, his freshman year was "undistinguished," but he got his act together as a sophomore and attracted the attention of Professor Alfred Wong, for whom he worked during the summer months until graduation. Impressed by Wickes, Wong advocated for his student's admission to the Princeton physics doctoral program, from which Wong himself had received his PhD. Wong's recommendation did the trick and Wickes headed across the U.S. to New Jersey.

Wickes was not particularly interested in astronomy but subsequently realized that his fascination with such topics as Mach's Principle was cosmological in nature. He had finished his prep work for the physics general exam when he heard about Dicke's search for a graduate student to work on what amounted to an observational cosmology project. To Wickes, it "*sounded like an intriguing challenge and* [gave] *the opportunity to work with a luminary like Dicke.*" So, it was natural for him knock on Dicke's door. Wickes describes his initial meeting with Dicke who "*being the digging-tiny-signals-out-of-noise genius that he was, came up himself with the general idea for the rotating double-aperture interferometer using a chopping wheel to detect fringe visibility. When I first met him, he sketched out the idea for me on a blackboard, and I went from there.*" Dicke was busy with other ongoing projects and as department chair, so he asked his postdoc Paul Boynton—who would join the University of Washington physics faculty the following year—to get Wickes going on the project. When Boynton departed Princeton, Wickes was thereafter pretty much on his own, with occasional consultation with Dicke.

Wicke's realization of Dicke's blackboard sketch is shown in Figure 6.4. It was called an automatic interferometer (Wickes & Dicke 1973). You will recall that

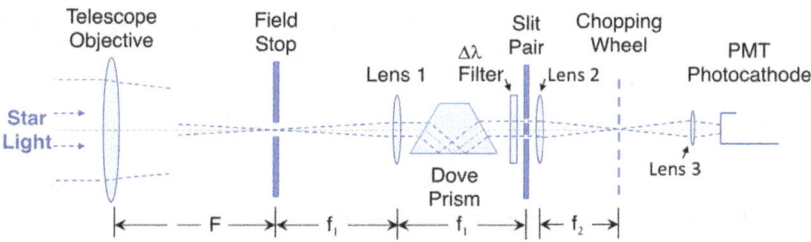

Figure 6.4. The Wickes–Dicke Automatic Interferometer with starlight entering at left and visibility signal detected at right. See the text for a description of the components. (Author's diagram adapted from Figure 1 of Wickes & Dicke 1973. © IOP Publishing. Reproduced with permission. All rights reserved.)

Anderson's double-star interferometer produced variations in fringe visibility as the "slit" pair (actually circular apertures) was rotated during visual inspection of a binary star. By noting the slit position angles at maximum and minimum visibility, the binary star geometry could be deduced. Wickes had fleshed out Dicke's concept to produce an interferometer that scanned through the double aperture position angle Ω and produced quantitative measurements of visibility as a function of Ω. This would have thrilled Anderson and Pease. From left in Figure 6.4, starlight enters the telescope objective (shown schematically here as a lens) that converges the plane starlight wavefronts to a focus at which a small aperture field stop is located to limit the field of view to ~5 arcsec. Lens 1 then collimates the diverging beam after the field stop, and the resulting plane waves pass through a Dove prism[2] mounted in super-precision ball bearings to allow the prism to rotate the emerging collimated beam about the optical axis. The bearing assembly is belt-driven by a DC motor and locked at a selectable rate between 0.5 and 2.0 Hz by a servo sensing the voltage produced by a second motor driven by the first. The beam (rotating at twice the prism rate) then passes through an interference filter of bandwidth 100 nm centered on 550 or 725.5 nm wavelength, depending on the star of interest. The filter enables two-color fringe detection while also yielding a workable coherence length (see Appendix B.2.2). Rotation of the Dove prism relative to the fixed double apertures maintains a fixed alignment of the fringes with the chopping wheel spokes.

The spectral-bandwidth limited and collimated beam next encounters a two-aperture field stop to serve as the interferometric slit pair with separation S that projects onto the telescope entrance pupil to have an effective separation leveraged to $S \cdot F/f_1$. This yields a theoretical resolution limit for binary stars of $\lambda f_1/2SF$. The slit pair and Lens 2 are rigidly connected and translatable along the optical axis to provide a fine stellar image focus onto a glass "chopping wheel" (4800 1/8 inch radial opaque "spokes" on a 1 inch radius circle—a challenging pattern to make—that was provided by the David Sarnoff Research Center in Princeton) for which the rectangular-wave transmission pattern has the same spatial frequency as the interference fringes. The wheel is spun at 5 Hz by a synchronous motor so as to rotationally chop the beam at 24 Hz. Finally, Lens 3 reimages the double aperture

---

[2] The Dove prism is capitalized in recognition of its non-avian inventor Heinrich Wilhelm Dove (1803–1879).

onto the photocathode of a photomultiplier tube (PMT), effectively locking the image onto a fixed region of the photocathode to eliminate any variation in the detected intensity that might arise from the image moving due to the beam rotation or star image movement from seeing or telescope motion.

The photocathode yields a light intensity signal with a component, modulated at the chopping frequency, that is proportional to the fringe visibility. That signal can be bandpass filtered to isolate the visibility component, with a small bandwidth to allow for stellar motion due to telescope drift and atmospheric turbulence. Visibility variation due to atmospheric turbulence, again at low frequencies, will appear as noise on the successive measurements of visibility variation with Ω. That noise will average out over many rotations of the light beam, typically many hundred for each individual data set. (See Appendix B.5.2 for more on signal filtering.) This approach of contriving to move a small desired signal to a predetermined frequency where it can be isolated from much larger noise left behind at another frequency is classic Robert Dicke methodology.

Interferometers typically utilize narrow spectral bandpasses in order to produce a larger coherence region within which fringes form. This makes fringes far easier to find and maintain, but it comes at the sacrifice of light and produces a brighter limiting sensitivity than one might hope for. This shortcoming inspired Wickes to devise a clever and effective way of having a broad bandpass to improve sensitivity and signal-to-noise ratio to reach fainter stars and improve measurement accuracy. His achromatic interferometer (Wickes & Dicke 1974) innovation in Figure 6.5 modified the light path of Figure 6.4 starting from the Dove prism and extending through detection. The innovation derives from relocating the double-slit mask to a re-imaged telescope entrance pupil downstream from the chopping wheel, which behaves as a diffraction grating that feeds its various orders through the slits. As shown in the right side of Figure 6.5, the arrangement ends up placing sweet spots in

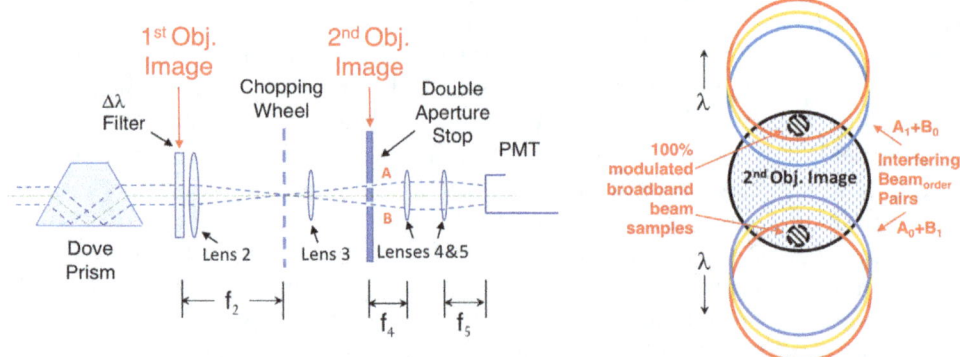

Figure 6.5. The Wickes–Dicke achromatic (broad-band) interferometer resulted from modifications to the configuration of Figure 6.4 that take advantage of the diffraction grating performance of relocating the fringe-forming double-aperture stop to a reimaged telescope entrance pupil in which two discrete regions of that pupil contain fully-modulated fringe samples from the A and B sides to be imaged with an appropriate mask onto the photomultiplier tube's photocathode. (Author's diagram adapted from Figures 5 and 7 of Wickes & Dicke 1974. © IOP Publishing. Reproduced with permission. All rights reserved.)

the pupil imaged on the detector that contain light across much of the spectrum from the zeroth- and first-orders of diffraction combined in interfered pairs of the A and B sides of the slit mask as indicated. Both beams possess all the modulation induced from the chopping wheel on the fringes. This yields a broad-band interferometer free of spectral dispersion and smearing that might have bedeviled the instrument had a less thoughtful way of utilizing a diffraction grating been incorporated. Wickes would use his achromatic interferometer for the majority of his observations.

He described his presentation of this radical redesign to his advisor whose concept was the basis of the original "automatic" device. *"I do recall that a few years later I was able to do a blackboard sketch for Bob* [Dicke], *showing my notion that the interferometer could be made achromatic by moving the double aperture behind the chopping wheel. This makes the wheel act as a diffraction grating, such that the separation of the images of the two apertures on the primary mirror is proportional to the light wavelength. As the wheel turns, the images go light and dark, from all wavelengths together. The ideal situation is limited because different wavelengths don't pass through quite the same atmospheric paths, which diminishes the net fringe visibility. So, I continued to use an optical bandpass filter to limit the spread-out aperture size rather than the optical bandwidth."*

Wickes developed a six-parameter model to fit the fringe modulation signal. Two parameters of that fit are the angular separation ρ and position angle θ of a binary star as defined in Appendix B.3.2, and a third yields the brightness ratio of the two stars. Also shown is the effective rotation of the slits with respect to the sky. This results in each increment of Ω, the orientation of the interferometer aperture pair on the sky, sampling a different component of the angular separation projected onto the orientation line joining the slits. See the 1973 paper by Wickes and Dicke for full details on the data analysis process (Wickes & Dicke 1973) and Appendices B.3.2 and B.4.2 for more on the interferometric measurement of binary stars. The signal generated by the interferometer inevitably contains noise arising from atmospheric seeing fluctuations. Additionally, systematic distortions in the curve result from focus variations as the aperture pair samples different parts of the telescope optics.

But, Mount Wilson was not his first stop. After testing everything on Princeton's 36 inch telescope as a preparation, Wickes headed for Kitt Peak where he would mount his interferometer on one of the national observatory's 36 inch telescopes. He reckoned that the observations would benefit from the much better seeing on a 6900 ft mountain in southwest Arizona compared with that provided by the sky over Princeton, New Jersey. Ironically, he found that the Kitt Peak telescope suffered from astigmatism to an extent that the biases resulting from sampling different regions of the telescope's primary mirror introduced variations in fringe visibility that required major modifications to the observing and reduction processes. Thenceforth, each measurement of a double star was modified to incorporate alternating observations of a bright single star as close by in the sky as possible. The latter provided an ongoing normalization of the double star data, correcting for the imperfect telescope optics.

Wickes then turned to the U.S. Naval Observatory in Washington, DC where he attached his interferometer to the 26 inch refracting telescope in 1972 March. Here he would find success. He was hosted by Charles E. Worley (1935–1997), one of the last and most-skilled visual observers of double stars (Mason et al. 2007). Following a very successful observing experience with Worley, Dicke gave Wickes the go-ahead to write up what he had done in terms of developing the instrument and its data analysis pipeline to fulfill the doctoral requirements. Newly married, Wickes had promised his wife that they would return to Southern California after the PhD was awarded. Dicke thwarted that plan by offering the freshly-minted Dr Wickes a postdoc job at Princeton to complete the achromatic redesign and "*to continue chasing μ Cas. So much for my promise to my new bride Susan that we would return to Southern California after I finished. Never did, actually.*"

Wickes' visit had made a great impression on Worley. The first of Antoine Labeyrie's speckle interferometry papers with measurements of double stars would appear the month after Wickes' USNO run (Gezari et al. 1972). In addition to diameters of several of Pease's stars, measurements of the Capella and β Cephei binary star systems appeared in that important paper. Those early measurements were primarily aimed at demonstrating the resolution and precision of the speckle technique, and they lacked careful calibration that would convert that precision into accuracy. As a result, no position angles were given for the double-star data. The speckle observers attempted to resolve μ Cas, the old binary system at the heart of Wickes' work, but were unable to breach the brightness gap of the components.

Over the next few years, Wickes' binary star results were coming out along with the speckle interferometry of Labeyrie's team. This piqued Charles Worley's interest in exploring the feasibility of such techniques as viable and powerful successors to visual micrometry. Although put off by the lack of reliable calibration of the early speckle data, Worley eventually decided to retire his micrometer and introduce a new binary star speckle program at the USNO. That effort was launched in the fall of 1990 and continues to this day. Figure 6.6 shows a photograph of Charles Worley visiting at Lowell Observatory when my group's new speckle camera was being tested there in 1985 June. Speckle interferometry would replace visual micrometry, the last application of human vision directly to astronomical measurements, as the

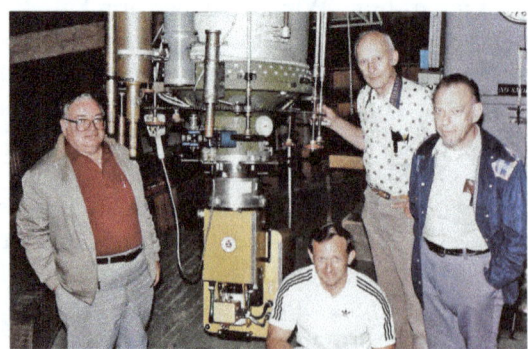

Figure 6.6. Charles Worley (left) at the Lowell 24 in refractor in 1985 June with the CHARA speckle camera at the telescope. Kneeling is Lowell's chief engineer Ralph Nye, while director Art Hoag and astronomer Otto Franz stand at right. (Author's photo.)

primary means of following the orbital motions of double and multiple stars on the plane of the sky. That transition began with Wickes' spring 1972 observing run at Charles Worley's big refractor. Wickes recalled to me his interaction with Worley as being "*a great honor to work with such an experienced double-star astronomer. He was very supportive, helpful, and patient. And he was quite pleased when I accidentally discovered that α Delphini was actually a binary star.*" In 1973, on the MWO 24 inch telescope with 35 cm aperture baseline—Wickes had used α Del as a reference single star for measurements of the nearby binary β Del with $\rho = 0.55$ arcsec. The following year, using the Mt. Wilson 60 inch telescope with a 100 cm baseline, he was astonished to see that the visibility versus aperture position angle curve of the supposed single star was unmistakably that of a binary. He ended up measuring both stars, using δ Del as the reference. For α Del, the data yielded $\rho = 0.241$ arcsec at that epoch.

There was one more stop before Bill Wickes first observed at Mount Wilson. Here is his description of that experiment. "*After the negative experience at the Kitt Peak reflector, and the positive one at the USNO on the 26-inch refractor, I tried another refractor—the 36 inch at Lick. That was a bust—after a short initial try, it became clear that the telescope tube was rather too "floppy." I couldn't seem to keep the star of interest in the small field aperture. Then the fog rolled in and stayed there for the rest of my four days of observing time.*"

He doesn't recall how he became aware of Mount Wilson, but its reputed good seeing and the "*excellent figure on the 60-inch telescope*" combined with the abundance of bright-moon time available for non-Carnegie astronomers had Wickes looking toward Southern California. Observatory Director Horace W. Babcock (1912–2003) first assigned observing time for the Princeton project in 1972 November. A list of eventual observing assignments appears in Table 6.1.

The 1972 activity using the first-generation interferometer was primarily concerned with instrument shakedown and performance testing. Wickes published three measurements of the binary system comprising φ UMa that reduced to mean values in θ and ρ of $58.51° \pm 0.25°$ and 0.2677 ± 0.0031 arcsec. His results from the Washington refractor earlier that year were formally somewhat better, with the mean of five data sets giving $58.58° \pm 0.18°$ and 0.2758 ± 0.0007 arcsec (Wickes & Dicke 1973).

Table 6.1. Wickes' MWO Observing Runs

Date	Telescope	Results	Paper
72 Nov	60 inch	Testing + φ UMa	Wickes & Dicke (1973)
73 Aug	24 inch	Testing + β Del	Wickes & Dicke (1974)
73 Oct	24 inch	λ Cas, λ Cyg, 85 Peg, μ Cas	Wickes & Dicke (1974)
74 Aug	60 inch	6 stars including μ Cas	Wickes (1975b)
74 Nov	60 inch	6 stars	Wickes (1975b)
74 Nov	60 inch	Hyades binaries: ADS 3135, 3210, 3475	Wickes (1975a)

The 1973 observing at the MWO 24 inch telescope utilized the second-generation achromatic instrument and provided new measurements of the binary stars β Delphini (with $\rho = 0.554$ arcsec), λ Cassiopeiae (0.530 arcsec), λ Cygni (0.857 arcsec), and 85 Pegasi (0.77 arcsec)—another moderately old system. The much sought-after μ Cassiopeiae (0.35 arcsec) was also bagged (Wickes & Dicke 1974).

The 60 inch telescope provided the 1974 results that comprised a mix of binaries having angular separations ranging from 0.19 to 0.74 arcsec with Δm up to 2 magnitudes (Wickes 1975a). Both 85 Peg and μ Cas were again measured. Table 6.2 compares the two epochs obtained for these old binaries.

The duplicity of 85 Peg had been discovered visually in 1878 by Sherburne W. Burnham (1838–1921). Its large magnitude difference and sub-arcsecond angular separation challenged all but the best visual micrometrists. The star's low metal abundance and evolutionary-model fits indicate that the system is nearly twice the age of the Sun (Bach et al. 2009). Wickes measured the magnitude difference between the A and B components to be 3.17 ± 0.17 magnitudes. The Δm of the μ Cas stars was much more challenging with Wickes finding a value of 5.5 ± 0.7 magnitudes, corresponding to a secondary with only 0.6% the brightness of the primary. Wickes' data were an invaluable addition because of their accuracy and especially in his determination in the brightness difference of the stars.

Like all who have worked on Mount Wilson, Bill Wickes has stories of his experiences. Here are three of them.

The rattlesnake incident—"One day about noon, as I wandered out of the double-double monastery doors, I heard a hissing noise. There was a rattlesnake between the outer and inner doors on one side. I called the night assistant, who showed up shortly and handed me a .22 rifle. The snake was outside on the dirt planting area next to the building, so I shot it. I was quite proud of hitting the small snake's moving head with one shot."

The baby incident—"I was sitting on the dome stairs in the 60-inch dome, talking with my wife on the phone hanging on the wall next to the stairs, when she informed me that we were expecting our first child. That would have been in November of 1974. That child, our son Ken, is now a 22-year veteran at Microsoft."

The dome light heresy—"I think I astonished other observers and night assistants by working away all night with lights on in the dome. Ambient and/or sky brightness just wasn't an issue for me. Often I had a little TV on too—I especially remember car commercials by Cal Worthington, who "would eat a bug" if you would buy one of his fine used cars."

Table 6.2. Measurements of Old Population Binaries

Star	Epoch	$\theta(°)$	ρ(arcsec)	Δm	Telescope
85 Peg HD 224930	1973.79	184.5 ± 0.3	0.77 ± 0.005	3.17 ± 0.17	24
	1974.65	194.0 ± 0.3	0.737 ± 0.009	3.0 ± 0.1	60
μ Cas HD 6582	1973.79	24.0 ± 3.0	0.35 ± 0.04	5.5 ± 0.7	24
	1974.65	333.9 ± 0.2	0.23 ± 0.01	4.5 ± 1	60

Wickes' astronomy career was advancing nicely during the years of his Mount Wilson visits. Following completion of his PhD in 1972 and the year as Dicke's postdoc, Wickes, shown in Figure 6.7 with his equipment at Mount Wilson, was given a one-year instructorship followed by a three-year appointment as an assistant professor. Knowing of Princeton's tradition of exporting its own graduates to other institutions, he would leave in 1978 for a tenure-track job at the University of Maryland where he would collaborate with Douglas Currie, whose own interferometry initiatives are described next, as well as with Carroll O. Alley (1927–2016), who led the Dicke-inspired Apollo program effort to place laser retro-reflectors on the Moon for tests of Einstein's General Relativity theory. Wickes described his attraction to Alley as *"another former Dicke protégé,"* while going on to say that he *"worked with Doug on the CCD camera for the Hubble telescope. Then, in 1980, Carroll and I were inspired by a talk given by John Wheeler about John's "delayed-choice quantum mechanics" Gedanken experiment to put a grad student on a Ph.D. project to translate Wheeler's concept into an actual experiment* (Alley et al. 1986)."

"There were mixed feelings in the physics community about whether the experiment was worthwhile, but Wheeler was an enthusiastic supporter, as were several other of my Princeton acquaintances." Bill Wickes was looking beyond double star interferometry for his future interests.

In the meantime, though, there was more to be done with his interferometer before heading off to College Park. Although the emphasis on μ Cas had waned, an interferometer to be built on Mount Wilson a generation after Wickes would measure the star's diameter and effective temperature to show that its very low metallicity led it to mimic normal stars of much later spectral type (Boyajian et al. 2008). But, that's jumping the gun.

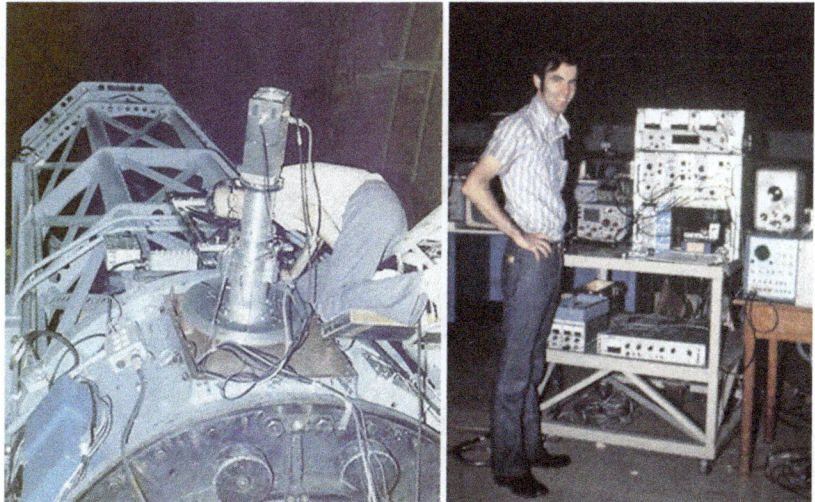

Figure 6.7. (Left) Bill Wickes installing his achromatic interferometer in 1975 at the bent Cassegrain focus of the 60 inch. (Right) Wickes with the interferometer's control and data-collection electronics. (Photos courtesy of William C. Wickes.)

Setting μ Cas aside, Wickes decided to obtain high-accuracy measurements of several Hyades star cluster binaries in order to better determine the masses of the component stars and see if the Hyades mass–luminosity relation differed from that of the Sun–Sirius group. This would benefit from a revised distance modulus for the cluster from a then new-application of the classical convergent point method. In the end, he concluded that a more accurate cluster distance would be required to determine if the masses of Hyades stars differed from stars near the Sun (Wickes 1975b).

I have already mentioned the contemporaneity of Wickes interferometer observations with speckle interferometry. In a sense, the two may be seen, as did Charles Worley, as contenders for a modern method of observing double stars to obtain masses of greater accuracy and free of subjective biases inherent in classical visual method. It is important to note (as described in Appendix B.3.2) that the limiting resolution for binaries using Michelson-type interferometry at an aperture-projected slit spacing S is $\lambda/2S$ while that provided by speckle interferometry exploiting a telescope's full aperture A is given by the Rayleigh Criterion of $1.22\lambda/A$ (Appendix B.7). Thus, if both techniques are employed on telescopes of comparable apertures, then A and S are approximately equal and Michelson interferometry has a factor of 2.44 advantage in limiting resolution over speckle interferometry. At the MWO 60 inch, Wickes could have resolved binaries with angular separations approaching 0.038 arcsec while speckle interferometry would fail to resolve binaries closer than 0.092 arcsec. In reality, though, the few speckle observers active at the time were using large telescopes like the Palomar 200 inch (Labeyrie and his collaborators) or the Kitt Peak 4 m telescope (me and my colleagues). Thus, the speckle teams enjoyed more or less the same limiting resolution that Wickes had at the much smaller 60 inch telescope. In practice, Wickes did not exploit that high-resolution capability.

The other attribute for double-star observations that relates to aperture is measurement accuracy. Indeed, because the angular separation eventually enters into the mass determination via Kepler's third law, the uncertainty in the angular separation has a factor of three input to the system's total mass. Here again, accuracy also approximately scales with limiting resolution, or more precisely with the highest spatial frequency components to which the technique is sensitive. One might thus expect Wickes's results to be of similar quality to those his speckle competition was extracting from large reflectors.

In Figure 6.8, Wickes measurements of 85 Peg and three Hyades binaries are shown plotted against high-accuracy orbits incorporating HIPPARCOS spacecraft parallax measurements to improve masses (Soderhjelm 1999). Also shown are speckle observations that I began accumulating in 1975 and continued for many years with students and colleagues at Georgia State, although I only include here speckle data up to about 1985. Plotted against the ADS 3475 orbit are visual interferometer measurements by M. Maggini and W. S. Finsen from which obvious conclusions are described in the figure caption. Due to crowding of points, I only indicate calculated positions for the 85 Peg observations and for two instances along the ADS 3475 orbit. It is nonetheless apparent from casual inspection that the speckle and Wickes interferometer data were essentially equivalent in performance.

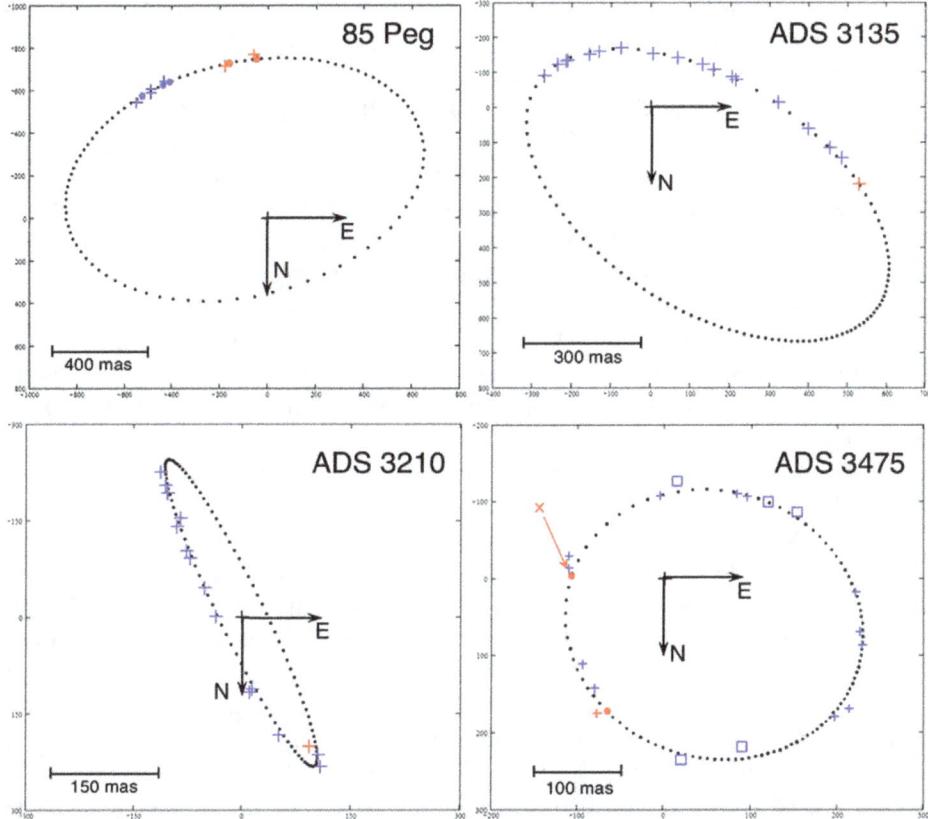

Figure 6.8. Four double star systems observed with the Wickes–Dicke interferometer on Mount Wilson telescopes. In these plots, the brighter "primary" star is considered fixed at the location of the direction vectors while the relative position of the fainter "secondary" star shows the motion of the two components about the center of mass. Wickes' measurements are shown as red + signs and CHARA speckle interferometry measures prior to 1990 with blue + signs. Their respective locations as predicted by the orbit used here are shown as black points. The measurement of 85 Peg closest to $q = 180°$ was from the 24 inch telescope while the second was from the 60 inch. They are of comparable accuracy. The remaining three systems, indicated by their Aitken Double Star numbers, are all Hyades cluster stars with orbital periods (clockwise from upper right) of 89.7, 16.3, and 27.4 yr. For ADS 3475, the blue squares are visual interferometry measures by W. S. Finsen between 1953.0 and 1962.1. The single red X is a 1923.94 visual interferometry result by M. Maggini with an arrow connecting it to its position as predicted by a modern orbit. Compare that with Wickes' result adjacent to its predicted location. The diagrams indicate that Wickes' results are essentially of equivalent accuracy as the speckle data. (Author's diagram.)

Of course, if Wickes had been using 4–5 m-class telescopes, he would have won the competition hands down. On top of that, speckle observations at the time could not reach Δ-magnitudes like that of μ Cas. Wickes' work in binary star interferometry ended with the Hyades. Although Princeton gave the interferometer to the University of Pittsburgh's Allegheny Observatory for use at the 30-inch Thaw refractor, where it might have embarked on a long-term and highly productive program, no more results were forthcoming.

Astronomy was still a relatively small field in the mid-1970s, and single-author papers were not uncommon. An individual could make a big difference in the future of a field. Had Wickes decided to continue in double-star interferometry, that method might have supplanted speckle interferometry as the usurper of the micrometer. As Wickes explains, "*My work in the Gravity Group was always a lone wolf project, at least until Dave Wilkinson and I started working together on primeval galaxies light, and I did a little project on galaxies with Jim Peebles. The motivation for my interferometry was always about cosmology, not interferometry or binary stars in general.*"

A chain of events would take him out of astronomy altogether. Bill Wickes described those events. "*Shortly after my marriage and my taking the post-doc position in Dicke's gravity group, David Wilkinson bought an HP-35 calculator. I was fascinated with it. Susan got a teaching position in a public school, so we thought we were rich. We made a deal where I would buy an HP-45 calculator and she would buy a set of crystal and china—about the same cost, and neither of us really understood why the other wanted their prizes. I do remember working out binary star ephemerides on the HP-45 by successive approximations.*" Before leaving Princeton, he had upgraded to an HP-65. When he got to Maryland, he found that Alley had an HP-41C, which had a bug that enabled hacking its OS and adding custom code to it. This unplanned feature intrigued Wickes who couldn't resist the opportunities it offered. He ended up writing a best-selling book about it (Wickes 1980). Hewlett–Packard soon invited him to give a talk at their Corvallis, Oregon campus "*which in turn led to a job offer. After much agonizing, I accepted. Not sure I can articulate exactly why—maybe it had something to do with HP offering me a salary considerably larger than I was earning at UMD, to do professionally what I had been doing for fun.*"

Wickes retired from HP in 2010 and still lives in Corvallis. He has gone back to Princeton from time-to-time to participate in milestone celebrations involving his mentors, including the *General Relativity at 100* conference at the Institute for Advanced Study in 2015 November, to which he was invited thanks to his collaborations with Dicke (research) and John A. Wheeler (teaching), the two instigators of the rebirth of gravitation experiments and theory originating at the university in the 1950s and 1960s. At this writing, he is looking forward to the celebration of Peebles' 2019 Nobel Prize in Physics, an event postponed by the 2020 COVID-19 crisis. Peebles' 1966 paper on the primeval helium abundance (Peebles 1966) had helped motivate the interest of Dicke and Wickes in interferometry of μ Cas from Mount Wilson.

Although Wickes foray into binary star work on Mount Wilson did not carry on beyond 1974, he elegantly demonstrated how technical advances in computing, control systems, and detectors could remove limitations from the Michelson–Pease era and lift their techniques to new heights of performance. Mount Wilson would continue to attract practitioners of high angular resolution astronomy, and, half a century later, interferometry has never been stronger on Hale's mountain.

6.3 Douglas G. Currie's Amplitude Interferometry

Doug Currie embarked on a career in theoretical physics in 1962 with the completion of his doctoral research at the University of Rochester on *The*

Hamiltonian Description of Interactions for Classical, Relativistic Particles. Currie had gone to Rochester after majoring in engineering physics at Cornell. He recalls that during his dissertation defense a Polish theoretician asked him to describe laser radiation. So, he did, but the examiner told him he was wrong and to "Do it again." After a second attempt, Currie got the same response—"you're wrong, do it again." At that point, another of the examiners piped up and declared Currie's explanation to be correct. It was time to move on. This was a "small world" event in that Doug Currie's defender was none other than Emil Wolf, who, a year or so earlier, had given a similar competency endorsement to the would-be interferometrist Willet Beavers.

Currie's trajectory following grad school took him to Princeton and Maryland for postdocs and then to Northeastern University as an assistant professor. By 1968, Currie was back at the University of Maryland as an associate professor of physics. He's been there ever since—now as Professor Emeritus of Physics.

With his experiences in practical and theoretical physics, it isn't entirely surprising to see a 1968 paper by him on "Conservation of Momentum and Angular Momentum in Relativistic Classical Particle Mechanics" followed by "Satellite Geodesy Using Laser Range Measurements Only." The latter publication signaled his transition to the newly emerged field of lunar laser ranging, which was being pursued by colleagues—including the afore-mentioned Carroll Alley— in UMD's physics and astronomy department. (The concept originally came out of Dicke's group at Princeton in the late 1950s as a potential means of detecting a change in the gravitation constant; Bender et al. 1973—an interesting connection between Bill Wickes and Doug Currie.) Currie would take a prominent role in developing both the theory and technology to allow laser light from Earth-based telescopes to illuminate corner-cube retro-reflectors placed on the moon by Apollo astronauts and measure the round-trip light travel time and distance from the Earth to the moon to an accuracy of about an inch. Such measurements have broad applications in orbital dynamics and lunar interior structure, as well as constraining fundamental aspects of physics. When I spoke with him in 2020 March, Currie had come full circle and was once again immersed in lunar laser ranging in his role as the lead scientist on the international effort toward next-generation laser retro-reflectors funded by NASA in 2019 July. NASA will have a 10 fold gain in distance measurement accuracy once the next-gen devices are placed on Mare Crisium as early as 2022. New vistas will open, including new tests of Einstein's General Relativity theory. Stay tuned.

A departure from lunar laser ranging that is relevant to this book is Currie's 1974 paper, written with Maryland graduate students Stephen Knapp and Kurt Liewer, on "Four Stellar-Diameter Measurements by a New Technique: Amplitude Interferometry (Currie et al. 1974)." This new research direction was the result of Currie's invitation to participate in the pivotal 1967 Woods Hole Summer Study we've already described. That opportunity arose from his theoretical exploration of the nature of laser radiation and the effects of the atmosphere on its propagation, and Currie initially had no particular motivation to get involved in the practical aspects of interferometry. But, once at the summer study, he found the emphasis to be highly theoretical and felt that an initiative toward a practical interferometer was

Figure 6.9. Doug Currie (at right) and his then graduate student Kurt Liewer sit at the observers' console of the 60 inch telescope in 1973 October with the amplitude interferometer mounted at the telescope's bent Cassegrain focus. (Photo courtesy of Douglas G. Currie.)

needed. That led Currie to pivot from a theoretical to an experimental approach, and his amplitude interferometer was the result.

Currie would soon find himself, his graduate students, and his interferometer at Mount Wilson and Mount Palomar where the 60, 100, and 200 inch telescopes were made available to him at the two mountaintops then under joint management as the Hale Observatories (Figure 6.9). For a span of seven or so years, he would typically come out from Maryland two or three times each year to tease diameters from interferometer observations of resolvable stars. Currie recalls that "those were the days when, if you needed to ship equipment, you could just check in at the airline counter with a 500 pound box and declare it to be your luggage. You didn't pay a thing!" Figure 6.10 shows Doug Currie and Bill Wickes at the Palomar 200 in telescope in the late 1970s.

He also vividly remembers a brush with personal and professional disaster he had in setting up his massive interferometer on the 60 inch telescope. Like the 100 inch, the smaller telescope has a Cassegrain focus fed by a tertiary mirror rather than passing the light from the secondary through a hole in the primary. In doing an afternoon alignment of the optics prior to that night's observing, Currie was up on a ladder at the 60 inch attempting to rotate the interferometer to the appropriate

Figure 6.10. Bill Wickes and Doug Currie atop the Palomar Observatory 200 inch telescope dome during a visit related to the development of the future Hubble Space Telescope (Photograph courtesy of William C. Wickes)

position when "all of a sudden, it came off in my hands." All 450 pounds of it! Fate smiled upon him as his arm, which might have been caught between his teetering interferometer and the telescope's instrument mounting adapter, happened to be safely in a gap. Currie was unscathed as his equipment headed to the steel floor and hit it so hard that the mountain superintendent heard the noise from his house and rushed up to the 60 inch telescope dome. On inspection after the accident, it turns out that the mounting plate, which Currie had always assumed to be of cast steel, was actually an early aluminum alloy that had crystalized and finally disintegrated under load. It was a trap waiting to be sprung, and Currie might have been its victim. Rather miraculously, his arm was intact and his instrument survived the fall. Although the 60 inch telescope was disabled due to the failed mounting bracket, Currie was bumped up to the 100 inch that night.

Observing on Mount Wilson had no further hair-raising experiences for Currie or his students, and he grew impressed with the very good seeing conditions he found there and not so much at Mount Palomar. There were times, he recalls, that the images at the 100 inch telescope were limited by the telescope's aberrations and not by the atmosphere—a condition one rarely encounters at any telescope.

His interferometer's beam-combining method was different from that previously used in that it did not produce a two-dimensional image across which fringes were seen. Instead, the interference occurred in mixing two collimated beams and scanning back and forth about the zero path length position, while letting the interfering amplitudes modulate the net intensity as they move in and out of phase.

In the arrangement he presented at Woods Hole, Currie, for illustration's sake, inserted a "random phase modulator" in one of the beams to demonstrate the concept, but the actual instrument allowed atmospheric turbulence variations across the telescope entrance pupil to accomplish this necessity. Doug Currie's amplitude interferometer is shown schematically in Figure 6.11 and uncovered in his lab at UMD in Figure 6.12. The instrument featured a reflective two-element system that injects a collimated beam into a Kösters prism after it is interrupted by a double-slit mask that projects the slit apertures and their spacing onto the telescope's entrance pupil by the factor f_{tel}/f_{int}. The prism, named for German metrologist Wilhelm Kösters (1876–1950) is a double-image device turned into an interferometer by virtue of a thin coating of aluminum applied on one of two identical right-angle prisms at their juncture. Interference occurs at the beam splitter and light sent to the detectors has reflected and transmitted components from the two beams entering from the slit pair. This is the methodology used in several modern interferometers, including the CHARA Array, and considerable detail about the reduction and analysis of such data is given in Appendix B.5.2.

Kösters prisms had been investigated as early as 1953 for the purpose of testing telescopes and other applications by J. B. Saunders of the National Bureau of Standards (Saunders 1953). Contemporaneous with Currie's application of these prisms to his interferometer, Paul H. Knappenberger, Jr explored their use in various interferometers, including the Michelson–Pease beam, for measuring diameters and double stars (Knappenberger 1966). But, it was Currie who produced

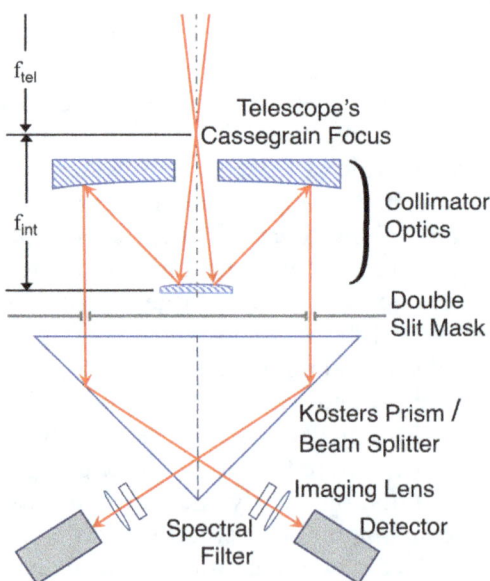

Figure 6.11. The light path and components of Currie's amplitude interferometer are shown here. (Author's diagram adapted from Figure 1 of Currie et al. 1974. © IOP Publishing. Reproduced with permission. All rights reserved.)

Figure 6.12. The amplitude interferometer shown uncovered in Currie's lab at UMD in 1973 October. (Photo courtesy of Douglas G. Currie.)

the first astronomical measurements with these compact optical devices that would later be used in the crucial Fine Guidance Sensors of the Hubble Space Telescope.[3]

Each program star was observed from Mount Wilson as well as from Mount Palomar with two goals in mind. The first was to compare the external accuracy measured through repeatability with the internal error bars derived statistically. The other was to measure the diameter of each star at different wavelengths, which will contribute to scatter in comparison with other measurements if not taken into consideration. These stars all had diameters measured by Francis Pease using the 20 foot interferometer beam, and thus we compare the two data sets in Table 6.3 and, except for α Ori, in Figure 6.13. Also included are the same subsequent measurements from "modern" interferometers as were considered in Section 3.3. Pease did not provide formal error estimates for his measurements, but his results as well as the amplitude interferometry diameters are generally in good agreement with later observations from longer-baseline interferometers.

The one exception is the measure from amplitude interferometer of the red supergiant star α Her. Pease was apparently ambivalent about his result. He at first called it "preliminary" in Year Book volume 24 but included it in his 1931 review article, apparently without reservation. In Section 3.3, we noted that Kuiper had seen a subsequent but never published manuscript, in which Pease omitted α Her. Perhaps this was because the star is rather faint—and varies in brightness by about

---

[3] See https://stsci.edu/hst/instrumentation/fgs.

Table 6.3. Amplitude Interferometer Diameters

Star	Pease (20 foot) Θ_{UD} (mas)	Ref.	Currie (Amp. Int.) Θ_{UD} (mas)	Ref.	Modern Value Θ_{UD} (mas)	Ref.
α Ori	47,34	1	47,57,57,44,(62)	2,7	49.4 ± 0.24 (800 nm)	5
			52.6 ± 0.5	3,7,8		
α Tau	20	1	24 ± 5	2,7	18.5 ± 0.5	5
α Boo	20	1	26 ± 7	2,8	19.5 ± 0.3	6
β Peg	21	1	21 ± 6	2,7	16.0 ± 0.2	5
α Her	30	1,9	58 ± 9	4,7,8	32.7 ± 2.6	10

Notes.
References
1. Pease (1931)—20 foot Int (Mt. Wilson) λ = 575 nm.
2. Currie et al. (1974)—Amp. Int during 1972.
3. Liewer (1979)— Amp Int during 1972–1976 (mean of 5 observations)
4. Knapp et al. (1975)—Amp Int.
5. Mozurkewich et al. (1991)—Mark III Int. (Mt. Wilson) λ = 450 nm.
6. Mozurkewich et al. (2003)—Mark III Int. (Mt. Wilson) λ = 550 nm.
7. From Mount Palomar 200 in telescope.
8. From Mount Wilson 100 in telescope.
9. Pease probably intended to withdraw this result (see Section 3.3).
10. Mean of speckle and interferometry measurements as discussed in text.

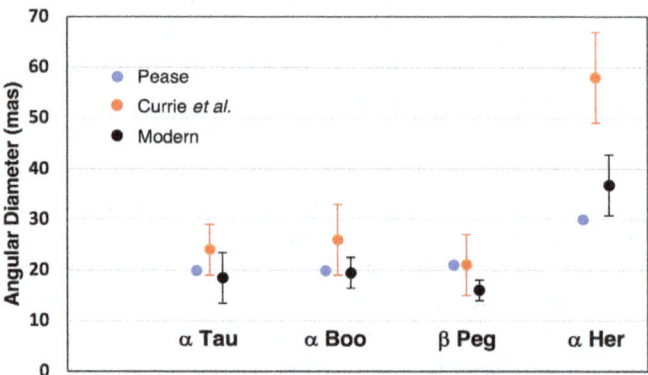

Figure 6.13. Diameters of stars observed by Pease and by Currie et al. Error bars: Pease provided none; "modern" value for α Her is the standard deviation of the mean for four measurements (Author's diagram.)

one magnitude—compared with the stars comprising the diameter collection. It also has a visual companion that might have added contaminating background light to wash out fringes. Despite his ambivalence, Pease's result is consistent with modern values.

The star had also been measured using speckle interferometry by Labeyrie and his collaborators at the 200 inch telescope in 1971. They obtained a diameter of 31 ± 3 mas (Gezari et al. 1972). Another speckle measurement came along in 1975 when S. P. Worden used the Kitt Peak 4 m telescope to get 30 ± 1 mas at λ = 510 nm (Worden 1975). Fifteen years later, Jim Benson and his colleagues used the University of Wyoming's Infrared Michelson Array (IRMA) outside Laramie to determine a value of 32.2 ± 0.8 mas at 2.2 μm, corresponding to an effective temperature of 3210 ± 40 K (Benson et al. 1991). They subsequently improved their diameter value to 33.0 ± 0.5 mas, increasing the effective temperature by 10 K (Dyck et al. 1993). Lastly, the Mark III interferometer on Mount Wilson found, from observations taken sometime prior to 1992 December when the interferometer was decommissioned, a diameter of 36.8 ± 0.6 mas at 550 nm (Mozurkewich et al. 2003). The mean of these four heterogeneous measurements is what is used in Table 6.3 while the amplitude interferometer result is about 2σ larger. It is unfortunate that α Her, like Betelgeuse, is over-resolved by the currently operational long-baseline interferometers.

The amplitude interferometry diameter, nearly twice that of the mutually-consistent results from other methods, leads to a temperature of 2450 K, a value shown to be discordant with stellar atmospheres model fits to the observed spectral energy distribution of the star by Takashi Tsuji (1978). With the remaining measurements from Currie's group being quite consistent with those of other observers and techniques, it is natural to wonder what might have gone awry with α Her. Worden asked this question and suggested that the discrepancy might have resulted from unknown systematic errors, a strong wavelength dependence of diameter, or a large time variation of diameter. Labeyrie pointed to possible biases in the statistical treatment of the conversion of the atmospherically modulated signal to visibilities (Labeyrie 1978). As for the wavelength dependence of diameter, note that the mean value from techniques other than amplitude interferometry derives from four measurements taken at wavelengths ranging from 510 nm to 2.2 μm. There is variation, but it does not seem relevant to the much larger offset of the amplitude interferometer result.

There is also the potentially problematic multiplicity of α Her, which is a triple star system comprised of a visual binary whose components are α^1 Her—the M5Ib supergiant whose diameter we have been considering—and α^2 Her. The latter is a spectroscopic binary comprised of α^2 Her A and α^2 Her B (Moravveji et al. 2013). This close pair has an orbital period of 52 days (Deutsch 1956) and would have an angular separation of about 1 mas, which is unresolved at the limiting resolution available to Currie. So, that subsystem is not a possible contributor to any bias to Currie's measurements.

A poorly-constrained orbit solution for the visual pair suggests an orbital period of 3600 yr and a semimajor axis of 4.7 arcsec (Hartkopf et al. 2001). Applying that orbit, one finds that the predicted angular separation during the summer of 1973, when the amplitude interferometry data were acquired, was 4.6 arcsec. Having a magnitude difference between α^1 Her and α^2 Her of 1.92 magnitudes, the secondary contributes an additional 17% to the flux of the primary star, with the amount actually detected significantly diminished due to the nearly 5 arcsec separation. Whatever contamination from the light of the secondary was present would likely have been at the few-percent level.

Figure 6.14 shows the individual visibility measurements from the amplitude interferometer and the best-ft visibility curve (in red) to them. The visibility curve for the mean of the other α Her diameter measurements is shown in blue. The two diameter curves are quite distinctly separated. Also shown in the figure are the visibility curves for the visual double in which the primary star's diameter is resolved while the secondary's is not. The two highly modulated curves correspond to the two distinctly different diameters under scrutiny. For clarity, the relative orientation of the interferometer apertures with the position angle vector for the binary is set in the figure at 15° to avoid blurring by the high-frequency fringe modulation of a fully-resolved binary star. The effect of the binary is to depress visibility below the diameter-only curve. If we assume that the smaller diameter is the correct value, then the amplitude interferometer visibilities should all still fall within the blue modulation band in Figure 6.14. They do not, especially for values of $d/\lambda > 20 \times 10^5$. It seems, then, that the companion is not responsible for the larger diameter. Does that settle the matter? No.

One other observational circumstance came up in an email exchange with Doug Currie that should have occurred to me. The darkest time of the month is reserved on large telescopes for those observing faint, extended objects like galaxies for which scattered light from a bright moon contaminates that from the galaxy. So, those who

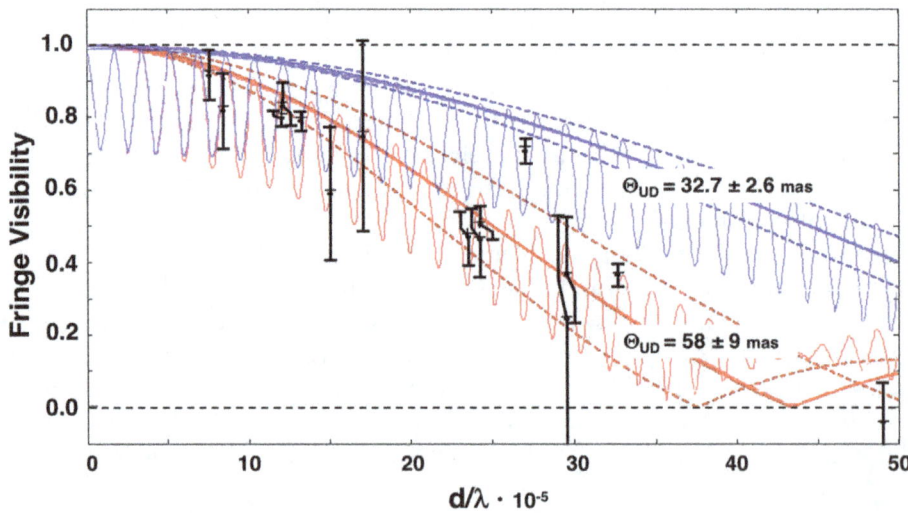

Figure 6.14. Visibilities of α^1 Her and the corresponding fit for Θ_{UD} diameter (in red) along with visibility curve predicted by other values (in blue). Also shown is the modulation that would arise from α^2 Her if it were in the field of view of the observation. See text for more discussion. Note that the abscissa is the ratio of baseline to wavelength, a dimensionless factor in the visibility function for a single star's diameter as shown in Appendix B.4.1. (Author's diagram adapted from Figure 1 of Knapp et al. 1975. © IOP Publishing. Reproduced with permission. All rights reserved.)

observe stars are given bright time surrounding full moon, and Currie was no exception. Interferometric methods, such as used by the early Mount Wilson interferometrists, that measure visibility by imaging fringes across a seeing disk must have high magnification to resolve the fringes either by eye in the early days or by a pixelated detector today. The same thing goes for speckle interferometry. Currie's amplitude interferometer did not require high magnification, and his field of view was quite large. He was always given bright time, so moonlight was always additive to the starlight. Being about two magnitudes fainter than his next faintest target star, α Her was more susceptible to this contamination than the others. Diffuse moonlight, which has an effective visual magnitude of +20 per $arcsec^2$ (Elias 1994), acts like a fully resolved object and lifts the zero visibility level to reduce the measured visibility of the star. A quantitative calculation can be made if one can reproduce the phase of the moon and its location on the sky for every observation, but those numbers are not easily accessible for Currie's measurements from the 70s. Nor is the size of his field of view, which was arcminutes across rather than arcsec.

This suggests that the Maryland astronomers' results for α Her might be explainable in terms of the companion and contaminating moonlight. And, of course, they got it right for three other stars as well as for Betelgeuse, which we turn to next. Those stars are all much brighter than α Her, and they have no comparable companions. To look for contemporaneous evidence that something was going on in the supergiant's very extended atmosphere that produced a larger diameter, I checked to see if the AAVSO brightness estimates for α Her might show something suggestive as they did for Betelgeuse in Section 4.2. They don't. So, there we'll leave it until the next chapter when we review the science from the Berkeley Infrared Spatial Interferometer (ISI).

Another supergiant with interesting things going on is Betelgeuse, and the amplitude interferometry for the star merits a few words. The diameter values listed in Table 6.3 are shown graphically in Figure 6.15 where error bars would be obscured by the filled markers for each diameter. Pease felt that his diameters variations, already discussed in Section 4.2, were real, and the Maryland group took a conservative approach to estimating their errors. The excursion to the largest observed diameter indicated by the open circle in the figure was given an uncertainty significantly larger than the other amplitude interferometer results, while that for the filled square is the mean of five measurements extending out to 1976. The standard deviation from the mean of is only ±0.5 mas. Here again, we find suggestions of activity in the enormous extended envelope of the supergiant. As with α Her, the Berkeley ISI will weigh in on Betelgeuse as well.

Additional diameter measurements were reported in meeting abstracts (Liewer et al. 1977) and a doctoral dissertation (Braunstein 1978) but never published in the refereed literature. Doug Currie told me that some were, once again, discordant with the French speckle measurements without an explanation being found. With students leaving for new jobs and Currie himself being pulled in a variety of research directions, there was little manpower available to get to the bottom of the mystery. And, so it remains.

Doug Currie has fond memories of his many days and nights on Mount Wilson. Rather than just send his students out to collect data, he went with them on every

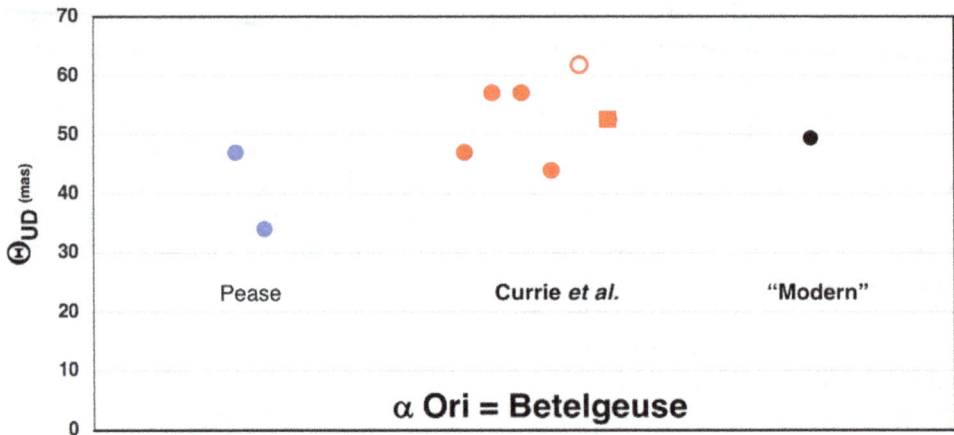

Figure 6.15. Diameter measurements of Betelgeuse by Pease and by Currie et al. Error estimates, when given, are smaller than symbol sizes. The open red circle was considered uncertain while the filled square is the mean of five observations during 1972–1976. (Author's diagram.)

observing run. Turns out a friend from Currie's student days was the Carnegie Observatories astronomer Arthur H. Vaughan, Jr. (1934–2015). Vaughan had been a fellow physics student with Currie at Cornell and likewise went on to Rochester where he received his PhD in 1964. His specialty was the design of astronomical optical systems. His first postdoctoral job was at Carnegie Observatories where he would stay for a number of years before launching a second career at the Jet Propulsion Laboratory. Vaughn naturally acted as Currie's liaison with management.

Art Vaughan was, like Pease, an unsung hero of both mountains—Wilson and Palomar. As recalled to me by Doug Currie, Art knew the telescopes and instruments inside and out and had an instinct as to the source of most any problem when it arose. He played a similar role when Carnegie established its southern hemisphere observatory at Las Campanas in the Chilean Andes. During his second career at JPL, Vaughan led the effort to design the optics for the Hubble Space Telescope Wide Field and Planetary Camera 2. He and his colleagues found themselves charged with including corrective optics to fix the notorious spherical aberration problem inherent in the HST primary mirror. Their work was splendidly and famously successful.

One of Art Vaughan's earlier optical designs was for a special photometer to measure the brightnesses of stars in the Fraunhofer's H and K lines of singly ionized Calcium. Emission in those lines is an indicator of activity in the chromospheres of stars, and the H–K Photometer was used for an important long-term survey for cycles of activity in stars analogous to the Sun's 22 yr cycle. Vaughan's Mount Wilson colleague Olin C. Wilson (1909–1994) was the principal user of the photometer and was often found on Mount Wilson observing stars for their chromospheric activity—which brings us back to Doug Currie.

Astronomers working on the mountain and staying at the Monastery would gather for dinner before heading off to their respective telescopes for the night. Currie laughs when he recalls O. C. Wilson teasingly wondering why in the world

Currie was given time on the 100 inch telescope when only a tiny fraction of the light collected by the big mirror was allowed to pass through the slits of his interferometer. "What a waste," Wilson would opine. Currie's immediate reply was that Wilson only used the narrow H and K line regions and therefore tossed out most of the spectrum. "Oh, that's not the same thing at all," Wilson would respond. This mutual harassment would be re-enacted each time their paths crossed on Mount Wilson. Bill Wickes has his own variant of wasted photon tragedies: "*My thing,*" Wickes told me, "*was to contemplate the fate of some unlucky photon who travelled millions of light-years across the universe only to plonk into the back of the secondary mirror on some telescope and never be detected.*"

With the success of Currie's interferometer mounted at the Cassegrain foci of large-aperture telescopes (see Figure 6.16), he, like most interferometrists, got the itch for higher resolution. The quality of the seeing on Mount Wilson enticed him to explore the feasibility of a long-baseline, two-telescope interferometer placed along a slope on the mountain so as to be parallel to the Earth's spin axis. This would avoid having a long delay line. Sound familiar? As you will recall from Section 5.3, this is precisely what E. H. Synge had proposed in his 1930 paper that no one read. Currie first explored the perfect 22° slope of the Haleakala shield volcano on Maui where the Air Force has telescope facilities that suffer from winds frequently scooping clouds up out of the caldera to shut down the summit facilities. He was told that Haleakala was therefore not viable until he showed during a 6 month sequence of video data that, while the summit suffered cloud capping, the slope did not. Even so, there was no progress in funding an interferometer there. Next, he looked around the grounds on Mount Wilson and succeeded in finding a workable location, but he was

Figure 6.16. The amplitude interferometer mounted at the north Cassegrain port of the 100 inch telescope in 1972 June, that same location utilized by Anderson, Merrill, and Pease in their interferometer work half-a-century earlier. (Photo courtesy of Douglas G. Currie.)

similarly unable to excite much interest at the funding agencies. As it would turn out, his efforts would be among those of others during this period that would prime the pump for the support that would eventually come.

References

Alley, C. O., Jakubowicz, O. G., Steggerda, C. A., & Wickes, W. C. 1986, in Proc. Int. Symp., Foundations of Quantum Mechanics in the Light of New Technology, ed. T. Yajima (Tokyo: Physical Society of Japan), 36

Bach, K., Lee, J., Demarque, P., & Kim, Y. C. 2009, ApJ, 703, 362

Bates, R. H. T. 1982, PhR, 90, 203

Bender, P. L., Currie, D. G., Dicke, R. H., et al. 1973, Sci, 182, 229

Benson, J. A., Dyck, H. M., Mason, W. L., et al. 1991, AJ, 102, 2091

Blazit, A. 1977, ApJ, 214, L79

Blazit, A., Bonneau, D., Keochlin, L., et al. 1977, ApJ, 217, L55

Boyajian, T. S., McAlister, H. A., Baines, E. K., et al. 2008, ApJ, 683, 424

Braunstein, R. H. 1978, PhD Thesis, Dissertation Abstracts International 39-09 B 4406

Currie, D. G., Knapp, S. L., & Liewer, K. M. 1974, ApJ, 187, 131

Dainty, J. C. 1984, in Laser Speckle and Related Phenomena, ed. J. C. Dainty (2nd enlarged ed; Berlin: Spriner), 255

Deutsch, A. J. 1956, ApJ, 123, 210

Dyck, H. M., Benson, J. A., & Ridgway, S. T. 1993, PASP, 105, 610

Elias, J. 1994, CTIO/NOAO Newsl. No. 37, https://www.noao.edu/noao/noaonews/mar94/art20.html

Gezari, D. Y., Labeyrie, A., & Stachnik, R. V. 1972, ApJ, 173, L1

Hartkopf, W. I., Mason, B. D., & Worley, C. E. 2001, Sixth Catalog of Orbits of Visual Binary Stars, http://astro.gsu.edu/wds/orb6.html

Knapp, S. L., Currie, D. G., & Liewer, K. M. 1975, ApJ, 198, 561

Knappenberger, P. H. 1966, Design and Applications of Optical Interferometers in Astronomy (Charlottesville, VA: Pub Leander McCormick Obs, Charlottesville UVa Press)

Labeyrie, A. 1970, A&A, 6, 85

Labeyrie, A. 1975, ApJ, 196, L71

Labeyrie, A. 1978, ARA&A, 16, 77

Labeyrie, A., Schumacher, G., Dugue, M., Thom, C., & Bourlon, P. 1986, A&A, 162, 359

Liewer, K. M., Braunstein, R. H., Currie, D. G., & Knapp, S. L. 1977, BAAS, 9, 598

Liewer, K. M. 1979, in Proc. IAU Coll. 50, High Angular Resolution Stellar Interferometry, ed. J. Davis, & W. Tango (Sydney: Chatterton Astron Dept.), 8-13

Mason, B. D., Hartkopf, W. I., Corbin, T. E., & Douglass, G. G. 2007, in Binary Stars as Critical Tools and Tests in Contemporary Astrophysics, ed. W. I. Hartkopf, E. F. Guinan, & P. Harmanec (Cambridge: Cambridge Univ. Press), 28

McAlister, H. A. 1976, PASP, 88, 317

Moravveji, E., Guinan, E. F., Khosroshahi, H., & Wasatonic, R. 2013, AJ, 146, 148

Mozurkewich, D., Johnston, K. J., Simon, R. S., et al. 1991, AJ, 101, 2207

Mozurkewich, D., Armstrong, D., Hindsley, R. B., et al. 2003, AJ, 126, 2502

Pease, F. G. 1931, ErNW, 10, 84

Peebles, P. J. E. 1966, PhRvL, 16, 410

Saunders, J. B. 1953, JRNBS, 58, 27

Scott, N. J. 2018, Proc. SPIE, 10701, 112

Soderhjelm, S. 1999, A&A, 341, 121

Texereau, J. 1963, ApOpt, 2, 23

Tsuji, T. 1978, A&A, 62, 29

Wickes, W. C. 1975a, AJ, 80, 1059

Wickes, W. C. 1975b, AJ, 80, 655

Wickes, W. C. 1980, Synthetic Programming on the HP-41C Calculator (College Park, MD: Larken Pub), 92

Wickes, W. C., & Dicke, R. H. 1973, AJ, 78, 757

Wickes, W. C., & Dicke, R. H. 1974, AJ, 79, 1433

Worden, S. P. 1975, ApJ, 201, L69

Chapter 7

Closing Out the 20th Century

7.1 Seeing Double on Mount Wilson (1985–2007)

My first trip to Mount Wilson, in the spring of 1985, was the result of a collaboration with Michael Shara who was then at the Space Telescope Science Institute (STScI). The Hubble Space Telescope (HST) launch was five years away, and the Institute staff was deep into planning the details of telescope operations. Absolutely fundamental to the telescope's success was the ability of the Fine Guidance Sensors (FGS) to accurately point and lock onto targets anywhere in the sky. This required that the FGS acquire two guide stars for any position, and great effort was being invested in a Guide Star Catalog (GSC). The catch was that those two stars had to be single, point sources. If one turned out to have a nearby companion with a magnitude difference less than one magnitude, that FGS could not achieve lock, and no observation of the target object would be possible. An imperative quickly arose for estimating the fraction of the GSC stars that would be binaries, and the STScI decided to quickly mount a survey to check guide star candidates for companions and to improve the known frequency of occurrence statistics for binaries. This would tell them the redundancy requirement for the GSC to ensure there are no blind spots in the sky.

I had been doing binary star speckle interferometry for a decade by then and had developed a small group at Georgia State University that would evolve into the Center for High Angular Resolution Astronomy (CHARA). We had our own speckle camera (Figure 7.1). based on an intensified CCD read out at video rates to grab speckle images for post processing and extraction of the observables for binary stars. (See Appendix B.7 for details on speckle interferometry.) Using the 4 m aperture telescope at Kitt Peak, we could measure binaries down to angular separations of about 35 mas. We were a sort of astronomical "have gun, will travel" group, and Mike Shara invited us to join in on a very quick survey of binaries to improve the statistics. He would get time for us at a variety of telescopes. How could I turn that down?

 © IOP Publishing Ltd 2020

Figure 7.1. The author with the GSU/CHARA speckle camera installed at the north bent Cassegrain port of the 100 inch telescope, 1993 December. (Courtesy of William I. Hartkopf.)

The first observing run Mike secured was at Mount Wilson on the 100 inch during 1985 March 20–27, a period in which Mount Wilson was in a Shakespearean Ides of March gloom due to its impending closing scheduled immediately after our observing run. We were thus the last users of the 100 inch telescope after 80 years of the Observatory's operation by the Carnegie Institution of Washington. I would be joined by CHARA postdoc Don Hutter (now retired from a career at the U.S. Naval Observatory) as well as by Mike Shara (today Curator and Professor of Astrophysics, American Museum of Natural History).

In planning this observing run, we decided to take a shot at a group of stars that had already been selected for the GSC. Big mistake—those stars were generally too faint for us and we had to change course thereafter to a statistical survey extrapolating from brighter stars into the GSC domain. So that first experience at Mount Wilson was a bit of bust as far as the HST GSC was concerned.

But it was a wonderful experience from the visiting astronomer perspective. By then, there was a philosophical acceptance of the impending closure and everyone was in a peaceful calm as the remaining days ticked by. The Monastery cook—at the time a restaurant-level chef—had a well-supplied larder that needed emptying. He pulled out all the stops, and we dined spectacularly off such delicacies as venison and giant prawns. During the nights, long-time Night Assistant Jim Frazier controlled the 100 inch from its original operating station with the grace and expertise of a concert pianist. At the appropriate hour after midnight, we'd adjourn to the Galley for night lunch before going back into the dome to complete the night. At each dawn, we battened down the hatches and walked down the ridge to the Monastery to our respective bathroom-down-the-hall sleeping rooms.

The heritage of the Observatory was palpable, and I fell in love with the place. I never then dreamed that I would oversee the development of a major interferometer on that mountain or that I would become the entire Observatory's principal caretaker for a dozen years.

We took our camera to other telescopes for the STScI GSC project and obtained a significant amount of data as a result. Two years after the Mount Wilson run—and three years prior to the HST launch—we published conclusions drawn from the effort (Shara et al. 1987). The switch to brighter stars inspired us to do an ambitious survey of objects from the Yale Bright Star Catalogue (BSC). By 1992, we had collected data for BSC stars at the Kitt Peak and Cerro Tololo 4 m telescopes and the 3.6 m Canada–France–Hawaii Telescope for 23% of the catalog's 9110 stars, discovering 75 new, bright binaries in the process (McAlister et al. 1993). How many people can lay claim to having discovered naked-eye stars even if those stars were unknown only because of their proximity to even brighter stars?! Had we observed the entire BSC, we likely would have found an additional 250 or so new systems. This productive outcome resulted from a faulty decision in 1985 March at Mount Wilson.

As for Mark Twain, the reports of Mount Wilson Observatory's death were greatly exaggerated. As mentioned in Section 3.4, the Mount Wilson Institute (MWI), established by Arthur H. Vaughn and C. Robert Ferguson, was incorporated as a 501(c)(3) non-profit corporation on 1987 April 7. An agreement for MWI's operation of the Observatory on behalf of the Carnegie Institution of Washington went into effect on 1989 January 5, with Vaughan serving as the Observatory's new director (Vaughan 2006). Mount Wilson was back in business. In 1992, Robert Jastrow, who had just retired from a professorship at Dartmouth College, succeeded Vaughan to become the MWI board chairman and the Observatory's director. Jastrow embarked on a successful fundraising effort to modernize the 100 inch under an "if you build it they will come" business plan based on that telescope having an aperture (and site) still suitable for many astronomers' science goals. High-angular resolution was to be the telescope's new theme. My speckle group comprised the first such astronomers who came.

The level of effort on 100 inch upgrades was substantial. The original DC solenoid relays and switches were bypassed (but not removed), and a new microprocessor-based control system was installed. Significant wiring replacement was carried out. Similarly, the original clock drive was set aside and a modern servo-controlled system installed. The telescope is famous for its mercury flotation bearings, which occasionally had memorable leaks over the years, and it was imperative that the mercury system be secured and enclosed to reduce the mercury vapor in the dome. A double-walled system was installed for this purpose along with mercury vapor detectors and alarms.

In parallel with all this, an adaptive optics systems was being fabricated that would, for a brief period, be the finest in operation. Bob Jastrow had secured some funding to pay for observer expenses to utilize this newly-equipped 100 inch with the hope that subscribers would continue indefinitely into the future and bring their own funding to offset observatory operating expenses.

As a result of all this effort, the telescope we would use in 1993 December would be very different from the one we had used eight years earlier. We were initially assigned time on 46 nights spread over nine observing runs extending to 1995 December during which we had 32 usable nights not lost to clouds, poor seeing, or the occasional instrument glitch. Altogether we gathered 1328 measurements of 975 binary star systems (Hartkopf et al. 1997). Those data were calibrated primarily by reference to stars observed near in time at the Kitt Peak 4 m telescope for which a double-slit mask served to calibrate position angle and angular separation. However, beginning with a run in 1995 June, we had devised a slit mask that fit over the full aperture of the telescope (see Figure 7.2 for the ultimate version of that mask) where 75 years earlier Mr Kimple had perched next to the 20 foot Michelson–

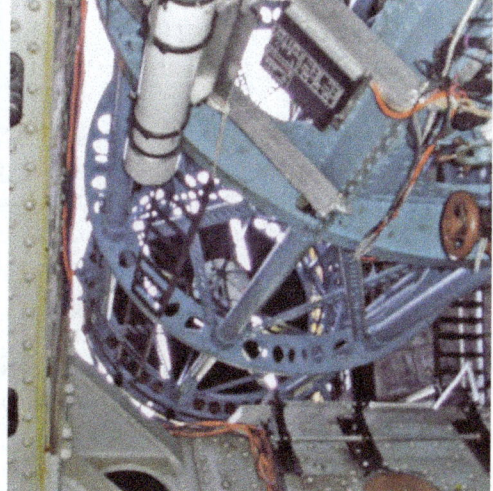

Figure 7.2. (Left) A full-aperture double-slit mask was fabricated using a rigid metal slit frame pin-registered to the Cassegrain secondary mirror cage with opaque material attached to cover the pupil outside the slits. A bright single star viewed through the mask produces Young's fringes in the focal plan whose angular spacing is only dependent on the telescope focal length, the slit spacing, and the wavelength of the observation. The orientation with respect to north can be accurately measured as well by switching off the telescope drive to establish an east–west drift. (Courtesy of William I. Hartkopf.) (Right) The daylight sky illuminates the mask in this view from the other end of the telescope. (Author's photo.)

Pease Interferometer. For the first time since those early days, the 100 inch was once again producing interference fringes!

The first night of our five-night observing run kick-off was 1993 December 1. Bill Hartkopf and I were the observers. Here is a synopsis from our log that is still in Bill's possession:

1993 Dec 1—Rain.

1993 Dec 2—Poor Seeing (Santa Ana winds) 10 arcsec! Curse you Turbo!

1993 Dec 3—Seeing 1.5 then 1.3 arcsec. Clouds came to shut us down.

1993 Dec 4—Seeing 1.4–1.7 arcsec. Closed dome for 4 hr, then seeing worse.

1993 Dec 5—Seeing 0.7–0.8 arcsec. Experimented with slit mask.

So much for the vaunted Mount Wilson seeing, which was mostly done in by one of the late-in-the-year Santa Ana winds that normally afflict the Southern California coast in the fall and can greatly increase the ferocity of wildfires. It was bad luck to get one of these in December. "Turbo," by the way, was our imaginary Greek god of seeing, who we only ever cursed and never praised. Just to tantalize us with what we might have experienced throughout, Turbo smiled on us that final night.

The first star we successfully observed and published from that grand-reopening of the 100 inch telescope was 72 Pegasi = HD 221673 = ADS 16836 = Bu 720, a binary star system discovered in late 1878 September by Sherburne Wesley Burnham (1838–1921). He was famous for being Clerk of the U.S. District Court of Chicago by day and double-star astronomer by night, before joining the staff of George Ellery Hale's Yerkes Observatory in 1897 while still serving the court. The year of his discovery of the duplicity of 72 Peg, Burnham was awarded an honorary A.M. degree by Yale University in recognition of his achievements in double-star astronomy (Aitken 1921). The Washington Double Star Catalog contains 1542 double or multiples stars discovered by Burnham during 1871–1910 (Mason et al. 2019).

The nearly equally-bright K giant stellar components of Bu 720 were oriented at $\theta = 127.7°$ and $\rho = 0.40$ arcsec at its discovery. When we observed it on 1993.9167, the system had revolved to $\theta = 89.7°$ and $\rho = 0.531$ arcsec. The orbital period is long, nearly 500 years, so the secondary had moved less than 40° in position angle since its discovery. In 2006, the star was observed by the Palomar Testbed Interferometer (PTI), a descendent of the Mark III Interferometer to be discussed in the next section. The PTI data were analyzed a few years later in a search for substellar companions, and, with a high degree of confidence, Bu 720 was found to have a brown dwarf as a third component orbiting one of the visual giant components with a period of 4.2 years (Muterspaugh, et al. 2010).

By 1995, CHARA had become pre-occupied with long-baseline interferometry, regrettably leaving diminished resources for its twenty-year-old binary star speckle interferometry program. We were also no longer receiving observing time at the national observatories, and we had already decided in 1993 to locate the GSU speckle camera at Mount Wilson as a matter of convenience. As a result, CHARA speckle observing continued at the 100 inch telescope for a total of 22 observing runs, the last of which was in 1998 October. That follow-on effort yielded another 893 binary star measurements along with 28 more garnered from the earlier observing (Hartkopf et al. 2000). Our speckle activities concluded soon after that 1998 run. Bill Hartkopf, who had taken the reigns of our speckle effort, would leave GSU in 1999 to join our former graduate student and postdoc Brian Mason at the U.S. Naval Observatory (USNO) in Washington, DC where Brian had succeeded Charles Worley as manager of the USNO's double star program.

Sadly for me, CHARA was done with binary star speckle interferometry, and our speckle camera was stored on the mountain with a hope that it might resume observations at the 100 inch at some time in the future. That wasn't to be, and, at the request of the Smithsonian Institution, it was given to their collection of historic scientific instruments in 2016.

However, CHARA staff did use the Mount Wilson Institute's wonderful new adaptive optics (AO) system developed under the technical leadership of J. Christopher Shelton (1995). Nils Turner, who had received his PhD with us and stayed on with CHARA when work commenced on our interferometric array, participated in the implementation of Shelton's system and became expert in its use. Nils and Theo ten Brummelaar, then CHARA's associate director, mounted a program of AO observations with two primary goals—accurately measure brightness differences between close double stars and search for faint companions to certain classes of nearby stars. Under Theo's initiative, CHARA had obtained observing time in 1995 on the U.S. Air Force 1.5 m telescope at the Starfire Optical Range in New Mexico. That telescope was equipped with a complex laser-guide star AO system that Theo and Nils used for differential photometry (ten Brummelaar et al. 1996). The timing was perfect as the MWI natural guide star AO system was then coming on line and became accessible in 1996 (ten Brummelaar et al. 1998). Theo's team's first 100 inch AO compilation included results for 36 binary stars (ten Brummelaar et al. 2000). In addition to measuring θ and ρ values for systems with separations as small as 0.11 arcsec, they calculated differential magnitudes in the U, B, V, R, and I photometric bandpasses from which could then be determined spectral types for the individual components from the (B–V), (V–R), and (V–I) color indices as well as estimates of effective temperatures and absolute magnitudes. The differential magnitudes were in good agreement with values from other techniques including those from the HIPPARCOS mission. Their observations continued through mid-1999, culminating in a search for faint companions to nearby Sun-like stars. Five such components were detected including newly discovered companions to μ Her A, HR 7123, and HD 190067 (Turner et al. 2001). Their program would continue at the Air Force's Advanced Electro-Optical System 3.6 m telescope on Maui as, at this juncture, funding realities forced MWI to cease AO operations on the telescope. This experience with adaptive optics at several different facilities not only produced a body of scientific results, it helped to prepare us for the recently completed retrofitting of AO to each of the 1 m telescopes comprising the CHARA Array.

Speckle interferometry may have been down, but it wasn't yet out. Another round of binary star observations would resume at the 100 inch when our old friends at the USNO conducted their own speckle program at Mount Wilson. Brian Mason and Bill Hartkopf (see Figure 7.3) secured observing periods in 2006 July and in 2007 April and October.

Figure 7.3. USNO double star experts Bill Hartkopf and Brian Mason seated at their speckle camera control center set up on the observing floor of the 100 inch telescope on 2006 July 24 with the author dropping in to say hello. (Courtesy of Susan J. McAlister.)

As they set up their speckle camera during the first day of their July observing run, they discovered with much dismay that the detector "had suffered a mechanical failure." Our camera, stored on the mountain and unused for eight years, provided a rescue option. Its detector was pulled out of the crate and turned on. It worked.

Fabricating a new cable was all that was needed for it to substitute for the newer, and more sensitive, failed sensor. They obtained 525 data series that night, and, with their own camera back in service, 1015 more data sets were collected in the two 2007 trips to Mount Wilson. These data yielded 607 new double-star measurements along with 222 stars found to be single within specific angular separation and magnitude difference limits (Hartkopf & Mason 2009). They would later dig another 10 measurements out of these three observing runs (Mason et al. 2010, 2011).

Hartkopf and Mason plotted the accumulated data for their stars that had existing orbit solutions, specifically looking for any divergence from those orbits. This flagged 35 systems for which it was worthwhile to modify orbital elements. A fine example of this is given in Figure 7.4 for the binary star system Cou 798, designated in the Washington Double Star Catalog as WDS 15347 + 2655. In that figure, two orbits are plotted—the dashed orbit being the one then thought to approximate the combination of visual (green + signs) and speckle (blue filled circles) measurements available at the time the orbit was calculated. The blue star symbol is the new Mount Wilson measure and displays a whopping departure from the initial orbit. Their resulting revised orbit—the solid ellipse—has a period of 216 ± 62 yr, a large uncertainty in this still preliminary solution due to the system

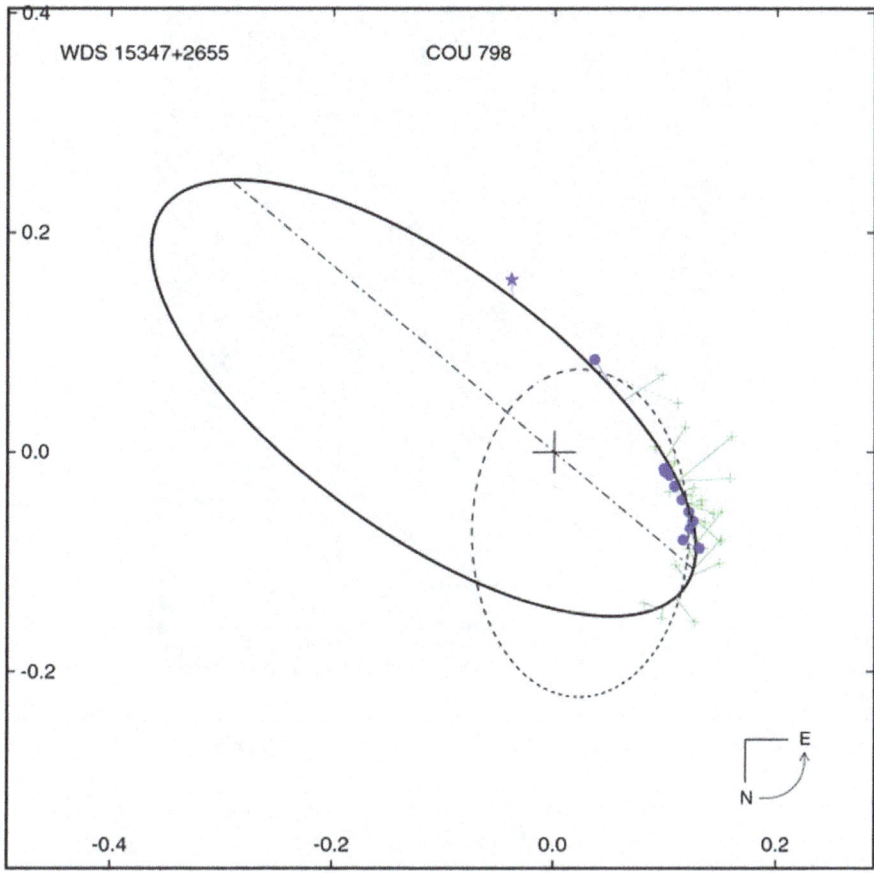

Figure 7.4. What a difference one measurement can make! See text for discussion. (Author adaptation from Figure 5 of reference Hartkopf & Mason 2009. © IOP Publishing. Reproduced with permission. All rights reserved.)

having only completed a small arc of its motion. Just a glance at the figure shows how off the mark was that earlier orbit. A definitive orbit will only arrive after a complete revolution of the pair of stars has been observed. Such is the case for thousands of visual binaries like Cou 798.

The namesake of this binary was its discoverer Paul Couteau (1923–2014), a consummate visual micrometer observer who discovered 2700 new pairs while accumulating nearly 26,000 measurements, mostly with the 50 cm and 76 cm refractors of the Nice Observatory. Couteau is shown in Figure 7.5 with Brian Mason in 2009. Couteau's goal was to survey the declination zone +17° to +52° while his Paris Observatory colleague Paul Muller (1910–2000) extended that effort to the pole, adding another 705 new binaries and 13,000 measurements. These two men were among the last to devote their careers to the visual study of double stars (Argyle et al. 2019).

Funding limitations have prevented the USNO speckle group from revisiting the 100 inch, and the telescope is, for the foreseeable future, no longer being routinely

Figure 7.5. Brian Mason (U.S. Naval Observatory) and Paul Couteau chat during an annual meeting of the CHARA collaboration at the Nice Observatory on 2009 March 17. (Author's photo.)

used for scientific purposes. During my tenure as the Observatory's director, a new visual observing capability was added to the 100 inch, so that this splendid telescope can play the capstone outreach role for Mount Wilson Observatory. Prior to this, one could only observe visually through the coudé focus, which gives poor image quality, or from the Cassegrain observing platform. The latter presents a precariously dangerous and uninsurable prospect.

Today, visitors standing below the 100 inch (see Figure 7.6) can look through this once world's largest telescope, with its magnificent light gathering power, to collect photons on their retinas from star clusters, nebulae, galaxies, and, yes—even double stars. This new capability utilizes a periscopic arrangement of mirrors and lenses to relay the focus at the north Cassegrain observing port, free of image degradation, to a safe and easily accessible position below the 100 inch mirror enclosure. This addition to the century-old telescope, which has provided a new income stream for the Mount Wilson Institute, was made possible through optical and mechanical design work by CHARA Senior Research Scientist Laszlo Sturmann, who designed and fabricated much of the electro-mechanical upgrades to CHARA Array facilities. Sturmann is to CHARA as Pease was to Mount Wilson Observatory. In parallel with the optical work, upgrades to the telescope's pointing and tracking systems by Gail Gant and Bill Leflang, who have devoted thousands of hours improving Observatory facilities over the years, have added greatly to the operational efficiency of the 100 inch.

Figure 7.6. Don Nicholson, son of famed MWO astronomer Seth Nicholson, looks on as CHARA's Laszlo Sturmann inserts an eyepiece into his new optical system for visual observations at the 100 inch on 2012 October 5. (Author's photo.)

I had always dreamed that the telescope could be used in a continuing program of binary star speckle interferometry. The problem was—you guessed it—money. The telescope is expensive to operate, but if we could assemble a group of volunteers, the operating costs would boil down to utilities and repairs. Of course, the latter could be crippling if a major failure occurred on the vintage telescope. But, some risks are worth taking. I thought there must be enthusiastic and talented amateur astronomers in the LA area who would be thrilled to join in on such a program. From early on, my colleagues and I had realized that the amazing advance in commercial technology available to amateurs could permit them to undertake binary star speckle interferometry programs. While still a graduate student, Nils Turner contributed a paper on this topic at a 1992 meeting we organized in Pine Mountain, Georgia (Turner et al. 1992) and a few years later as a postdoc added a nice amateur speckle sidebar to an article I wrote for *Sky and Telescope* magazine in 1996 (McAlister 1996). He has followed this up with other publications (e.g., Turner & Argyle 2012). Brian Mason and Bill Hartkopf have also been highly supportive of amateur involvement in the field.

I first became aware of this concept becoming a reality in 2008 when Russell M. Genet contacted me and others about the feasibility of such an undertaking. Genet subsequently produced a portable speckle camera and obtained time on the Kitt Peak 2.1 m telescope to test and refine the instrument and observing technique (Genet 2015; Genet et al. 2015). Russ later brought his equipment to Mount Wilson for experiments and workshops at both the 60 and 100 inch telescopes (Genet et al. 2016). He participated in the more recent activity I'm about to describe.

An unanticipated benefit of the new public-viewing focus is the accessibility it provides to advanced amateurs in possession of sophisticated observational and analysis tools for acquiring speckle data, reducing it, and then publishing it. Their compact speckle optical assemblies and imaging cameras insert directly into Laszlo Sturmann's eyepiece holder. One such visit took place in 2019 June when a group of eight students and their teacher from the Paso Robles High School (PRHS) in San Luis Obispo County, California (about 220 miles northwest of Mount Wilson) came to the mountain. Their trip was under the auspices of the Astrometry Field Research Program carried out by their school and organized by their teacher Jon-Paul Ewing in collaboration with the Institute for Student Astronomical Research (InSTAR).

In addition to their immersion in a specific research topic at Mount Wilson, the students were trained in real time to operate the telescope and speckle camera system, which debuted on their observing night of 2019 June 11. The success of this "engineering run" resulted from a group of highly experienced double star observers —Rick Wasson, Reed Estrada, and Chris Estrada—who had assembled the speckle instrumentation. Equally important was their data pipeline that the students used under the guidance of these experts. Another key player was Tom Meneghini, the Mount Wilson Institute's Executive Director, who hosted the group at the Observatory and operated the telescope for them.

The day following their successful observing session, the PRHS students reduced their data to extract new measures of the binary stars Cou 1757 and Hu 1268, the latter discovered in 1905 by William J. Hussey (1862–1926). The experience culminated with the students publishing the results of their speckle observations at the 100 inch in the *Journal of Double Star Observations* (Gates et al. 2020) along with a description of this "new opportunity" written by their expert instructors (Wasson et al. 2020). Figure 7.7 provide two glimpses from the PRHS visit to Mount Wilson.

Such experiences will hopefully continue, and this high-resolution specialty will produce scientifically valuable results at the 100 inch telescope while inspiring a few young people to become scientists and making even more good science citizens. My fingers are crossed.

7.2 The Mark III Interferometer (1987–1992)

Although it only operated from 1987–1992, the Mark III stellar interferometer had an enduring influence on the field. Its scientific output included more than 100 stellar angular diameters, the orbits and masses of 16 resolved spectroscopic binaries, the demonstration of high-precision stellar astrometry (the determination of stellar positions and motions), and the characterization of atmospheric seeing over Mount Wilson. Beyond that impressive body of work, key technologies developed by the Mark III team were subsequently adopted by follow-on interferometers such as the Navy Precision Optical Interferometer (NPOI) outside Flagstaff, the Palomar Testbed Interferometer (PTI) adjacent to Hale's 200 inch telescope dome, and the CHARA Array on Mount Wilson. Equally important to those technologies was

Figure 7.7. (Above) At their observing station set up just below the 100 inch primary mirror enclosure, a PRHS student is shown by Jimmy Ray how to keep an object centered in the small field of view using the telescope's hand paddle for small pointing adjustments. Richard Harshaw, Russ Genet and Reed Estrada oversaw details of data acquisition. (Below) Rick Wasson walks a student through the autocorrelation analysis of speckle data while other students get a head start on writing their JDSO paper. (Photos courtesy of Reed Estrada.)

the training the Mark III afforded to young interferometrists who would go on to participate in the development of other ground- and space-based interferometers.

As its name implies, this was the third in a series of interferometers originating at the Massachusetts Institute of Technology and deployed at Mount Wilson. The Mark I interferometer (Shao & Staelin 1979) was a prototype instrument with a 1.5 m baseline developed by MIT graduate student Michael Shao. His supervisor was electrical engineering and computer science professor David H. Staelin (1938–2011), who had entered MIT as an undergraduate in 1956 and never left. Mike Shao's 1978 doctoral dissertation on "A Long Baseline Optical Interferometer for Astrometry" launched a career in interferometry with a broad and abiding impact. That instrument's primary goal was to demonstrate fringe tracking—stabilizing the interferometric fringes to a fraction of a wavelength. This it did admirably in 1979 March, servo-locking onto fringes and maintaining equal path lengths to ±0.1 μm (Shao & Staelin 1980). The target star was Polaris, using light over the broad spectral band of 0.4–0.9 μm. This was a vitally important achievement, as fringe-tracking would be one of the essential elements of any future interferometer. Measurement of star position, i.e., stellar astrometry, is achieved interferometrically by accurately measuring the position of the fringe center as opposed to its visibility. This requires very accurate determination of the optical delay in the path length compensation process and places severe constraints on the stability of an interferometer. The Mark I's second notable achievement took place in 1979 June when it simultaneously tracked fringes at two separate wavelengths (Shao 1981).

The next phase was undertaken in early 1980 by the Mark II interferometer (Shao 1981); this time in a partnership between MIT and the Naval Research Laboratory, which had a mission-oriented interest in accurate stellar astrometry. Since well before the Civil War, the U.S. Naval Observatory had undertaken transit circle measurements of star positions to support the navigation of ships at sea. The USNO was particularly interested in new techniques for improving the accuracy of those positions, and interferometry offered intriguing possibilities. It was quite natural for the Navy to become involved and support the MIT astrometric interferometry work.

Unlike the Mark I, which stared at Polaris as its only target, the Mark II was designed to look at stars within 15° of the zenith. The new goal was to achieve day-to-day, 1 mas repeatability in measuring the angular separation between stars within 1.5° of each other on the sky. This precision was relaxed to 20–50 mas for stars 10°–20° apart. A secondary goal was to identify the main sources of instrumental error in measuring fringe positions. The Mark II employed a 3.2 m baseline between 5 cm aperture siderostats akin to those shown in Figure 7.9. This allowed the use of a two-color method developed by Shao & Staelin (1977) for eliminating atmospherically-induced angular displacements between stars. Optical delay compensation was done in vacuum in order to avoid dispersion in long air paths, and the retro-reflector position of the delay line was measured by a laser-interferometer metrology system to an accuracy of ±10 nm. In a 1981 preliminary report on the goals of the Mark II (Shao 1981), it was noted that these techniques could be utilized in a future space-based interferometer, which would become the ultimate goal of Shao's group. For the near-term, the performance limits of the Mark II would feed into the design of the Mark III instrument, which would be longer-lived than its progenitors and capable of carrying out a significant science program, while also continuing technology development toward a major astrometric interferometer.

By the late summer of 1983, the Mark II was performing at the 20 mas level with a series of 1 s integrations. But, observations carried out over periods of hours showed significant excursions, which were attributed to mechanical and thermal instabilities. It was concluded that *the long-term drift of the instrument is a serious problem.* This situation would be solved in the Mark III by incorporating laser-interferometry-based metrology *to create, in effect, an astrometric instrument that is mechanically and thermally perfect. All known errors for wide angle (>15°) astrometry at the milliarcsecond will be monitored or corrected (Shao 1986).* While observations of five bright stars carried out in 1984 September showed some sets of residuals in delay measurements at the few-μm level, they were intermixed with other sets having residuals as large as 20 μm. The nature of these excursions suggested a combination of thermal drift effects and ball bearing imperfections in the siderostat mounts, thus pointing to improvements that would be made in the Mark III. The final report on the Mark II expressed confidence that the large deviations in delay could be reduced to achieve accuracies of ±0.1 μm—a factor of 200 improvement in performance—translating to ±10 mas accuracy in star positions. The authors were optimistic that this goal would be achieved and exceeded by the Mark III interferometer (Shao et al. 1987).

This was a challenge that the MIT–NRL team would tackle in its third iteration, this time employing longer baselines and multiple telescopes. The new instrument—the last this group would place on Mount Wilson—would be a major extrapolation of the two prototypes but was not intended as their ultimate astrometric interferometer. At the time the Mark II had fulfilled its mission, Shao was by then at the Smithsonian Astrophysical Observatory (SAO) in Cambridge, Massachusetts. On the basis of the considerable observing experience already in hand, Shao and his SAO colleagues Peter Nisenson and Robert Noyes rhetorically asked at the 1984 January Tucson meeting of the American Astronomical Society if Mount Wilson was the best interferometer site in the world (Nisenson et al. 1984). The Mark III and, much later, the CHARA Array would provide prodigious amounts of data relevant to that question, which I attempt to answer in Appendix A. Another of Staelin's MIT students, M. Mark Colavita, joined Shao at SAO in 1985 after finishing his dissertation on the atmospheric limits of two-color interferometry. Shao and Colavita formed a powerful and complementary partnership that would lead to two ever-grander successors to the Mark III and develop the new technology for a space-borne interferometer. If you run a search on NASA's Astrophysics Data System, you'll them both among the authors of each of 121 published papers and abstracts to date.

Ground-based optical interferometry is a small field in which the players know each other and will from time-to-time change teams. One such interferometrist is Donald J. Hutter, who came to work as a postdoc in my speckle interferometry group at Georgia State in 1984 March after completing his doctorate at the Indiana University. His research topic there involved an X-ray and optical survey of Markarian Galaxies, which seems pretty distant from binary star speckle interferometry except that we then had ambitions—never to be fulfilled—of high-resolution studies of galaxies with compact bright nuclei. Don, whom you encountered briefly in the previous section, retooled as an interferometrist and went to work at the Naval Research Lab in the fall of 1985 after his postdoc with us. NRL put him to work on the Mark III and assigned him the responsibility for procuring, assembling, and integrating the optical-fiber system that distributed the laser-interferometer metrology system for monitoring the delay line positions and the siderostat pointing. Don did much of the design, parts specification, assembly, and alignment of the siderostat metrology optics as well, and assisted in the beam combiner assembly.

When the integration phase started in 1986, Don found himself traveling out from Maryland to Mount Wilson and spending two out of every six weeks on the mountain. Once the metrology components were in place, Don participated in the night-time observing. He recalls their state-of-the-art nine-track tape drives to which data were written by a control computer with "*a 20 Mbyte disk drive the size of a manhole cover.*" Observing was done from a trailer (see Figure 7.8) that was situated such that if you came out of the door in the middle of a dark night and made the wrong turn you could, in just a few steps, rudely discover that you'd walked off the mountain's eastern escarpment.

Mike Shao and Mark Colavita were both heavily involved with the installation of the Mark III, which started in early 1986. Don Hutter remembers that they joined in with him in installing cables, a task that first involved laying them out in the proper lengths in the parking lot by the 100 inch dome. It was a very buggy day, and "*every once in a while, Mike would stop and run around in circles to get away from the bugs, and then get back to work*" until the job got done.

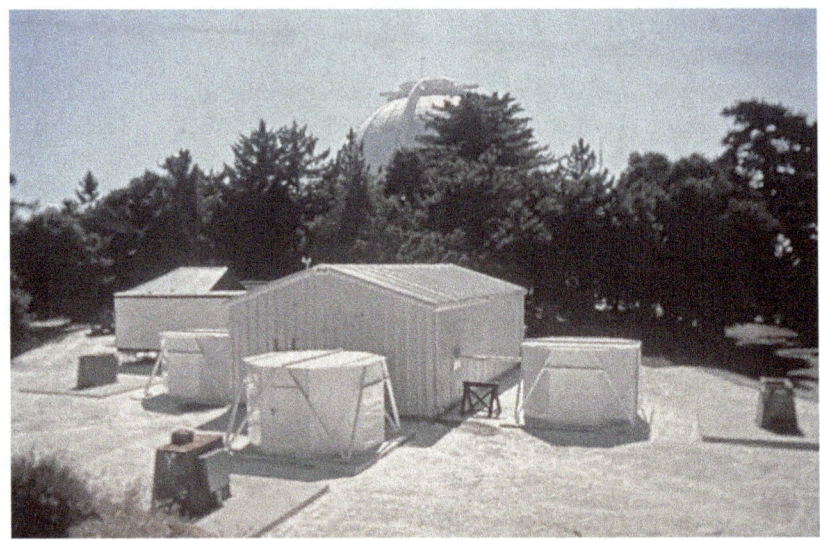

Figure 7.8. The Mark III Interferometer site to the east of the 100 inch telescope dome. Three siderostat enclosures flank the north, east, and south sides of the central beam-combining laboratory while the control trailer sits in the left background of this view. The trapezoidal plinths next to each enclosure were intended for possible longer baselines but were never used. In 1988, a variable baseline providing siderostat spacings from 3 to 32 m was added. (Courtesy of Donald J. Hutter.)

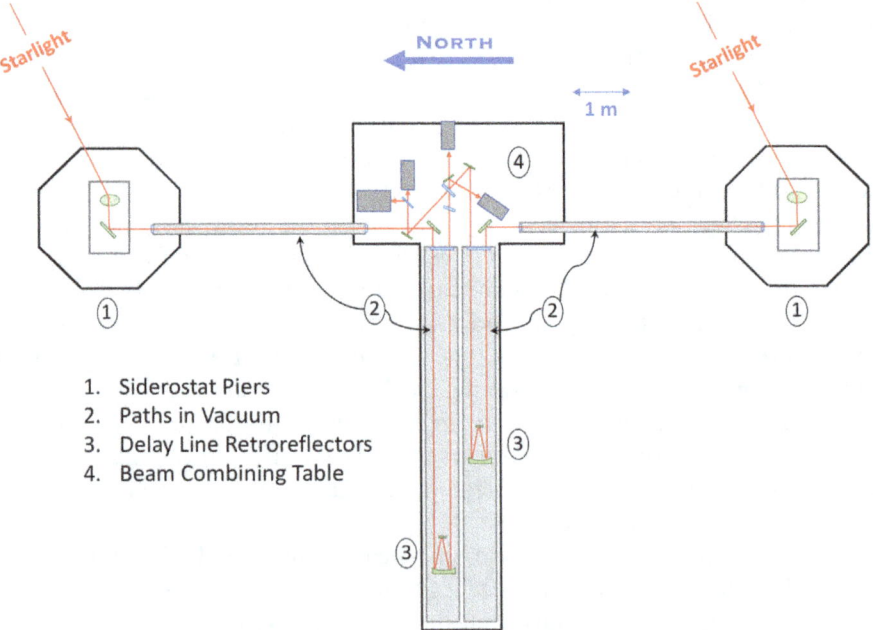

Figure 7.9. General layout of the N–S astrometric telescopes and the beam-combining table of the Mark III interferometer. (Author's adaptation of Figure 2 from Shao et al. 1988a. © ESO.)

Achieving its first fringes in 1986, the Mark III flew under the flag of an SAO–MIT–NRL–USNO partnership. It occupied the same site of its predecessors, a football-field's distance east of the 100 inch telescope dome, adjacent to the mountain's escarpment. This location had been used by Caltech physicists Gerry Neugebauer (1932–2014) and Robert B. Leighton (1919–1997) during 1965–1968 for one of the first surveys for infrared sources in the northern sky—the Two-Micron Sky Survey (TMSS)—which helped launch the new field of infrared astronomy by cataloging some 5000 extremely red objects (Neugebauer & Leighton 1969). Their highly compact *f/*1 telescope employed a 62 inch mirror that had been spun cast from epoxy resin and then coated with aluminum (Neugebauer et al. 1965). The enclosure for that historic experiment had been home to the Marks I and II instruments during their brief tenure on the mountain and afterwards suffered the ignominious fate or providing storage for paint and cleaning materials. The TMSS telescope itself, known as the Caltech Infrared Telescope, is on display at the National Air and Space Museum's Udvar-Hazy Center in Chantilly, VA. It is rumored an aperture of 62 inches was chosen by Neugebauer so that the TMSS telescope would be Mount Wilson's second-largest.

The general layout of the initial astrometric configuration of the Mark III is shown in Figure 7.9. The north and south siderostats rest on massive piers (1) inside temperature controlled enclosures as shown in Figure 7.10. Each siderostat directs its beam through an evacuated tube (2) toward the optical path-length-matching delay lines (3), also in vacuum. The beams are injected into the delay lines by two

Figure 7.10. A siderostat inside its enclosure is shown here. The temperature inside the enclosure is maintained at the night-time value during the day to avoid thermal shock upon opening for observing. (Courtesy of Donald J. Hutter.)

folding mirrors that preserve polarization vectors. A third astrometric siderostat, located 12 m east of the north–south astrometric baseline, was later installed to enable an east–south baseline. Also added were two imaging siderostats that could be positioned on any of 11 siderostat piers to achieve north–south baselines ranging from 3 to 32 m. With just two delay lines, the interferometer could only be used for pairwise combination, either in the astrometric or imaging mode.

We pick up from there by now referring to Figure 7.11 and following the beam paths in greater detail. The siderostat beams exit their transport tubes though optically flat windows (1) after which folding mirrors (2), which are precisely pointed by piezo-electric transducers (PZTs), direct them into the delay lines. Figure 7.12 shows details of the delay line system.

After the south beam emerges from path length adjustment by its delay line, it is intercepted by an annular wedge (3) that passes most of the beam uninterrupted, but also a secondary beam split off by 45 arcsec. A folding mirror then turns these beams toward the beam splitter (5). Also arriving at that beam splitter is the path-length matched north beam that has passed through a glass compensator (4) of equal thickness to the beam splitter (5). This ensures that the N and S beams suffer the same dispersion in traveling through glass prior to their interference at the coating on the NW face of the beam splitter. At this point, the reflected component of the S beam interferes with the transmitted component of the N beam and vice versa, following which the outputs of the two sides of the beam splitter are directed by folding mirrors to two fiber feed assemblies for fringe detection by photomultiplier tubes (8). The slide-in mirror (6) can be moved out of the beam to feed an alignment telescope (7). The secondary beam from the wedge (3) is picked off by an annular mirror (9) to feed an angle tracker camera (10) to ensure interference between beams.

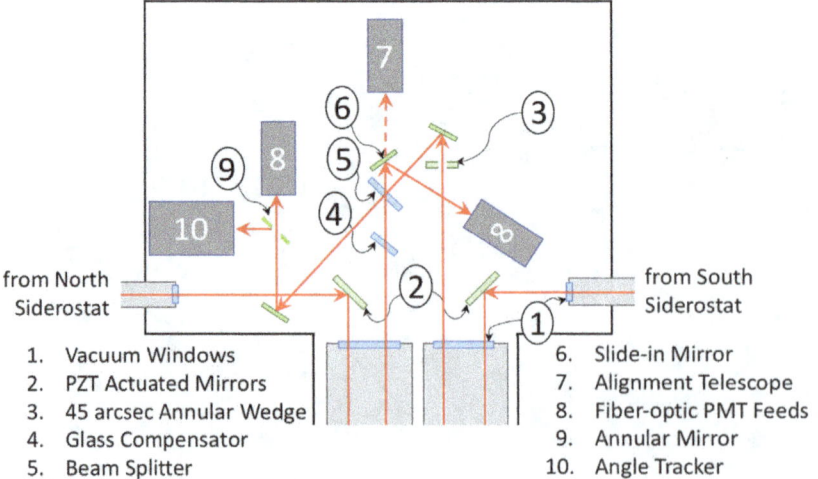

1. Vacuum Windows
2. PZT Actuated Mirrors
3. 45 arcsec Annular Wedge
4. Glass Compensator
5. Beam Splitter
6. Slide-in Mirror
7. Alignment Telescope
8. Fiber-optic PMT Feeds
9. Annular Mirror
10. Angle Tracker

Figure 7.11. Some details from the Mark III beam combination table. See text for a description. (Author's adaptation of Figure 2 from Shao et al. 1988a. © ESO.)

Figure 7.12. (Top) A path-length compensating retro-reflector cart moves on rails inside a cylindrical vacuum chamber (middle) in order to match the path lengths that common wavefront light intercepted by each siderostat must travel before being interfered. (Bottom) Light from a siderostat enters the delay line from one side of the window, and exits from the other side after being returned by the retroflector cart. (Courtesy of Donald J. Hutter.)

Figure 7.13. As this closeup of a portion of the central optical systems of the Mark III makes clear, navigating the busy traffic on the beam combining table of an optical interferometer requires a roadmap and/or a knowledgeable guide. (Courtesy of Donald J. Hutter.)

More details of the Mark III optical layout, including the star tracker, delay lines, fringe detection, siderostat control, laser metrology, etc, are given in the project's main technical publication (Shao et al. 1988a). Figure 7.13 shows the fully-packed optical table of the Mark III.

In parallel with the project's primary astrometric goals, which we will examine later in this section, observations were secured of single stars for their diameters as well as for binary stars for their orbits, masses, and distances. While interferometric astrometry was new to Mount Wilson, interferometry for stellar masses and diameters at that site went back to Anderson, Merrill, and Pease. Evidently there was good karma on the mountain for that work, as the Mark III quickly settled down to scientific business. The first star diameter results came from a single night, 1986 November 21, when α CMi (Procyon), β Gem (Pollux), α Per (Mirfak), and α UMa (Dubhe) were resolved at 701 nm with β Tau serving as the calibrator star (Shao et al. 1988b). These initial Mark III diameters had accuracies ranging from ±1.7% to ±10%.

Two additional stars were observed: α Gem (Castor), a visual binary with additional companions too complex for visibility calibration, and α Leo (Regulus). Regulus was subsequently modeled by the CHARA Array to show its oblateness from rapid rotation that resulted in long- and short-axis diameters of 0.771 ± 0.032 and 0.651 ± 0.016 mas (McAlister 2005). These are well below the Mark III's resolution limit of 4 mas at its 12 m baseline, so its non-resolution is understandable. Still, this limit gave the Mark III considerable sensitivity to resolving giant and supergiant stars down to a limiting visual magnitude of about +7.

A second set of diameters, for 24 giants and supergiants observed at 674 nm, followed on 29 nights from late-July to mid-October 1987 (Hutter et al. 1989). The diameters of a third of these stars had accuracies of ±5% or better while three were in the range ±10% to ±12%. Comparison of the new diameters with ones already measured by other interferometers as well as with relations found between a star's color and its surface brightness—the so-called Barnes–Evans relation (1976)—showed generally acceptable agreement except for a few discrepancies as large as a factor of two.

These and subsequent diameter determinations are summarized in Table 7.1. Of these eight investigations, half were targeted at either individual stars or groups of special interest. The measurements for Arcturus and Mira relate back to the 20 foot interferometer results in Table 3.1. In the case of Mira, the Mark III visibilities were reproduced by fitting uniform elliptical disks at three phases over the star's cycle of variability. The models showed that two such disks of different major axes gave a reasonable fit to the observations (Quirrenbach et al. 1992).

In addition to Mira's variable size, the surface brightness is not spherically symmetric, and the position angle of the asymmetry varies with phase. The bottom line is that the star cannot be represented by a single expression of its diameter. That said, Pease's final word in 1931 on the star's diameter as being 47 mas is consistent with the Mark III's model of the star. For Arcturus, the comparison of the Mark III (Quirrenbach et al. 1996) and 20 foot diameters can be more simply stated—they agree to within one standard deviation of the Mark III error estimate.

The final diameter measurements by the Mark III were published 11 years after the facility's closure. Some 15,000 visibilities obtained using siderostat spacings from 3 to 32 m provided by a variable baseline capability installed in 1988 resulted in 220 angular diameters of 85 stars (Mozurkewich et al. 2003). Those data were recorded at several wavelengths between 451 and 800 nm on 133 nights between 1988 September 17 and 1990 October 15. A large fraction of the diameters have uncertainties at the ±1% level, and the agreements with results from other interferometers were close to 1σ (1 standard deviation). By the time that publication appeared, the Navy Prototype Optical Interferometer had arrived on the scene and

Table 7.1. Summary of Stellar Diameter Publications from the Mark III

No. of Stars	Stellar Type	Reference
4	Non-specific	Shao et al. (1988b)
24	Giants & Supergiants	Hutter et al. (1989)
12	Giants	Mozurkewich et al. (1991)
1	o Ceti (Mira)	Quirrenbach et al. (1992)
15	Giants & Supergiants in and out of TiO bands	Quirrenbach et al. (1993)
3	Carbon Stars	Quirrenbach et al. (1994)
1	α Boo (Arcturus)	Quirrenbach et al. (1996)
85	Cumulative analysis of most Mark III observations	Mozurkewich et al. (2003)

secured diameter measures for 39 stars in common with the Mark III (Nordgren et al. 1999). Of those stars, 51% agreed to within 1σ, 89% within 2σ, and 95% were within 3σ. Enough said. The Mark III results were well calibrated and have formal error estimates likely representative of the true uncertainties.

We now come to the binary stars observed by the Mark III interferometer. The targets were all spectroscopic binaries whose resolution would lead to component masses using the same principles as Merrill employed in 1920 for Capella and thereby demonstrating the power of interferometrically resolving a double-lined spectroscopic binary (see Appendix B.3.4). Indeed, Capella was measured with very high accuracy by the Mark III, as we shall see shortly.

Table 7.2 lists 16 spectroscopic binary star systems resolved by the Mark III. A few of these stars had previously been resolved by speckle interferometry, but the higher resolution of a long-baseline interferometer really pays off in terms of accuracy of measurement. We saw in Section 2.3 that the visual interferometry of Capella by John Anderson and Paul Merrill during 1919–1921 led to mean residuals $\overline{\Delta\theta}= +0.16° \pm 1.61°$ in position angle and $\overline{\Delta\rho}= -0.4 \pm 0.6$ mas in angular separation compared with $\overline{\Delta\theta}= +0.67° \pm 1.59°$ and $\overline{\Delta\rho}= -1.4 \pm 1.2$ mas for my own speckle observations. They had observed by eye whereas I had a more objective detection process, but those 1920 instruments had higher resolution than I did albeit my telescope was larger. Recall from Chapter 6 that the limiting resolution of a Michelson interferometer is $\lambda/2D$ compared with $1.22\lambda/D$ for speckle interferometry (see Appendices B.3.2 and B.7). Anderson stated that the slits in his interferometer projected onto the primary mirror with spacings ranging from 120 to 200 cm (Anderson 1920) compared with the aperture of the Kitt Peak 4-m telescope, which, due to a mask over the primary mirror's turned-down edge, is actually 3.9 m. Thus, the Mount Wilson measurements had a factor $2.44(D_{\text{Anderson}}/D_{\text{speckle}}) = 2.44(2.0/3.9) = 1.25$ higher resolution than I did. So, we should expect the Mark III, with $D = 12$ m, to do very well, indeed.

I need to point out that, unlike a Michelson interferometer on a telescope pointed at a star, modern interferometric arrays do not directly measure the binary star observables θ and ρ. Instead, they measure fringe visibilities along a projected baseline dependent upon the geometric relationships of the telescopes in the plane of the ground and the object's position on the sky. The latter, of course, changes due to Earth rotation. Having recorded the various geometric and time parameters along with the visibilities, an orbit can be calculated, and the visibility residuals from that orbit can be used to calculate astrometric errors relatable to θ and ρ. With that caveat, the mean "residuals" and their standard deviations to the Mark III orbit of Capella (Hummel et al. 1994a) are $\overline{\Delta\theta}= +0.39° \pm 0.58°$ and $\overline{\Delta\rho}= -0.02 \pm 0.28$ mas.

Looking only at $\sigma_{\Delta\rho} = \pm0.28$ mas compared with ±0.6 and ±1.2 mas for the Anderson interferometer and my speckle residuals, respectively, we see that the Mark III residual has factors of 2.1 and 4.3 improvement over its predecessors. Nine of those residuals are from baselines between 3 and 8 while 3 others are from baselines exceeding 22 m. If we only include the remaining 29 that are close to 12 m, we see an improvement with $\sigma_{\Delta\rho} = \pm0.24$ mas, implying Mark III advantage factors

Table 7.2. Summary of Binary Star Results from the Mark III

Star Name	Spectral Types	Orbital Period (days)	Semi-major Axis (mas)	Orbital Inclination (°)	M_1 (M_{sun})	M_2 (M_{sun})	Distance (parsecs)	Reference
β Ari	A5 V	107	36.1 ± 0.3	44.7 ± 1.3	2.34 ± 0.10	1.34 ± 0.07	18.87 ± 0.61	Pan et al. (1990)
φ Cyg	K0 III + K0 III	434	23.7 ± 0.4	78.4 ± 0.4	2.54 ± 0.09	2.44 ± 0.08	80.8 ± 1.8	Armstrong et al. (1992a)
γ Equ	A9 Vp	98.8	11.987 ± 0.078	151.5 ± 1.1	2.13 ± 0.29	1.86 ± 0.21	55.3 ± 2.3	Armstrong et al. (1992b)
α And	B8 IVp + A3 V	96.7	96.6960 ± 0.0013	105.66 ± 0.22	3.8	1.8	—	Pan et al. (1992a)
θ² Tau	A7 III	—	18.6 ± 0.2	46.2 ± 1	2.0	1.6	43.7 ± 1.9	Pan et al. (1992b)
η And	G8 III-IV + G8 III-IV	115.72 ± 0.01	10.37 ± 0.03	30.4 ± 0.4	2.59 ± 0.30	2.34 ± 0.22	76 ± 2	Hummel et al. (1993)
α Aur	K0 III + G1 III	104.022 ± 0.002	57.46 ± 0.05	137.18 ± 0.05	2.69 ± 0.06	2.56 ± 0.04	13.3 ± 0.1	Hummel et al. (1994a) (radii also measured)
π And	B5 V + B5 V	143.53 ± 0.06	6.69 ± 0.05	103.0 ± 0.2	—	—	—	Hummel et al. (1995)
θ Aql	B9.5 III + B9.5 III	17.122 ± 0.001	3.2 ± 0.1	143.5 ± 3.0	3.6 ± 0.8	2.9 ± 0.6	75.4 ± 6.0	Hummel et al. (1995)
β Aur	A1m IV + A1m IV	3.9600 ± 0.0001	3.3 ± 0.1	76.0 ± 0.4	2.41 ± 0.03	2.32 ± 0.03	24.9 ± 0.5	Hummel et al. (1995)
ζ¹ Uma (Mizar A)	A2 Vp + A2 Vp	20.5377 ± 0.0003	9.64 ± 0.05	59.7 ± 0.2	2.51 ± 0.08	2.55 ± 0.07	26.1 ± 0.3	Hummel et al. (1995)
93 Leo	G5 III + A7 V	76.69	7.5 ± 0.1	50.1 ± 0.5	2.25 ± 0.29	1.97 ± 0.15	72.7 ± 2.7	Hummel et al. (1995)
113 Her	G8 III + A0.5	245.52 ± 0.08	10.1 ± 0.1	40.2 ± 0.6	3.6 ± 0.8	3.6 ± 0.8	76.2 ± 1.8	Hummel et al. (1995)
β Tri	A5 IV	31.387 ± 0.001	8.02 ± 0.05	129.9 ± 0.2	(3.7 ± 0.4)	1.6 ± 0.2	—	Hummel et al. (1995)
δ Tri	G0 V + G9-K4 V	10.0200 ± 0.0001	9.80 ± 0.06	167 ± 34	—	—	—	Hummel et al. (1995)
ζ Aur	K4 Ib + B5 V	972	16.2 ± 0.1	87.3 ± 1.0	5.8 ± 0.2	4.8 ± 0.2	261 ± 3	Bennett et al. (1996)

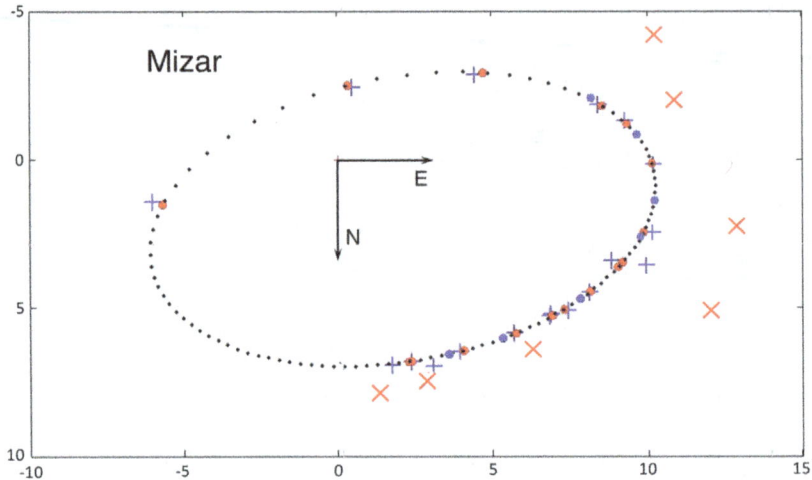

Figure 7.14. Francis Pease's observations from 1925–27 shown as red X symbols are plotted against the 12-m baseline observations and orbit derived from Mark III observations from 1990–92 shown as blue + symbols. Filled circles of the opposite colors show the predicted positions. (Author's diagram)

of 2.5 and 5.0 over its competition at that time. This is the long-way around to saying that, yes, longer baselines do give higher-accuracy results.

Figure 7.14. Francis Pease's observations from 1925–1927 shown as red X symbols are platted against the 12 m baseline observations and orbit derived from Mark III observations from 1990–1992 shown as blue + symbols. Filled circles of the opposite colors show the predicted positions. (Author's diagram.)

The only other binary system in Table 7.2 in common with the Anderson–Merrill–Pease era is ζ^1 UMa—Mizar. In Figure 7.14, I've plotted the Mark III orbit overlaid with its equivalent measurements (Hummel et al. 1995) after culling out those made at shorter baselines, which, as would be expected, show larger deviations from the orbit. Pease's results are also plotted. In Section 3.3 we saw that Henry Norris Russell combined Pease's observations with the available spectroscopically-determined orbital elements for Mizar to obtain the geometric elements unique to visual orbits (see Appendix B.3.3). Pease's values of ρ led to an orbital semi-major axis of 11.5 mas or about 20% larger than that seen in Table 7.2. A glance at Figure 7.14 shows why—Pease's angular separations were systematically too large. His values of θ were in better agreement. Subject to the caveat mentioned above, the Mark III mean residuals to the orbit derived from them are $\overline{\Delta\rho} = 0.052 \pm 0.251$ mas and $\overline{\rho\Delta\theta} = 0.005 \pm 0.061$ mas. Mizar was also observed with the Mark III's successor NPOI, which has significantly longer baselines, making only minor modifications to the Mark III orbital elements (Hummel et al. 1998). But, the NPOI data were used to construct a lovely image of the unresolved and fully detached components of the binary (Hummel et al. 1997).

Although I do not normally include results from unrefereed journals, a preliminary orbit for the Hyades binary θ^2 Tauri (Pan et al. 1992b) was used in a new solution of the spectroscopic orbit to determine the distance of the Hyades star cluster (Torres et al. 1997). The resulting distance to the binary of 47.6 pc is in excellent agreement with the mean cluster distance of 47.34 pc determined by Hubble Space Telescope observations of binaries (McArthur et al. 2011) as well as with 46.45 from HIPPARCOS (van Leeuwen 2009). Long one of the more vexing challenges to astrometry, the problem of determining the distance to this key rung in the cosmic distance ladder was finally accurately determined from space astrometry.

These few examples of their binary star work show that the Mark III scientists fully exploited their interferometer to extend a large cantilever out into the classical gap separating the "visual" and spectroscopic binaries.

Stellar diameters and the orbits of binaries were sidelines to the Mark III's original rationale—astrometry. The goal starting with the Mark I was to develop a new means for measuring star positions with accuracies exceeding that available from traditional instruments such as meridian circles. Recall that after climbing up the experience curve of the Mark II, the anticipation was that the Mark III would achieve star position accuracies of ±10 mas or better. The first relevant Mark III results derived from observations obtained in 1986 of 10 stars that were observed for over 90° in hour angle. The internal errors in right ascension (RA) and declination (Dec) were found to be ±6 and ±22 mas, respectively (Mozurkewich et al. 1988). Two years later, the Mark III group observed a dozen stars from the Fifth Fundamental Catalogue (FK5) of the positions of 1,535 stars that had just been updated and so were representative of the best that classical astrometry had to offer (Shao et al. 1990).

I calculated the mean errors for FK5 positions on the sky to be ±17 mas in RA and ±23 mas in Dec to give a combined uncertainty of ±29 mas (Fricke et al. 1988). By that time, the third astrometric siderostat had been added to the Mark III. It could feed into the delay lines in lieu of the north siderostat, thereby providing an E–S baseline intended to improve the RA measurement precision. Other measures by then undertaken were application of the two-color atmosphere correction, improved thermal control, and a new central fringe detection algorithm. As it turned out, the precision in RA was unchanged while that in Dec reduced to ±10 mas. Among these 12 stars, only half were within 50 mas of their FK5 positions indicating that the real external accuracy was affected by unknown systematic effects that persisted in the Mark III and FK5 results in spite of the great efforts in thermal and vibrational control on the part of the interferometrists and the even more complex situation underlying the multitude of input sources to the FK5. The issue of accuracy versus precision was subsequently addressed from additional observations of 11 of the stars in the sample and an argument was made that the displacement was due to FK5 positional errors (Hummel et al. 1994b). Nevertheless, under the metric of the precision of its star position measurements, the Mark III had attained its ±10 mas goal while surpassing the best ground-based positional catalogs, demonstrating interferometry's viability for wide-angle astrometry.

However, this competition was in the process of being made irrelevant by the mind-boggling success of astrometry from space. The Leander McCormick Observatory at the University of Virginia (where I went for grad school) had been engaged for most of the 20th century in measuring parallaxes of stars. In 1940,

Observatory director Samuel Alfred Mitchell (1874–1960) reported with pride of accomplishment that, since 1916, his observatory had taken 42,000 photographic plates of stars whose parallactic displacement on the sky arising from the Earth's revolution about the Sun revealed the distances of 1350 stars. The mean error in those trigonometric parallaxes was something like ±10 mas (Mitchell 1940). Hold those numbers in your head for a minute—1350 stars/±10 mas/24 yr.

In 2018, the science team for the Gaia astrometry mission of the European Space Agency (ESA) published the second data release from their spacecraft, which measures positions, parallaxes, and angular motions across the sky (Lindegren et al. 2018). Gaia was launched in 2013, so this release came from the first four years or so of observations, which will continue into 2023. Here are Gaia's numbers from that release—1,693,000,000 (yes, that's 1.69 billion!) stars/±0.04 to ±0.7 mas (for brightest to faintest stars)/4 yr. Those represent multiplicative gains in improvement in comparison with the McCormick work by—1,250,000/250 − 14/6. In the final release after 2023, those numbers will be even more impressive (except the time factor will drop to 2.5, but who cares). Sure, Gaia cost money—about a billion dollars, in fact—but the cost per parallax is only 59¢. That would have been about 4¢ per star in Mitchell's day, and I'll bet those McCormick parallaxes weren't that cheap.

From almost every angle you approach the comparison of ground- versus space-based astrometry, the ground loses the competition. The Earth's turbulent atmosphere is such an impediment to making positional measurements through it that there is no longer much reason for ground-based astrometric programs. There are important exceptions to that blanket statement. One is that, for the present at least, the brightest stars are too bright for missions like Gaia and yet they are among the most interesting objects for examination. Additionally, space missions are typically survey efforts of finite duration in time. They cannot react to new discoveries nor can they go back and revisit objects at will to, for example, confirm a suspected astrometric perturbation indicative of a planet orbiting a star. This could change if an observatory-class space interferometer were orbited, but the massive cost associated with such spacecraft makes it quite unlikely for the foreseeable future. This leaves room for ground-based astrometric efforts in important niches.

Of course, Gaia was not the first space astrometry mission. From 1989 to 1993, ESA flew HIPPARCOS. By the way, Gaia was also originally an acronym representing Global Astrometric Interferometer for Astrophysics, but the name became irrelevant when interferometry was dropped as the mission's observational technique. Rather than come up with a new acronym and name, it was decided to keep Gaia but lose the acronym.

The point is that space astrometry was already in planning during the 1980s while the Mark I and Mark II were in development, and in orbit while the Mark III was operational. True to its name, HIPPARCOS was intended to produce a very large catalog of stellar parallaxes, which it achieved with great success—it was this mission that forever changed the way astrometry would be done.

The Mark III had great momentum and was part of a continuing chain of projects that would give the Navy a dedicated interferometer facility from which to carry out its own unique mission. Furthermore, the Mark III was anticipating that an interferometer would one day exploit from space the very high resolution that interferometry has to offer. Indeed, some members of the Mark III group, including its leader Mike Shao, would venture in that direction by relocating to Caltech's Jet Propulsion Laboratory (JPL) while others would remain with the Navy to produce and operate its Navy Prototype Optical Interferometer (NPOI; Armstrong et al. 1998)—later rechristened the Navy Precision Optical Interferometer—for which Don Hutter, who had relocated from NRL to the USNO in 1989 served as director for 13 years prior to his retirement in 2017. The JPL interferometry group would build the Palomar Testbed Interferometer (PTI; Colavita et al. 1999) and link the two 10 m Keck telescopes in Hawaii as the Keck Interferometer (KI; Millan-Gabet 2007) for which Mark Colavita would serve as system architect. They would also design the delay line carts and build the associated control system for the CHARA Array. More than any other interferometer yet built, the Mark III heritage propagated into successor projects as a consequence of the reliability and performance designed into it by its originators.

The ultimate focus of Mike Shao's group at JPL would become technology development for the Space Interferometry Mission (SIM) with the primary goal of detecting extrasolar planets and additional key projects on determining the masses of the most and least luminous stars, measuring parallaxes of stars across the Milky Way Galaxy, and mapping the distribution of Dark Matter in the Local Group of galaxies. The Space Interferometry Mission, which was reduced in scope and cost to SIM-Lite, involved NASA, JPL, and Northrop Grumman. NASA awarded initial contracts totaling $200M in 1998. The JPL group was charged with completing proof-of-performance demonstrations in eight new technology areas critical to SIM's performance. They successfully completed that task, but various delays took place to postpone the launch by a decade over its originally proposed date.

But SIM was not to be. The National Academy of Sciences' decadal surveys of astronomy and astrophysics have long been roadmaps for federal funding of astronomy. The 2000 survey report[1] gave strong support to SIM, but when the NAS Astro2010 review[2] was released for the 2010–2020 decade, it did not recommend continued funding for SIM. NASA promptly terminated the project. Two years later, funding was also ended for the Keck Interferometer. The Palomar Testbed Interferometer, originally built as a technology stepping stone to linking the Keck Telescopes coherently, had been shuttered in 2008, after more than a decade of scientific productivity.

The mercurial nature of funding thus ended a remarkable chain of developments that all traced back to Mike Shao's Mark I interferometer on Mount Wilson. But, the hardware DNA of the Mark III interferometer persists in several currently operational interferometers and many of the scientists who worked on the Mark III and its descendant facilities are today providing leadership at the handful of working interferometers and large telescope projects around the world.

---

[1] Available at https://www.nap.edu/catalog/9839/astronomy-and-astrophysics-in-the-new-millennium.

[2] Available at https://www.nap.edu/catalog/12951/new-worlds-new-horizons-in-astronomy-and-astrophysics.

References

Aitken, R. G. 1921, PASP, 33, 85

Anderson, J. 1920, ApJ, 51, 263

Argyle, B., Swan, M., & James, A. 2019, An Anthology of Visual Double Stars (Cambridge: Cambridge Univ. Press), 30

Armstrong, J. T., Hummel, C. A., Quirrenbach, A., et al. 1992a, AJ, 104, 2217

Armstrong, J. T., Mozurkewich, D., Vivekanand, M., et al. 1992b, AJ, 104, 241

Armstrong, J. T., Mozurkewich, D., Rickard, L. J., et al. 1998, ApJ, 496, 550

Barnes, T. G., & Evans, D. S. 1976, MNRAS, 174, 489

Bennett, P. D., Harper, G. M., Brown, A., & Hummel, C. A. 1996, ApJ, 471, 454

Colavita, M. M., Wallace, J. K., Hines, B. E., et al. 1999, ApJ, 510, 505

Fricke, W., Schwan, H., Lederle, T., et al. 1988, VeARI, 32, 1

Gates, E., Hughes, A., McNerney, M., et al. 2020, JDSO, 16, 163

Genet, R. M. 2015, JDSO, 11, 183

Genet, R. M., Smith, T. C., Clark, R. K., Wren, D., & Mathis, H. 2015, JDSO, 11, 226

Genet, R. M., Row, D., Meneghini, T., et al. 2016, JDSO, 12, 263

Hartkopf, W. I., McAlister, H. A., Mason, B. D., et al. 1997, AJ, 114, 1639

Hartkopf, W. I., Mason, B. D., McAlister, H. A., et al. 2000, AJ, 119, 3084

Hartkopf, W. I., & Mason, B. D. 2009, AJ, 138, 813

Hummel, C. A., Armstrong, J. T., Quirrenbach, A., et al. 1993, AJ, 106, 2486

Hummel, C. A., Armstrong, J. T., Quirrenbach, A., et al. 1994a, AJ, 107, 1859

Hummel, C. A., Mozurkewich, D., Elias, N. M., et al. 1994b, AJ, 108, 326

Hummel, C. A., Armstrong, J. T., Buscher, D. F., et al. 1995, AJ, 110, 376

Hummel, C. A., Mozurkewich, D., Armstrong, J. T., et al. 1998, AJ, 116, 2536

Hummel, C. A., & Benson, J. A. 1997, in IAU Symp. 189, Fundamental Stellar Properties: The Interaction Between Observation and Theory, ed. T. R. Bedding (Australia: Univ Sydney School of Physics)

Hutter, D. J., Johnston, K. J., Mozurkewich, D., et al. 1989, ApJ, 340, 1103

Lindegren, L., Hernandez, J., Bombrun, A., et al. 2018, A&A, 616, A2

Mason, B. D., Hartkopf, W. I., Raghavan, D., et al. 2011, AJ, 142, 176

Mason, B. D., Wycoff, G. L., Hartkopf, W. I., Douglass, G. G., & Worley, C. E. 2019, VizieR Online Data Catalog: The Washington Double Star Catalog

Mason, B. D., Hartkopf, W. I., & McAlister, H. A. 2010, AJ, 140, 242

McAlister, H. A., Mason, B. D., Hartkopf, W. I., & Shara, M. M. 1993, AJ, 106, 1639

McAlister, H. A. 1996, S&T, 92, 28

McAlister, H. A., ten Brummelaar, T. A., & Gies, D. R. 2005, ApJ, 628, 439

McArthur, B. E., Benedict, G. F., Harrison, T. E., & van Altena, W. 2011, AJ, 141, 172

Millan-Gabet, R. 2007, AdSpR, 40, 659

Mitchell, S. A. 1940, PMcCO, 8, 1

Mozurkewich, D., Hutter, D. J., Johnston, K. J., et al. 1988, AJ, 95, 1269

Mozurkewich, D., Johnston, K. J., Simon, R. S., et al. 1991, AJ, 101, 2207

Mozurkewich, D., Armstrong, J. T., Hindsley, R. B., et al. 2003, AJ, 126, 2502

Muterspaugh, M. W., Lane, B. F., Kulkarni, S. R., et al. 2010, AJ, 140, 1657

Neugebauer, G., & Leighton, R. B. 1969, Two-micron Sky Survey. A Preliminary Catalogue (Washington, DC: NASA)

Neugebauer, G., Martz, D. E., & Leighton, R. B. 1965, ApJ, 142, 399

Nisenson, P., Noyes, R., & Shao, M. 1984, BAAS, 16, 908

Nordgren, T. E., Germain, M. E., Benson, J. A., et al. 1999, AJ, 118, 3032

Pan, X., Shao, M., Colavita, M. M., et al. 1990, ApJ, 356, 641

Pan, X., Shao, M., Colavita, M. M., et al. 1992a, ApJ, 384, 624

Pan, X., Shao, M., & Colavita, M. 1992b, in ASP Conf. Ser. 32, IAU Coll. 135, Complementary Approaches to Double and Multiple Star Research, ed. H. A. McAlister, & W. I. Hartkopf (San Francisco, CA: ASP), 502

Quirrenbach, A., Mozurkewich, D., Armstrong, J. T., et al. 1992, A&A, 259, L19

Quirrenbach, A., Mozurkewich, D., Armstrong, J. T., Buscher, D. F., & Hummel, C. A. 1993, ApJ, 406, 215

Quirrenbach, A., Mozurkewich, D., Hummel, C. A., Buscher, D. F., & Armstrong, J. T. 1994, A&A, 285, 541

Quirrenbach, A., Mozurkewich, D., Buscher, D. F., Hummel, C. A., & Armstrong, J. T. 1996, A&A, 312, 160

Shao, M. 1981, in Proc. Conf. in Sunspot, New Mexico, ed. R. B. Dunn (Sunspot, NM: Sacramento Peak Nat Obs), 240

Shao, M., Colavita, M., Staelin, D. H., et al. 1987, AJ, 93, 1280

Shao, M., Colavita, M. M., Hines, B. E., et al. 1988a, A&A, 193, 357

Shao, M., Colavita, M. M., Hines, B. E., et al. 1990, AJ, 100, 1701

Shao, M., Colavita, M., Staelin, D., Simon, R., & Johnston, K. 1986, in IAU Symp. 109, ed. H. K. Eichhorn, & R. J. Leacock (Dordrecht: Reidel), 331

Shao, M., & Staelin, D. H. 1977, JOSA, 67, 81

Shao, M., & Staelin, D. H. 1979, BAAS, 11, 396

Shao, M., & Staelin, D. H. 1980, ApOpt, 19, 1519

Shao, M., Colavita, M. M., Hines, B. E., et al. 1988b, ApJ, 327, 905

Shara, M. M., Doxsey, R., Wells, E. N., & McAlister, H. A. 1987, PASP, 99, 223

Shelton, J. C., Schneider, T., McKenna, D., & Baliunas, S. L. 1995, Proc. SPIE, 2534, 72

ten Brummelaar, T. A., Mason, B. D., Bagnuolo, W. G., et al. 1996, AJ, 112, 1180

ten Brummelaar, T. A., Hartkopf, W. I., McAlister, H. A., et al. 1998, Proc. SPIE., 3533, 391

ten Brummelaar, T. A., Mason, B. D., McAlister, H. A., et al. 2000, AJ, 119, 2403

Torres, G., Stefanik, R. P., & Latham, D. W. 1997, ApJ, 485, 167

Turner, N. H., ten Brummelaar, T. A., McAlister, H. A., et al. 2001, AJ, 121, 3254

Turner, N. H., & Argyle, R. W. 2012, in Observing and Measuring Visual Double Stars, ed. R. W. Argyle (New York: Springer), 261

Turner, N. H., Barry, D. J., & McAlister, H. A. 1992, in ASP Conf. Ser., 32, IAU Coll. 135, Complementary Approaches to Double and Multiple Star Research, 577

van Leeuwen, F. 2009, A&A, 497, 209

Vaughan, A. H. 2006, Notes on the History of the Formation of the Mount Wilson Institute (Pasadena, CA: Mount Wilson Institution Internal Publication)

Wasson, R., Estrada, R., Estrada, C., et al. 2020, JDSO, 16, 151

Chapter 8

Into the 21st Century

8.1 The Infrared Spatial Interferometer (1988–Present)

One cannot write about the Berkeley Infrared Spatial Interferometer (ISI) without first writing about its creator. Charles Hard Townes (1915–2015) (Boyd 2015) was a scientist, teacher, academic administrator, and advisor to presidents whose contributions to his nation and to the world are remarkable and diverse. Most people approaching their mid-70s are well settled into retirement—but not Charlie Townes. Instead, he and his talented group of graduate students and postdocs in Berkeley's physics department embarked on a research initiative that would provide new views of red giant and supergiant stars and their circumstellar environments. The spectral window they were opening to interferometry was especially suited to studying the spatial structures of dust produced by these stars and blown out in expanding shells by the strong winds of these mass-losing behemoths. Much would be learned about the environment of that Mount Wilson perennial—Betelgeuse—which would have delighted the interferometrists of the 1920s.

One of my fond memories of "Professor Townes," as he is often referred to, derived from sitting on the front row of his audience when, after fumbling in his pockets, he looked up and asked if someone had a laser pointer he could borrow. I jumped up and loaned him mine. For the duration of his talk, I sat musing at my having the privilege of my little laser being in the hands of the person whose brilliance made lasers possible. Townes' 1964 Nobel Prize in Physics, which he shared with two Soviet scientists, was "for fundamental work in the field of quantum electronics, which has led to the construction of oscillators and amplifiers based on the maser-laser principle."

During WWII, Townes' research centered around microwave radiation. While sitting on a Washington, DC park bench in 1951, an idea came to him that would be called "microwave amplification by stimulated emission of radiation." Townes' revelatory "maser" morphed into the "laser" when, together with his brother-in-law and fellow physicist Arthur L. Schawlow (1921–1999), they showed how the maser process could work for visible light (Townes & Schawlow 1955). Schawlow joined Townes as a Nobel physics laureate in 1981 for his work on laser spectroscopy.

Born and raised in Greenville, South Carolina, Townes always retained his Southern lilt and mannerly ways. He attended Furman University in Greenville and graduated *summa cum laude* in 1935 with a BS in physics and a BA in modern languages. He remains a favorite son of his hometown as evidenced in Figure 8.1. The next year he earned a MA degree in physics at Duke as a jumpstart to entering Caltech's physics doctoral program in 1936. Townes' 1939 PhD was awarded for research on the "Concentration of the Heavy Isotope of Carbon and Measurement of its Nuclear Spin." His professional journey for the ensuing 76 years led sequentially to Bell Telephone Laboratories where he worked through the war on radar development, next to Columbia University where the maser inspiration occurred, and then MIT where he served as provost for six years.

In 1967, Townes was appointed as University Professor at the University of California, Berkeley where he would spend the rest of his life. He never retired to emeritus status; instead he became a Professor in the Graduate School at Berkeley, or "one of the PIGS" as he would joke. At Berkeley, Townes' interests turned from quantum electronics to astrophysics and astronomical instrumentation. After participating in the then new and thriving radio astronomy searches for interstellar

Figure 8.1. Sculpture of Charles H. Townes in downtown Greenville, SC depicting the famous inspired moment of the maser concept. (Courtesy of Carol M. Highsmith Archive, Library of Congress.)

molecules, his attention retargeted to infrared astronomy—in particular to the mid-infrared (mid-IR) N spectral band covering the 8–14 microns (µm) wavelength region that provides a window of atmospheric transparency where dust formation around cool, evolved stars is vividly revealed. This would be his final research focus, and it would result in the first mid-IR astronomical interferometer.

Although I met Charlie Townes a number of times and spoke with him on the phone even more often, I never asked him about the origins of ISI and why he chose Mount Wilson for its location. So, I turned to several of Townes' colleagues and former students who contributed mightily to ISI's success. Walt Fitelson joined Townes' group as an electronics tech soon after Townes came to Berkeley. By the time ISI materialized, Fitelson was the electrical/electronics engineer for the project. All agree that without Walt, ISI would likely have shut down years ago, and he remains key to any refurbishment that might occur in the future. Just as I designated Laszlo Sturmann as CHARA's Francis Pease in Section 7.1, ISI's Pease is Walt Fitelson, shown engaged in his work in Figure 8.2.

After completing his dissertation research at the University of British Columbia on far infrared applications of interferometry of the spectrum of hydrogen, Ed Wishnow took a postdoc at the Lawrence Livermore National Laboratory working on infrared cameras. When that program came to an end, he was loaned to Berkeley for more camera development work. Ed, who early on in his graduate training found "interferometry to be beautiful," wound up in Townes instrumentation group where it was inevitable that his interest and expertise in infrared detectors and

Figure 8.2. Walt Fitelson at work on the telescope delivery day in 1988 January to rewire quick-connect power cables following discovery that an electrician had incorrectly wired the site's electrical box. Note the receding snow. (Courtesy of Manfred Bester.)

interferometry made him a natural for ISI. As the years passed and Charlie Townes entered his nineties, Ed transitioned to become Townes' successor as ISI's lead scientist, which he continues to include among his other responsibilities at Berkeley's Space Sciences Laboratory (SSL). Townes' first postdoc to work with ISI was Manfred Bester, who had just gotten his PhD at the University of Cologne and was deeply involved with ISI's launch on Mount Wilson and oversaw operations during the ensuing golden years of the project. Although he still consults with SSL, Manfred is now retired and relocated to Seattle. My final ISI interviewee was John Monnier, who joined Townes group as a doctoral student, completing his PhD dissertation in 1999 on "Interferometry and Spectroscopy of Circumstellar Envelopes." John is now a professor of astronomy at the University of Michigan and a leader in interferometry internationally. He is also a valued colleague who brought imaging to the CHARA Array years before we had expected it.

Many others have contributed substantially to ISI's success over the years. In the project's early period, graduate student Ed Sutton and postdoc Bill Danchi played important roles. Danchi also oversaw the procurement of the third telescope, which was subsequently integrated into ISI by David Hale after he took over operations from Manfred Bester. That telescope then enabled grad student Ken Tatabe to produce ISI's first closure phase results to be described below. Another student, Everett Lipman developed an infrared guiding system for ISI that allowed observations of objects heavily obscured by dust that were too faint for visible guide cameras. While also contributing to ISI science, Peter Tuthill, a postdoc from Cambridge University, devised an aperture-masking technique used at single telescopes to permit diffraction-limited imaging. Tuthill, Monnier, and Danchi used that method at the 10 m Keck Telescope on Mauna Kea to dramatically image the rapidly-rotating spiral feature of the Wolf–Rayet star WR104 (Tuthill et al. 1999). I could go on and on.

As I'd anticipated, none of the gentlemen I interviewed speak of ISI without first talking at length and with much admiration of Charles Townes as a man who dearly loved science and mentored with the kind persuasiveness and by self-example as one might expect of a South Carolinian. Walt Fitelson, who worked for Townes for four decades, always found his boss "*to be exceptionally open, friendly, and respectful to everybody: students, staff, and fellow faculty members. Obviously he expected results, but never by demand, only by example and inspiration. He was able to get to the heart of a problem or issue immediately and make the right decision. On the other hand, he was very receptive to ideas from others, letting them to do it their way, even making mistakes. Over the years, a few students decided to shift out of astrophysics into biophysics and other fields, which he supported fully. He also accepted students who had left other groups to join ours—which was quite unusual then, at least at MIT and Columbia before and Berkeley then. On big issues, like the direction of research, he was famous for studying the problem very carefully until he was convinced he was right, no matter what anyone else said. For that, I would include his own career choices —the maser and then astrophysics, especially stellar spectroscopy, molecular astronomy, stellar interferometry, IR astronomy, etc.*"

Such recollections infuse this story of Townes and his mid-infrared interferometer. I only regret that I don't have more space in this book to adequately reflect all that I learned from them. For now, let's go back to Charlie Townes' introduction to Pasadena and how he likely chose Mount Wilson for his future interferometer.

George Ellery Hale was in the next-to-last year of his life when Townes arrived in Pasadena in 1936, but the two likely never met. Astrophysics as his research arena was more than 40 years into Townes' future. But, from his excursions with friends into the San Gabriel Mountains during his graduate study years, Townes must have been well acquainted with Mount Wilson. Walt Fitelson remembers Townes' stories of driving his old Model A Ford up into the mountains to enjoy the spectacular scenery from the completed stretches of the Angeles Crest Highway. The twists and turns of the Mount Wilson Toll Road could have brought Townes and his Caltech buddies right up to the Mount Wilson Hotel for lunch and a walk around the Observatory grounds to inspect the world's largest telescope.

Fast forward half a century to 1988 January. A caravan of two tractor-trailer trucks and smaller vehicles made its way along the narrow single-lane road to the central intersection of the Observatory and stopped across from the Pease 50 foot interferometer building. An area adjoining that structure had been designated as the site for ISI, and the chaparral had been cleared away to accommodate ISI's components. The truck trailers not only contained the two telescopes of the interferometer but served as their housings as well. A third would be the control room for the instrument.

Was it Mount Wilson's fine seeing that brought ISI to Mount Wilson? Townes had been a trustee of the Carnegie Institution since 1965 May (Haskins 1965). He was thus well aware of the pros and cons of the mountain and no doubt felt a comfortable familiarity there. Seeing is almost always the first consideration in astronomical site selection at visible wavelengths. Because the seeing is nearly a factor of two improved in the N band than in the visible, the ISI team could expect mostly very good seeing for their observations. Townes could have chosen Lick Observatory, which belongs to the University of California—his home institution. But, Lick had insufficient open land to accommodate an interferometer. That requirement trumps seeing although Mount Hamilton's location yields inherently poorer atmospheric stability in comparison with Mount Wilson. Hewlett–Packard Corporation co-founder William Hewlett had offered a fine site to Townes, but it was undeveloped and would have required significant resources for the basic infrastructure. Besides, Townes had envisioned that after ISI had demonstrated its scientific viability, its components would be shipped from Los Angeles harbor to a high and dry site with outstanding seeing like Mauna Kea, Hawaii or Las Campanas in the Chilean Andes. For now, though, Mount Wilson would do just fine.

The origins of ISI go back to the early 1970s, when Townes' group used the pair of auxiliary coelostats at Kitt Peak National Observatory's McMath Solar Telescope (later renamed the McMath–Pierce Solar Telescope) to provide them with a 5.5 m baseline for interferometry experiments at a wavelength of 10 µm. As mid-infrared interferometry had never been done, their initial concern was whether the atmospheric properties would be suitably benign. Using the planet Mercury at maximum western elongation, the group obtained the first measurements of fringe amplitude and phase ever made in the mid-infrared—the atmosphere would be cooperative (Johnson et al. 1974a, 1974b). This use of the McMath auxiliary telescopes was not ideal, but it initiated observations of circumstellar dust shells that would later be the pinnacle of ISI's contribution to stellar astrophysics. Among the objects observed were α Ori (Betelgeuse—of course!), VY CMa, α Sco (Antares), R Leo, and o Ceti (Mira) (Sutton et al. 1977, 1978). All of these stars were fully resolved except for R Leo. The astrometric accuracy of the technique was also shown to have a nightly precision of less than 0.1 arcsec with an inherent atmospherically-imposed limiting accuracy of 0.01 arcsec (Sutton et al. 1982).

In addition to these high-resolution studies of red giants and supergiants, Townes' team developed the approach to interferometry that they would take forward to the ISI. From the beginning, the McMath experiment was a prototype for a dedicated interferometer with a much longer baseline and higher resolution to provide structural information of the dust around red giants. Coming into the mid-infrared from radio astronomy work in interstellar molecules, it was natural that they would import radio techniques to the shorter wavelengths with the intention of not having to deal with complications such as optical delay lines. And so, they took advantage of the heterodyne principle to conceive of an N-band interferometer as a radio interferometer.

Heterodyning was first attempted early in the 20th century by Reginald A. Fessenden (1866–1932), a Canadian who conceived of using amplitude modulation of radio signals to carry audible sounds (Fry 1973). The method exploits the beat frequency phenomenon experienced in acoustics that arises when waves of frequencies f_1 and f_2 combine to form new frequencies that are the sum and the difference between the two mixed values. If, for electrical signals such as those produced by an N-band detector, the science signal of frequency f_N is mixed with a stable, locally-generated signal with a frequency f_L close to f_N, then the resulting value of $(f_N - f_L)$ can be made much lower than f_N. The other much higher-frequency heterodyne, $(f_N + f_L)$, can be removed by electronic filtering. It was inevitable, considering ISI's principal investigator, that a laser provides the necessary reference signal. Specifically, ISI uses a $^{13}CO_2$ laser emitting at a wavelength of 11.15 μm. Mixing this carbon dioxide laser in the N band produces a heterodyne with a bandwidth in the gigahertz regime rather than in the terahertz realm corresponding to 11.15 μm. The main downside of heterodyning results from the very narrow frequency of the local oscillator that only couples to the science wavelength with a similarly narrow frequency. This results in a very narrow wavelength region over which visibility information is obtained, limiting detection to a modest number of relatively bright objects.

Devices then available for mid-infrared detection were made from cadmium telluride and mercury telluride—typically referred to as HgCdTe or simply MCT detectors—and they remain the best detectors available for mid-infrared heterodyning. In this application, starlight is mixed with a "local oscillator" (in this case the output of a CO_2 laser) in a beam splitter just prior to detection by the MCT diode that then produces an electrical output signal of much lower frequency for further processing using conventional digital signal-processing methods.

Despite its sensitivity shortcomings, mid-infrared heterodyne interferometry has considerable advantages over the way interferometry is now done at visible and near-infrared wavelengths. Figure 8.3 compares the heterodyne approach with that employed by current optical interferometric arrays, such as the CHARA Array on Mount Wilson. A major overhead of such instruments involves correction required to match path lengths traveled by light collected by telescopes in a horizontal plane that must observe stars moving across the sky as the Earth rotates.

The resulting path length "delay" must be compensated to sub-micron accuracy in order for the light from the two telescopes to combine coherently and form

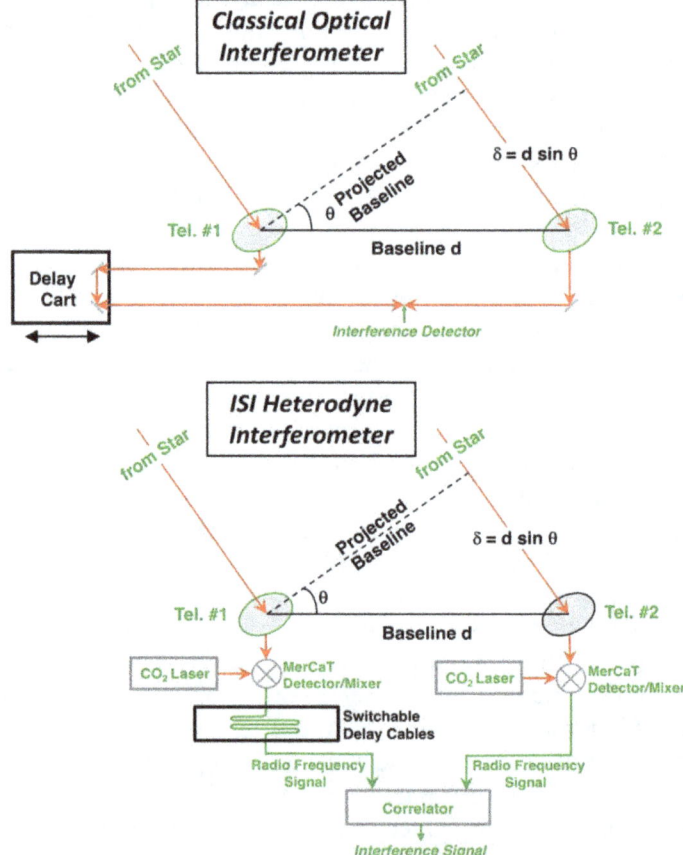

Figure 8.3. A comparison is shown here between two telescopes being used as a simple interferometer in (above) an optical system that uses starlight to produce interference fringes, such as is done by the Mark III and CHARA interferometers and (below) for ISI in which the infrared starlight is mixed with a local oscillator in the form of a CO_2 laser to produce a much lower radio-frequency electrical signal. Both interferometers must track the time varying "projected baseline" by equalizing paths arising from Earth rotation using subsystems outlined in black. The optical interferometer does this by moving retro-reflectors on precision rails that can be hundreds of meters in length while the heterodyne instrument switches in coiled cables of the appropriate length to insert the required "delay" amount. For more details see Section B.5.1. (Author diagram inspired by David Hale's image in Section VI of the excellent ISI system overview, adaption from http://isi.ssl.berkeley.edu/system_overview.htm)

fringes. Furthermore, light from each telescope must be transmitted through evacuated light pipes to avoid ground-level turbulence. Such handling requirements for starlight propagation have large budgetary impacts for construction and continued maintenance. For a heterodyne interferometer like ISI, these two subsystems—delay lines and light tubes—are dispensed with as the heterodyned signals are conveyed and delay-compensated using coax cables in a box. Extrapolating these advantages to shorter wavelength can, in principle, be accomplished using *frequency combs* produced by pulsing a mode-locked laser to produce a series of precisely-separated laser emission spikes within a Gaussian-like envelope to serve as local oscillators (Glindemann & Käufl 2006; Schliesser et al. 2012). This provides multiple local oscillators spanning a broader bandpass to improve sensitivity significantly over that of the monochromatic CO_2 laser pioneered by ISI. Near-term progress in this direction using laser combs and other advances in quantum optics could reinvigorate infrared heterodyne interferometry with substantial sensitivity gains and permit kilometric baselines to open up new science areas from the ground including imaging of exoplanet systems[1] (Ireland & Monnier 2014).

Planning for a standalone, two-telescope interferometer was initiated in 1983 (Danchi et al. 1988) while observations were still underway at Kitt Peak, and a Pfund-type telescope design was developed that brings light to a focus using only two large mirrors. This significantly reduces reflective losses in comparison with, for example, the altitude–azimuth configuration of the CHARA Array telescopes. This degree of efficiency helps offset the sensitivity reduction of the very narrow bandwidth of the heterodyne approach. As shown in Figure 8.4, the telescope consists of an 80 in diameter flat mirror that is steerable and feeds a fixed 65 in paraboloid that passes light back through a hole in the flat to a focus. Each of the pair is mounted on a truck trailer to simplify their occasional reconfiguration to better suit particular targets. In Figure 8.4, south is to the right, which gives access to the equatorial regions of the sky. Vignetting becomes serious for high northerly declinations unless the telescopes (and trailers) are rotated 180° in azimuth.

Following Mount Wilson's selection as their site, the group installed multiple 2 ft thick concrete pads for parking the telescope trailers that would provide a variety of baselines and orientations between 4 and 28 m. A longer baseline of 56 m was subsequently installed. At 10 μm, this permits formal limiting resolutions of from 0.5 to 0.07 arcsec as determined by the first null in visibility. In practice, higher resolution is attained by making measurements down the visibility curve without having to reach that null. Thus, ISI can resolve stars with diameters down to a few hundredths of an arcsecond. A third telescope would be added in 2003 that would permit phase measurement through the method of closure phase (see Appendix B.5.5). This allows what amounts to an imaging capability by detecting asymmetries not accessible with a pair of telescopes.

When Manfred Bester joined Townes group in 1986, he was given the tasks of writing the software for telescope pointing as well as for processing ISI data. He

---

[1] A workshop on Revisiting Infrared Heterodyne Interferometry for Astronomical Aperture Synthesis was to have been held in Grenoble, France in 2020 March. It was postponed due to the Covid-19 pandemic.

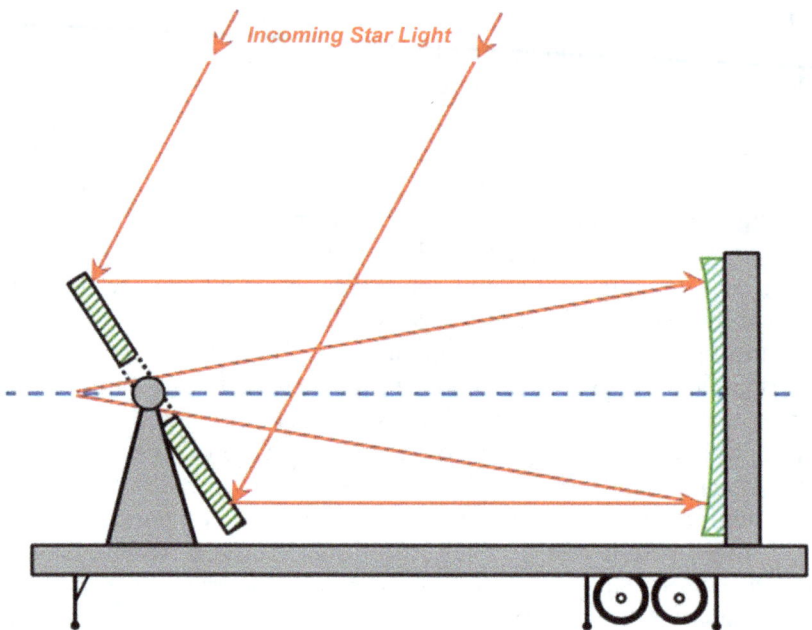

Figure 8.4. Schematic of an ISI telescope using the Pfund configuration of a steerable flat feeding a fixed paraboloid. (Author's diagram.)

recalls that they "*drove the trailers to MWO in January 1988, after having done some initial pointing testing outdoors in Berkeley. Our first fringes were seen on IRC +10216 on June 29, 1988.*" Those fringes, shown in Figure 8.5, were "*difficult to get … due to the unknown zero delay. The infrared detection systems were initially tuned up using soldering irons, so we knew we had good sensitivity. We also used coffee cans painted black and filled with hot water. Eventually we developed more accurate, thermostatically controlled hot body sources for this purpose.*"

"*Next we needed a very bright 10 micron point source to tune up on. Dr. Townes proposed to use the planet Mercury—we called it the "soldering iron in the sky." This is how he and his team had seen fringes with the interferometer prototype at the Kitt Peak solar telescope. The problem with Mercury is of course that it is always relatively close to the Sun, so we would have to be very careful to not do harm to the telescope optics and other components, not to mention to ourselves.*"

"*The safety procedure we came up with called for Dr. Townes to watch where the image of the Sun would fall, standing in between the large 65-in and 80-in mirrors of one of the telescopes. Peter McCullough, a graduate student at the time, was standing behind him, and I was at the computer controls inside the trailer. As we were carefully moving the telescopes and approaching Mercury, the image of the Sun—about 1" in diameter—crept up on Dr. Townes from behind, landed on his camouflage hat and set it on fire. Peter quickly extinguished the flames with his hands, and we aborted our quest to obtain fringes for the time being. That was a rather close call! After taking a*

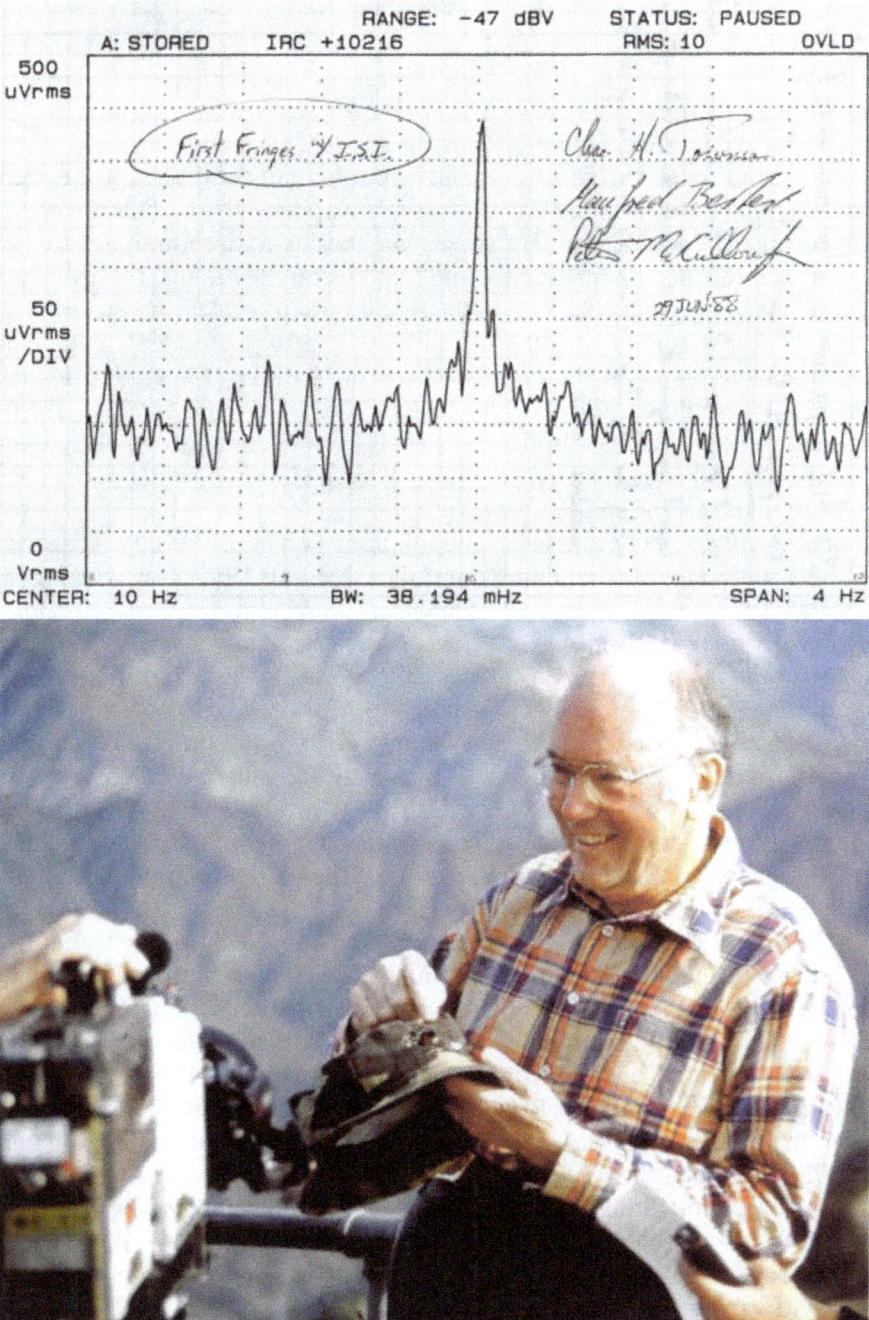

Figure 8.5. (Above) ISI's first fringes were obtained on 1988 June 29 as witnessed here by Charles Townes, Manfred Bester, and graduate student Peter McCullough. (Below) Townes showing off his "sunhat" during a filming on Mount Wilson by NHK–Japan television in 1990 December. (Photographs courtesy of Manfred Bester.)

deep breath and regrouping, we then managed to observe fringes on Mercury, and a short time later also on IRC +10216.”

“Eventually we were able to joke about this rather scary episode. Dr. Townes kept wearing his beloved 'sunhat' for many years and truly enjoyed showing off the 1-in hole that it now featured.” (See Figure 8.5.)

Townes was first and foremost a scientist, but he had footholds in the arenas of national security and arms control in such capacities as chairing the DoD Committee on the MX missile. He served on presidential committees for science advisement and for the first mission to land humans on the moon. (I heard him recall President Kennedy getting cold feet on the Apollo program, fearing the implications on national prestige should it fail.) It addition to being a trustee of the Carnegie Institution, Townes was on the boards of GM and Perkin-Elmer. His association with the giant automaker allowed him to have use of any GM vehicle from each current model year. Walt Fitelson recalls that Townes invariably selected vehicles with all the bells and whistles. One year he ordered a Corvette, which he let members of his group take out for spins.

Seeing no conflicts between science and religion, Townes believed that anchorages in both are essential to understanding our universe. In his 1995 book *Making Waves*, Townes states: *“To me science and religion are both universal, and basically very similar. In fact, to make the argument clear, I should like to adopt the rather extreme point of view that their differences are largely superficial, and that the two become almost indistinguishable if we look at the real nature of each.”* He points out that faith can send a scientist in search of an elusive answer to a question about the physical universe. Einstein's decades of effort in trying to unify his theory of gravity with quantum mechanics was fueled by faith. Similarly, revelation can result in scientific and not just religious breakthroughs (Townes 1995). Although unstated in his book, Townes' discovery of the maser/laser principle is one such revelation that altered our world.

Over the years, John Monnier has encountered those who thought that Townes was an evangelical fundamentalist. Monnier always hastens to correct that mis-impression, explaining that while his advisor would discuss religion if you asked him, he never brought it up on his own and certainly did not seek to convert those he encountered. Instead Townes wrote on the topic with his 1966 article on "The Convergence of Science and Religion" (Townes 1966) receiving considerable attention. In 2005, Townes was awarded the $1.5M Templeton Prize, given annually since 1972 by the John Templeton Foundation (see http://templeton.org) in recognition of an individual's *“exceptional contribution to affirming life's spiritual dimension, whether through insight, discovery, or practical works.”* Townes donated half of the prize money to his alma mater, Furman University, and gave much of the remainder to various faith-based organizations (Avasthi 2005).

Walt Fitelson recalls that Townes *“was generally a reserved, gentle person who was something of an amateur naturalist. One day, however, as a small group of us was walking out the gate of our site, a rattlesnake appeared in our path. Without a word, Townes took a few steps back into the site, grabbed a shovel and proceeded to kill the snake with one or two well placed blows. I have always thought that was partly due to him growing up on a farm."*

After obtaining first fringes from ISI, Townes' group settled into regular observations that would leave a productive legacy from this pioneering venture into mid-infrared interferometry. The group has produced some 80 publications in the refereed literature, conference proceedings, and as doctoral dissertations. We will look at a small sample of their results involving Betelgeuse, Mira, and the problematic α^1 Her. Supergiants like Betelgeuse and Antares are expected to have larger diameters in the infrared compared with the visible region of the spectrum due to significantly reduced limb darkening at the longer wavelengths (Scholz & Takeda 1987). A 1996 *Astrophysical Journal* paper (Bester et al. 1996) presented ISI diameters of 56.6 ± 1 and 44.4 ± 2 mas, respectively, for Betelgeuse and Antares. The extended dust surrounding these stars begins well away from the apparent stellar disk and so does not interfere with mid-infrared diameter measurements. Observed changes in the brightness of Betelgeuse were then attributed to temperature rather than diameter changes. This view evolved as observations of Betelgeuse continued and were expanded following the addition of the third telescope (Ravi et al. 2011) (see upcoming Figure 8.9). From visibilities obtained during the interval 2006–2009 (Townes et al. 2009), and subsequently expanded through 2010 (Lockwood et al. 2014), the ability to then measure closure phase allowed for the detection of asymmetries in the photosphere that might be due to bright spots. Such asymmetries changed from year to year. The ISI's diameter measurements for the period 1994–2009, shown in Figure 8.6, revealed a 16% change in the apparent diameter, which decreased from 56 to 47 mas. But, what we perceive as the diameter of Betelgeuse may be a mix of changes in effective temperature, emergence and disappearance of bright spots, and opacity variation of a possible cool shell above

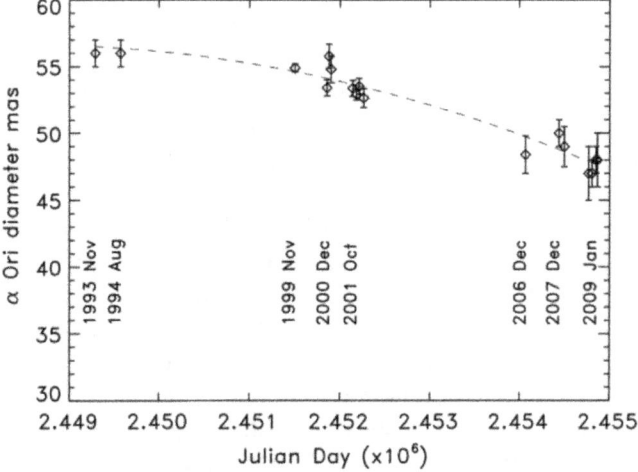

Figure 8.6. ISI's measurements of the apparent angular diameter of Betelgeuse are seen here to gradually diminish from 1993 to 2009. (Figure 1 from Townes et al. 2009. © IOP Publishing. Reproduced with permission. All rights reserved.)

the photosphere. Thus, as with Mount Wilson interferometry, Betelgeuse enters its second century of tantalizing the astronomers who observe it.

Another ISI program star familiar to us is α^1 Her. Having already observed the star with their prototype system at Kitt Peak, the ISI group included the M5 Ib–II star (Moravveji et al. 2013) in 1992 as part of their survey from Mount Wilson for dust around 13 bright late-spectral type stars (Danchi 1994). These stars fell into two groups, the first showing warm dust shells with inner radii of 3–5 times their central star's radius and the second having much larger inner radii. The latter category included α^1 Her along with α Ori and α Sco. This grouping was considered indicative of sporadic episodes of substantial dust production at intervals of a few decades.

This conclusion naturally spurred further observations of these stars in the hopes of catching such a dramatic event in the offing. That investment of observing effort paid off in the case α^1 Her, which was caught in the act of ejecting dust into its circumstellar environment (Tatebe et al. 2007). Visibilities obtained from 1993 to 2004 showed variation that could be well fit by a model in which an ejection of dust amounting to about 10^{-6} M_{sun} occurred in 1990. The shell expanded at an unexpectedly large rate of approximately 75 km s^{-1}, corresponding to angular expansion of about 70 mas yr^{-1}. The dust, of course, dissipated with this expansion, and no further episodes were seen by ISI over the ensuing 14 years of scrutiny.

ISI also contributed to the collection of diameters for the star measured at various wavelengths (Weiner et al. 2003a). Their value for α^1 Her of 39.32 ± 1.04 mas at 11 μm, corresponding to nearly 500 solar radii, is about 30% larger than those from the near-infrared due to the wavelength dependence of limb darkening.

The last sampling from ISI science we'll look at concerns o Ceti (Mira), the prototype for the Mira variables. Along with Betelgeuse, Mira's diameter was measured with high precision by Townes' graduate student Jonathon Weiner in late 1999 using ISI's longest baseline of 56 m (Weiner et al. 2000). The 11 μm diameter was then determined to be 47.82 ± 0.48 mas. This long baseline allowed visibilities to be measured all the way down to the first null thereby attaining 1% precision. Such exquisite measurements are shown from follow-on observations in Figure 8.7.

Wishing to see how the star's diameter behaves through its 11 month pulsation period, a series of 20 observations taken in the star's spectral continuum at 11.149 μm wavelength was accumulated from fall 1999 until late 2001 December (Weiner et al. 2003b).

Care was taken to ensure that the observed bandpass avoided spectral line features that might significantly affect the apparent diameter of the star and that dust was not a biasing factor. The resulting diameters are shown in Figure 8.8 plotted against date and pulsation phase. Weiner's mid-infrared diameters are nearly twice those from near-IR interferometry. This remarkable difference was subsequently explained as arising from the wavelength dependence of opacity arising from an extended layer of H_2O (Weiner 2004). Additionally, the star was seen to be 11% larger in 2001 than in the prior two cycles they observed, a result that could have arisen from an elongation of the star that was more resolved after a baseline change in 2001 June rotated the baseline orientation by 36.5°.

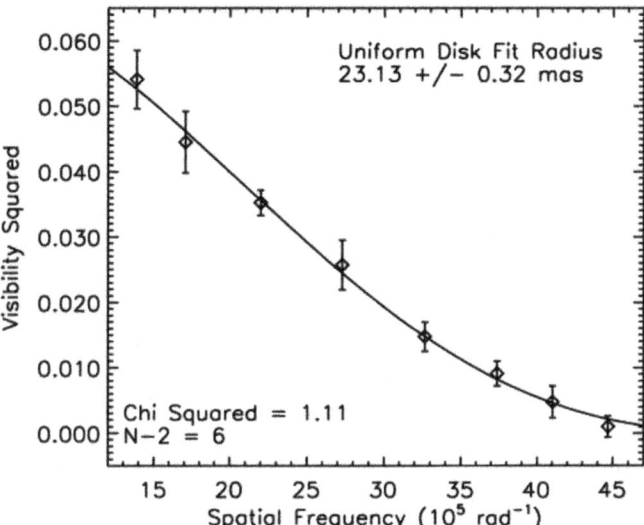

Figure 8.7. ISI's 1999 visibilities for Mira were measured all the way down to the first null resulting in a 1% precision diameter. (Figure 1 from Weiner et al. 2003a. © IOP Publishing. Reproduced with permission. All rights reserved.

When asked what they think are ISI's premier achievements, Ed Wishnow and John Monnier both replied that the Berkeley instruments' hallmarks are the characterization of the extensive circumstellar envelopes possessed by highly evolved red stars like Betelgeuse and the surprisingly short time scales for changes in their sizes and asymmetries. Wishnow points out that *"this work presaged the prominent and news-making dimming of Betelgeuse seen in the winter of 2019, followed by its brightening again during March–April 2020* (Overby 2020). *Generally people did not expect such rapid changes in the appearance of stars."*

Monnier, who was Weiner's contemporary at Berkeley, cited his fellow grad student's water vapor scrutiny of Mira as demonstrative of a lesson learned. Monnier had also pursued molecules discovered spectroscopically in these stars and was attracted to species with strong spectral features—molecules like silane (SiH_4) and ammonia (NH_3). He bypassed water due to its apparent anemic spectral presence, only to see Weiner hit the jackpot with that molecule.

Monnier now tells his own graduate students about this hiding-in-plain-sight circumstance arising from a massive and nearly stationary water vapor envelope lacking the rapid expansion typically seen in other cases. As a result, the wavelengths of re-emitted energy from the halo of water around Mira and the absorbed energy from water in the line of sight to the star's surface lay smack on top of each other rather than displaced spectrally due to the Doppler effect. They nearly cancel each other and give the impression of very little water around the star. To John Monnier, who admires Weiner's discovery, it was one of those things that got away.

It has been more than three decades since the first two ISI telescopes were driven up to Mount Wilson—a remarkable longevity for any astronomical instrument. The facility, shown in its final three-telescope form in Figure 8.9, has been highly

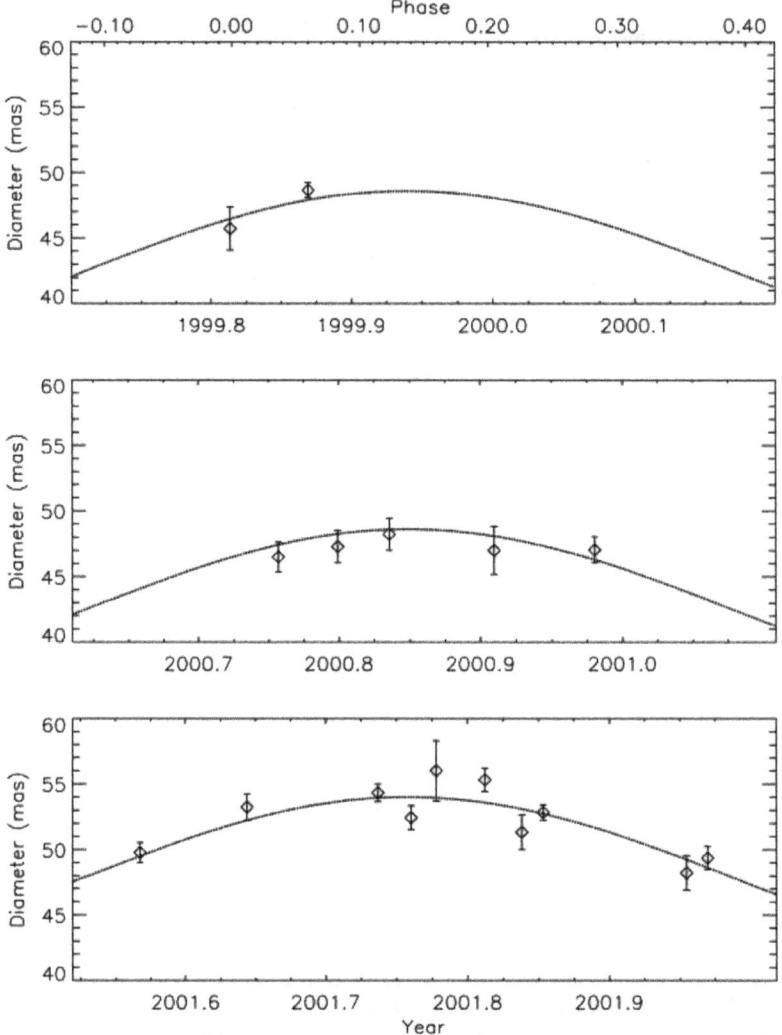

Figure 8.8. ISI Mira diameter variation with date and phase. (Figure 5 from Weiner et al. 2003b. © IOP Publishing. Reproduced with permission. All rights reserved.)

productive, not only scientifically but also in training many very talented scientists to ensure the continuation of mid-infrared interferometry. And yet, ISI may not be done. Its potential for high-spatial and high-spectral resolution has yet to be fully exploited. New technical upgrades such as laser frequency combs and high-speed digital signal sampling and spectrometer-correlators will greatly increase the detection bandwidth. This makes possible observations to fainter magnitudes, and hence, the number of objects accessible to ISI while improving the accuracy of results for the brighter objects it has been observing for years. Digital spectrometer-correlators will also provide visibility amplitudes and closure phases within and adjacent to specific molecular spectral lines allowing one to trace the formation of molecules as material streams away from stars.

Figure 8.9. The ISI as photographed from atop the Michelson 50 foot interferometer building shortly after the third telescope was added in 2003. The telescopes are in their shortest baseline configuration of 4–8–12 m E–W. The building just to the left of the telescopes is the master local oscillator which sends the CO_2 laser signal to each telescope via the periscopes emerging from the building's roof. (Photo courtesy of David D. S. Hale.)

As a dedicated facility, ISI also possesses a powerful capability as a testbed for technology development, and, for example, can contribute toward subsystem exploration for next-generation interferometers through experiments with fiber-optic linkage of telescopes, new frequency comb technologies, and various approaches to high spatial resolution mid-infrared spectroscopy. It also has unique capabilities relevant to Department of Defense goals for imaging geostationary satellites. With no other mid-infrared facility of its type in the United States, the Berkeley Infrared Spatial Interferometer maintains its uniqueness well into the future.

ISI is also a tribute to one of the 20th century's greatest scientists (see Figure 8.10) who compounded his founding knowledge of lasers with a deep love and curiosity for nature to produce a new high-resolution window on the universe.

8.2 The CHARA Array (1996–Present)

For years, Theo A. ten Brummelaar, now Director of the CHARA Array, would end each meeting of CHARA's technical management team with the facetious comment, *"It'll never work."* Obtaining first fringes is a signal achievement for an interferometer. In our case, that milestone (see Figure 8.11) was reached on 1999 November 23 as proof that the Array did indeed work! That key achievement did not come easily or quickly, and it would be another five years before the Array would settle into routine operation. By then, more than two decades had passed

Figure 8.10. Sporting his sunhat, Charlie Townes peers through the hole in the steerable flat mirror of one of the ISI telescopes. (Photo courtesy of Manfred Bester.)

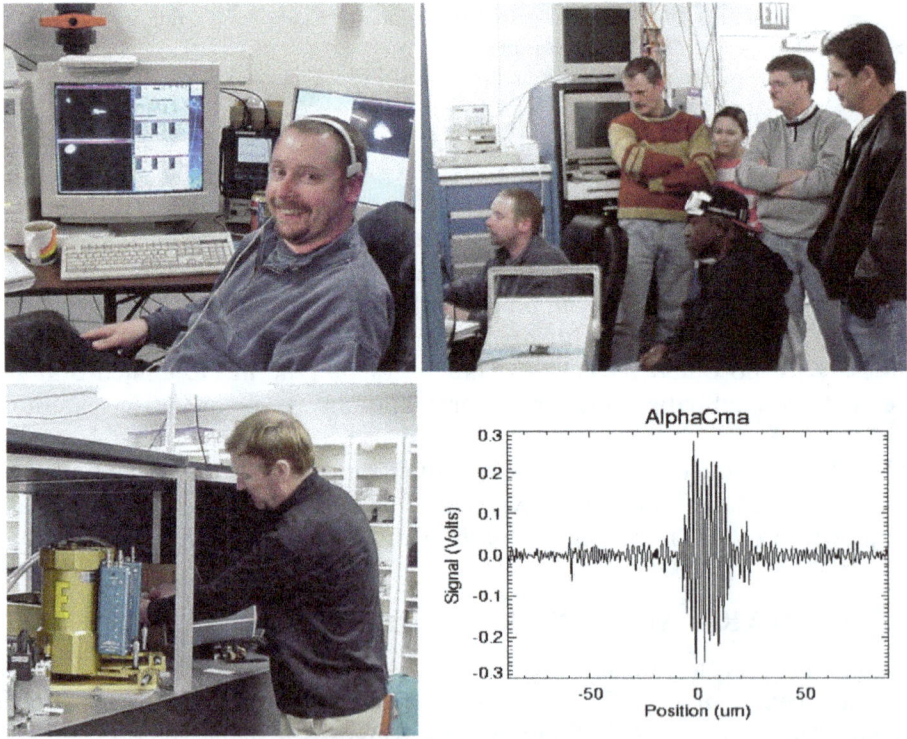

Figure 8.11. (Upper left) 1999 Nov 20: Theo ten Brummelaar is optimistic of finding fringes. (Upper right) That same night, Theo undertakes the search surrounded by anxious onlookers: Laszlo Sturmann, Joey Seymour (seated), Judit Sturmann, the author, and Eric Simison (CHARA's prime contractor). (Lower left) 1999 Nov 21: Steve Ridgway adjusts the detector dewar prior to the search to be continued that night, but weather would intervene. (Lower right) 1999 Nov 23: A scan of the first fringes from the CHARA Array obtained on the bright star Sirius. (Author's photos.)

since the Center for High Angular Resolution Astronomy (CHARA) was established at Georgia State University.

The stories of the origins of the CHARA Array and the somewhat unorthodox manner in which it was designed and built were told at the 2014 Montreal meeting of the SPIE in three back-to-back presentations by me, Theo, and our close colleague and core member of the CHARA technical management team Stephen T. Ridgway (McAlister et al. 2014; Ridgway et al. 2014; ten Brummelaar et al. 2014). Much of the next few paragraphs is paraphrased from my Montreal paper.

My yearning for very high angular resolution started early in my two-year postdoc at Kitt Peak National Observatory (KPNO) where C. Roger Lynds had initiated a project using a 35 mm film-based speckle camera developed by James B. Breckinridge and William G. Robinson (Breckinridge et al. 1979) not long after Labeyrie's inspired invention of speckle interferometry (see Section 6.1 and Appendix B.7). My job during 1975–1977 was to use this camera to initiate a binary star astrometry program at the brand new 4 m Mayall Telescope. This required developing observational procedures to accurately calibrate the data—high resolution without high accuracy would not make speckle interferometry very useful for measuring the orbital motions of binary stars. Resolution and accuracy could make speckle the best tool for the job.

After calibration was nailed down using a double-slit mask that had been designed and fabricated by Jim Breckinridge and Bill Robinson, I went to work on trying to resolve spectroscopic binaries for reasons I've described earlier in this book (see also Appendix B.3.4). Speckle interferometry's 35 mas limiting resolution at a 4 m telescope gives a cantilever into the discovery gap—actually "gorge" is more like it—separating the visual and spectroscopic binaries. At the same time, improvements in radial velocity techniques, starting with the pioneering work of Roger Griffin,[2] were extending a cantilever from the gorge's opposite side.

To a double-star devotee, the prospect of interferometrically constructing a four-lane superhighway across the divide was irresistible. KPNO astronomer Arthur A. Hoag (1921–1999), who was the Lowell Observatory director during 1977–1986, was very supportive of my work and was the first to encourage me to think about much higher resolution, pointing me to Richard H. Miller's seminal study described in Section 5.7.

My initial naïveté regarding the challenges of long-baseline interferometry was appropriately adjusted by going to meetings of scientists far more expert in the field. My first such meeting was IAU Colloquium No. 50 on "High Angular Resolution Stellar Interferometry" held at the University of Maryland in late summer 1978 (Figure 8.12). John Davis, whose Sydney University group was already at work planning the successor to their famously successful Narrabri Intensity Interferometer, chaired the scientific organizing committee for the colloquium while Doug Currie handled the local arrangements. That was my first indoctrination on the challenges of building an interferometer.

---

[2] See 2016, Celebrating Paper 250, ed. D. J. Stickland, R. W. Argyle, & S. J. Fossey, The Observatory, 136, 232.

Figure 8.12. This group photo from IAU Colloquium 50, held at the University of Maryland during 1978 Aug 30—Sep 1, shows many of the people who are mentioned in this book: 1—W. C. Wickes, 2—the author, 3—J. C. Dainty, 4—D. L. Fried, 5—P. Nisenson, 6—R. V. Stachnik, 7—C. H. Townes, 8—J. B. Breckinridge, 9—L. Koechlin, 10—C. O. Alley, 11—K. M. Liewer, 12—D. G. Currie, 13—J. Davis, 14—R. Hanbury Brown, and 15—S. T. Ridgway. That is not to say that others in this photo were not leaders in the field.

Ingemar Furenlid (1934–1994), another KPNO staff astronomer who joined me as a colleague at Georgia State in 1982, was a strong supporter of this drive toward unprecedented high resolution at optical wavelengths. By 1983 October, I was sufficiently buoyed to go to GSU Dean Clyde W. Faulkner (1936–2011) with a proposal to set up a research center intended to design build, fund, and then operate an interferometer with an estimated cost of $3—$4M. Paying special attention to the sound of its acronym, I picked Center for High Angular Resolution Astronomy as the proposed center's name. Dean Faulkner soon gave his approval and some start-up funds. Throughout all the ensuing years, the GSU College of Arts and Sciences and the upper administration of my University have been generous and supportive of CHARA's goals.

The speckle program I continued from Kitt Peak was doing well, and Bill Hartkopf was working with me as senior research associate. Don Hutter joined us a postdoc shortly after CHARA was established. The work had enjoyed support from the National Science Foundation (NSF) since 1978 with additional funds awarded in 1980 for a digital speckle camera to replace the KPNO film system. We also had resources from the Air Force Office of Scientific Research that funded observing and provided a state-of-the-art image processing system for the reduction and analysis of speckle data that included a VAX-11/750 computer. By the time we submitted a Phase A Feasibility Study proposal to the NSF in 1984 September, we could point to 32 published papers constituting 85% of all binary star speckle results at the time. After demonstrating our science creds, the proposal laid out mostly conservative science goals centered around stellar angular diameters and spectroscopic binaries.

Our baseline design concept was an unabashed borrowing of the Narrabri Intensity Interferometer's circular-track concept. The proposed diameter of 300 m would provide baselines ranging from 10 m to 300 m for two 1.5 m telescopes. The proposal sought funds for mostly engineering work on optical design, atmospheric effects, path-length compensation, and beam alignment. We also asked for support for a new senior research associate resident at GSU. The proposal resulted in a $262K grant from the NSF in 1985 December, and we spent the next three-and-a-half years on the feasibility effort.

The primary GSU contributors to the Phase A effort were William G. Bagnuolo, Jr, who we hired as the senior research associate, Bill Hartkopf, Ingemar Furenlid, and graduate students Donald Barry and Wean-Shun Tsay. We had entered into a partnership with the Georgia Tech Research Institute (GTRI) that brought Allen Garrison, Morris Hetzler, and David Roberts to the team. We entered into a consulting arrangement with United Technologies Optics and Applied Technology Laboratory (OATL) where Bill Robinson had relocated from KPNO. Bill, Richard Simkins, Robert Tyson, and James Pearson participated from OATL. Finally, Lowell Observatory astronomer Nat White, who had a scientific interest in interferometry, participated in connection with site selection to be described below. Professor Daniel M. Popper (1913–1999) of UCLA's astronomy department provided considerable scientific wisdom to the study as well as to our ensuing proposals.

The resulting Phase A Final Report was submitted to the NSF in 1989 March, and we put together a full-blown construction proposal asking for $5.3M for a $10.6M facility, the other half of which would be funded by GSU. The circular two-telescope array had evolved into a Y-shaped array with 200 m radial arms on which would be located seven fixed, 1 m aperture telescopes. We focused our attention on Lowell Observatory's Anderson Mesa station outside Flagstaff as the likely site.

Determining the fundamental stellar properties of mass, radius, luminosity, surface temperature, and shape remained the core of our scientific justification, but we expanded our goals to include measuring the radii of pulsating variable stars and probing the extended atmospheres and emission regions of hot, early-spectral-type stars such as Wolf–Rayet, P Cygni, and emission line B stars. For binary stars, we also included the detection of submotions due to low-mass unseen tertiary companions as well as imaging features in close binary systems including distorted photospheres, disks, streams, etc, pointing to the close 2.9 day system comprising Algol AB as a target with good prospects. We also considered imaging compact solar system objects, planetary nebulae, novae, and even detecting the relative motions of stars in globular clusters. The Array's ultimate sensitivity would determine if such objects were truly reachable.

Our scientific justification closed with a sentence contributed by Dan Popper that read: "*While it is often attractive (and certainly very painless) to invoke serendipity as a justification for a new scientific endeavor, history has indeed taught us that whenever a new technique enters a new realm of observational phase space, the most striking and productive results tend to be those not anticipated by even the most prescient thinkers.*" We shall see that the word "prescient" would apply to Dan Popper.

The proposal was well received by reviewers, but the NSF astronomy budget did not, at that point in time, have enough slack in it to meet our funding needs. So, the proposal morphed into a Phase B Preliminary Design Study for which CHARA was awarded $485K to put more flesh on the interferometry bones we had constructed. That effort was conducted from 1992 January until the spring of 1994 and was invaluable in defining the path forward. Its 300 page final report would form the basis for our proposal—at long last—to actually build an array.

But, there was a decision we had to make, as the limited funds that the NSF could invest in CHARA were insufficient for seven 1 m telescopes. We had two choices—stay with seven but decrease the aperture so the unit cost would go down, or keep the aperture but buy fewer telescopes. We decided on the latter path. We had specifically chosen 1 m primary-mirror diameters with the goal of eventually adding adaptive optics to fully correct that aperture and gain sensitivity to fainter targets such as the nuclei of active galaxies. The funds available could only afford five telescopes, so, in late 1994—a decade since I had set up CHARA—we happily went forward with that reduced number while continuing to seek other support for additional telescopes.

Georgia U.S. Senator Sam Nunn came to campus to make the announcement of the capital funding from the National Science Foundation. A press conference was held, and we got some nice publicity. A site for the facility had not yet been selected, and our press release simply said we would build the telescope array in the Southwest. Shortly thereafter, the University president's office got a letter from a member of the Georgia General Assembly asking why we weren't building this thing in Georgia. As I was CHARA's director, the letter was punted to me for response. I gulped, sat down, and wrote a reply to this gentleman attempting to explain the importance of good astronomical seeing, high elevation, and the large difference in the number of clear nights in our home state and the American Southwest. I was delighted when he replied to say that he understood and wished us good luck.

By then, we had been engaged in trying to find a home for our interferometer for years, realizing that most developed sites with good seeing don't have 35 more-or-less level acres to set aside for an optical interferometric telescope array. The path that led to Mount Wilson was a tortured one. It began with the identification of ten potential sites examined under our selection criteria of terrain, meteorological patterns, atmospheric seeing, night-sky illumination, geologic activity, and logistics. These sites included: Anderson Mesa and Mount Graham, Arizona; Sacramento Peak, Blue Mesa, and Magdalena Ridge, New Mexico; Mount Fowlkes (Flattop), Texas; Haleakala and Mauna Kea, Hawaii; and, Mount Pinos and Table Mountain, California. Site visits were initiated way back in 1986. A first elimination based on the above criteria left Blue Mesa, Mount Fowlkes, and Anderson Mesa in the running, and a final elimination pointed to Anderson Mesa as the best compromise. Our visits were all very graciously hosted by the institutions controlling each site.

In order to better evaluate the seeing conditions at Anderson Mesa, which had been developed by Lowell Observatory as a dark site away from the lights of Flagstaff, our graduate student Wean-Shun Tsay undertook a seeing monitoring program at the site. His results (Tsay et al. 1990) comprised the first astronomy doctorate awarded by GSU. They indicated a median seeing-disk size of 1.18 arcsec, not competitive with superb sites such as in Hawaii and Chile but deemed acceptable at the time.

So, by 1990, we were all set to find the funds to construct an interferometer on Anderson Mesa. However, as a result of the passage of time and the deliberate pace of obtaining major peer-reviewed funding, that site was not to be ours. Another well-known interferometer ended up on "our" site—the Navy Precision Optical Interferometer (NPOI; Armstrong et al. 1998).

Following the loss of Anderson Mesa, we were contacted in early 1992 by the University of New Mexico in regard to a new site they wished to develop on Mesa Negra, a beautifully flat mesa located on the Acoma Pueblo just off I-40 to the east of Grants, NM. An artist's concept of our facility on that site is shown in Figure 8.13. This location was an interferometrist's dream with plenty of flat real estate, but it was noisy geologically as a result of active underground lava flows. It also lacked any existing infrastructure, thereby driving up costs. For a variety of reasons, negotiations fell apart by 1995, and we were once again siteless.

With an NSF construction grant in hand, it was high time that a site was also in the bag. By early 1995, we were under discussions with the Kitt Peak and Mount Palomar directors whose sites had abundant infrastructure but were subject to space restrictions that had to be worked around.

And then, out of the blue, Bob Jastrow, director of Mount Wilson Observatory, called to ask if we had ever considered that historic mountaintop. Since we were then getting regular speckle observations at the 100 inch, the response was "sure, but we won't fit up there with all the existing structures." My impression was that there was no way we could be shoehorned into the complicated terrain with dozens of existing structures. Anticipating this, Dr Jastrow had already had his superintendent, Bob Cadman, rough in our array concept and found that it could indeed be

Figure 8.13. Artist Bill Pounds' concept of how the CHARA Array might have looked if constructed on Mesa Negra in New Mexico. (Copyrighted image courtesy of William Pounds.)

accommodated without having to demolish anything but an old garage. We jumped on it! A memorandum of understanding was signed on 1995 October 30 (Figure 8.14), and we immediately undertook the prerequisite planning and permitting work that would lead to construction. It was like coming home.

During these next few years, my most lasting contribution to CHARA's success was to bring a handful of truly outstanding people into the project. I've already mentioned Bill Bagnuolo who contributed greatly to our Phase A effort and, as we shall see, to making the case to another potential funding source when we needed an additional telescope. In 1994, I invited Stephen T. Ridgway, a Kitt Peak astronomer I got to know during my postdoc and whom I had grown to respect for his expertise in instrumentation in general and interferometry in particular. Steve has stayed with the project to this day, and it is hard to imagine it as having succeeded as it has if he had declined my invitation. His name will come up repeatedly below.

About that same time, John Davis made me aware of a truly exceptional student of his in Sydney who was clever in all regards and intensely so in making machines bow to computer control. Thus, in 1996, Theo came to work with CHARA as a postdoc with the intent to return to Australia after a few years. Theo never left and is today Director of the CHARA Array, which did indeed bow to his competence.

Figure 8.14. 1995 October 30: seated beneath the 100 inch signing the MoU to locate CHARA on Mount Wilson are GSU President Carl Patton and MWO Director Robert Jastrow; standing are GSU VP for Research Cleon Arrington, the author, and GSU Arts and Sciences Dean Ahmed Abdelal. (Courtesy of CHARA/Theo ten Brummelaar.)

Never one to disrespect a colleague who tells you that you'd be crazy not to hire someone, I followed that advice from Vanderbilt astronomer Doug Hall in regard to Laszlo Sturmann. Laszlo had been an engineer at Konkoly Observatory in Budapest before deciding to come to the U.S. for a PhD. Like Theo, Laszlo came to Atlanta as a postdoc in 1997 and is today a Senior Research Scientist with CHARA. Earlier in this chapter, I referred to him as CHARA's Francis Pease. Enough said, for now. In 1999, Judit Sturmann finished her physics doctorate at Vanderbilt and also came to work for CHARA. She has done a superb job overseeing the numerous and complex optical systems in the 10,000 ft^2 Beam Synthesis Facility building where starlight is turned into science. I had hired Bob Cadman away from the Mount Wilson Institute to be our first Site Manager. When Bob retired, I coaxed Larry Webster into joining us as Site Manager in 2008. Larry had come to work for Carnegie Observatories years before they ceased operations and then became senior observer at the 150 ft solar tower telescope operated by UCLA. In addition to being able to fix just about anything that breaks, Larry is deeply devoted to Mount Wilson Observatory and more knowledgeable about the Observatory than any living person. "Larry would know" has been the answer to innumerable questions about Mount Wilson's history and heritage as well as its physical plant.

While this essential group was still being assembled, we put together a team of consultants in order to fulfill the requirements of the National Environmental Policy Act (NEPA) as well as the U.S. Department of Agriculture National Forest Service (USFS) who owns the land beneath the Observatory. There were a few white-knuckle episodes along the way, but we maintained a very good relationship with Terry C. Ellis (1934–2004), the District Ranger for the Mount Wilson area. The NSF fulfilled its NEPA obligations and issued a "finding of no significant impact" on 1996 May 17. Ground was broken on July 13. Renowned Carnegie Observatories astronomer and relentless Mount Wilson Observatory supporter Allan R. Sandage (1926–2010) gave a stirring keynote address (Sandage 1997) at that kickoff of the Array (Figure 8.15). Thirteen years after CHARA's founding, we were finally off to the races.

Much work had been underway throughout our soap opera of a site search with multiple subsystems to be designed, parts ordered, and components fabricated. As we had to have six of almost every item, a computer-controlled milling machine was among the first items we bought for the GSU Physics and Astronomy machine shop. Charles Hopper, our shop manager in Atlanta, was responsible for machining countless parts for the Array over a number of years. Charles' craftsmanship coupled with the superb design work by Laszlo Sturmann fill the Beam Synthesis lab and adorn the telescopes.

Another propitious decision was the suggestion by Steve Ridgway that instead of contracting with one of the handful of telescope manufacturers in the U.S., we hire telescope designer Larry Barr, who had played key design roles in a number of large-aperture telescopes over the years. A telescope optimized for interferometry has particular need for stiffness, immunity to thermal expansion (to maintain beam collimation), and smooth tracking. Because we weren't exactly awash in funding, we also wanted a compact design to fit a 1 m telescope into a dome that normally

Figure 8.15. 1996 July 13: Allan Sandage delivers the keynote address at the Array's groundbreaking. (Courtesy of CHARA/Theo ten Brummelaar.)

housed nothing larger than half that size. Steve and Larry collaborated on a design detailed all the way down to shop drawings that we could bid out to any number of vendors with machining capabilities suited to the large parts for a telescope. Our own shop could not handle components of such size, and we contracted with M3 Engineering & Technology Corporation of Tucson to machine and assemble the telescope parts (see Figure 8.16). This worked out brilliantly and resulted in six telescopes of remarkable performance capability (see Figure 8.17).

The procurement of telescope optics was critical to a successful interferometer as well as a major expense for the project. We contracted with Telescope Engineering Company (TEC) of Golden, Colorado. The mirrors themselves would be fabricated by the Leningradskoye Optiko-Mechanichesckoye Obyedinenie (LOMO) in St. Petersburg, Russia, shortly after that country became a federation, which put us a bit out on a limb. The primary mirrors were fabricated from the glass-ceramic material Sitall, which has a very-low coefficient of thermal expansion. The secondaries, custom-matched to specific primaries were made from Zerodur, another low-thermal expansion glass-ceramic. Together they form a Mersenne-type afocal optical system comprising confocal primary and secondary paraboloidal

Figure 8.16. 1998 July 29: Steve Ridgway and Larry Barr sit in an assembly pit in which a CHARA telescope is being assembled for testing in Tucson by M3 Engineering. (Author's photo.)

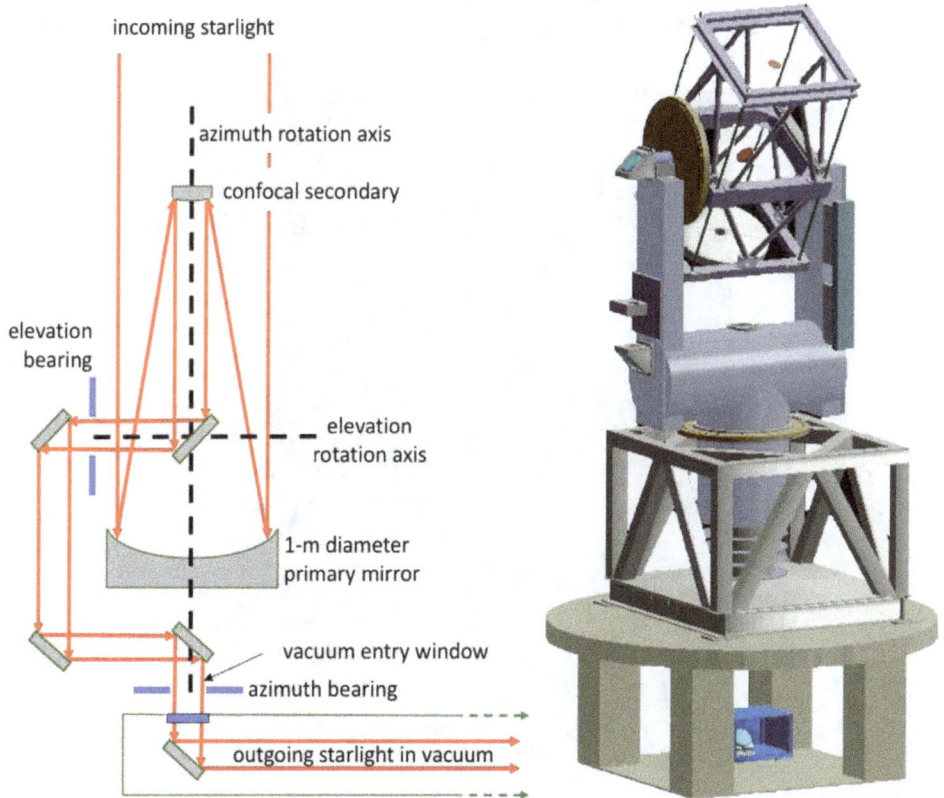

Figure 8.17. (Left) Schematic of the Mersenne optical system and coudé mounting arrangement of the CHARA light collecting telescopes. (Author's diagram.) (Right) A CAD rendering of the telescope and its alt/az mount. (Courtesy of Laszlo Sturmann.)

mirrors. This compresses the 1 m diameter beam collected by the primary to a 12.5 cm collimated beam off the secondary to be sent to the center of the array for path-length compensation. TEC eventually delivered at regular intervals seven matched mirrors that met or exceeded our $\lambda/30$ surface accuracy specification ($\lambda = 633$ nm). The pricing was so favorable that we could afford to order seven of these systems, with one in reserve as a potential seventh telescope. Figure 8.18 shows one of the primaries after being aluminized on Mount Wilson by Larry Webster.

I could go on and on about how we built the CHARA Array, but this is not a book solely about CHARA. That said, there was one more goal we wanted to achieve—getting our telescope number back up to six from the five funded by the NSF.

With N telescopes working together in all possible pairs, there are $N(N−1)/2$ unique baselines with differing orientations and lengths that sample different spatial frequencies in the target object's structure at any instant. Furthermore, the rotation of the Earth causes these orientations and baseline lengths to change continuously as their projections on the line of sight to a target star change. Radio astronomers have long taken advantage of this with a process called *aperture synthesis*, which permits the construction of images that can approach the resolution of a telescope the size of the whole array of telescopes. Aperture synthesis is extremely powerful, and its

Figure 8.18. 1998 November 24: CHARA Site Manager Bob Cadman inspects a newly aluminized primary mirror of one of CHARA's six telescopes. The mirror was coated by Larry Webster using the 100 inch mirror's aluminizing tank on the mountain. Larry would become CHARA's Site Manager upon Bob's retirement. (Author's photo.)

invention by the radio astronomers Martin Ryle and Antony Hewish resulted in their being awarded the 1974 Nobel Prize in Physics. For more about this topic see Appendix B.5.6.

The five telescopes for which we had funding would provide 10 such unique baselines. Armed with imaging simulations done by Bill Bagnuolo, I approached the W. M. Keck Foundation for a sixth telescope that would increase the 10 baselines of a five-telescope array to 15—a gain of 50% in imaging capability achieved at a bargain 20% increment in the cost of telescopes. Going from six to seven would net us only another 40% increase in baselines, so we accepted six telescopes as a sweet spot on the cost–benefit curve and set that as our goal. After a thorough review, the Keck Foundation gave us $500K toward a sixth telescope and its delay line with a commitment of another $1M when we got first fringes. As already mentioned, that happy event occurred on 1999 November 23, and we were funded for a six-element interferometric array with several years to go before it would be up and running.

All went well in spite of what could have gone wrong in building such a complex instrument. By late 2004 we were at the point where we could begin routine observing scheduling and science would start flowing—more than 20 years after I got the urge to scratch the very-high-resolution itch.

As shown in Figure 8.19, CHARA's telescopes are arranged non-redundantly on the three arms of a Y-shaped configuration providing 15 baselines from 34 to 331 m capable of measuring angular sizes from 40 mas down to 0.3 mas with a retinue of beam combiners operating from the visible to near-infrared wavelengths (0.5–2.2 μm). While the European Very Large Telescope Interferometer (VLTI) has much larger telescopes and higher sensitivity to faint objects, the CHARA Array has four times the resolving power of the VLTI, giving it access to a broad range of scientific investigations pertaining to stars. Before describing a sample of that science, let's look at the top-level elements of the Array in the order in which an incoming photon encounters them.

Each telescope is contained in an enclosure like that shown in Figure 8.20. The housing features a dome purchased from Ash Manufacturing atop a structure designed by Sea West Enterprises, CHARA's prime contractor, incorporating two concentric cylinders. Those can be raised and lowered in such a way as to seal the enclosure in the day and open it at night to allow ambient air to flow through and quickly bring the interior temperature into thermal equilibrium with the outside air. This decreases the local ill-effects that would result from turbulent eddies of hotter interior air boiling out of the dome slit. The heat output of electronics in the dome is minimized by housing computers, power supplies, pointing and slewing drive electronics, etc. in a separate temperature-controlled building adjacent to each enclosure. The approach works quite well.

When a lucky photon arrives at a telescope entrance pupil and manages to not "plonk onto the back of the secondary," which Bill Wickes philosophically described to me as the fate of an "unlucky" photon, it will be directed by a series of seven reflections down to a coudé box beneath the telescope where it will pass through an optical window and find itself speeding through evacuated tubes toward the central

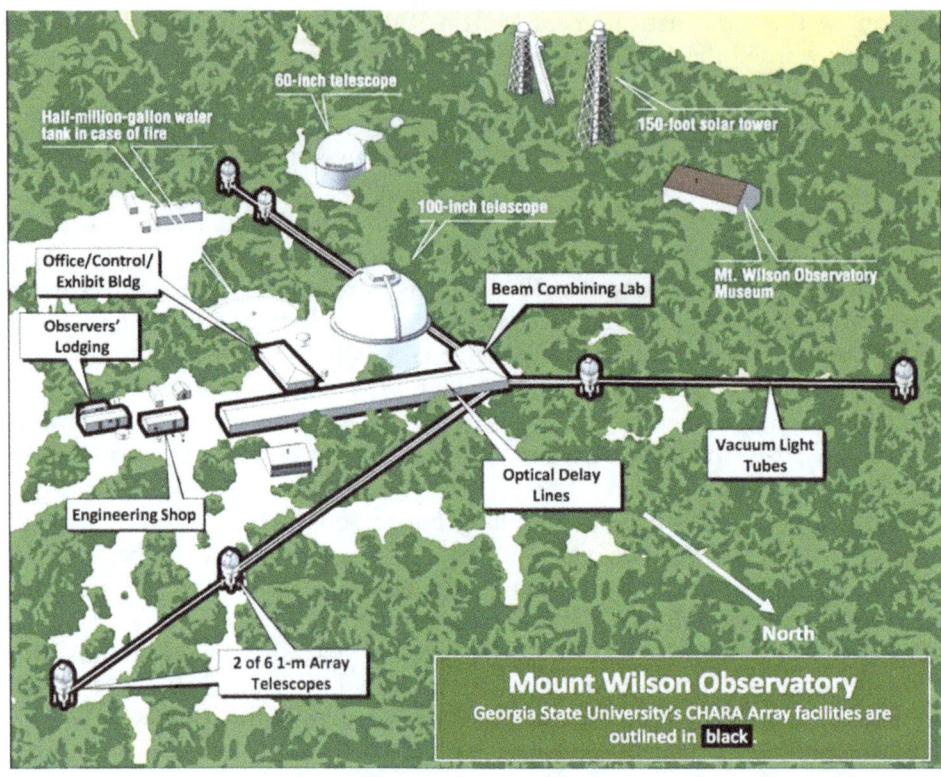

Figure 8.19. The layout of the CHARA Array on Mount Wilson. (Author's diagram.)

Figure 8.20. An open telescope enclosure features concentric cylinders that allow ambient air to convect upward from below to equalize the temperature inside the dome with the outside air. The block building at left houses the heat-producing control computers and electronics. (Courtesy of Steve Golden.)

Figure 8.21. (Left) CHARA Telescope S1 seen from the side fed light by the tertiary mirror. The optical assembly incorporates adaptive optics for correcting atmospheric turbulence across the telescope's 1 m aperture. (Photo courtesy of Steve Golden.) (Right) Evacuated tubes carry collimated beams of light from the two telescopes on the south arm of the CHARA Array through the trees to the Beam Synthesis Facility. (Author's photo.)

Figure 8.22. In this image from 2000 August, Susan McAlister stands behind the vacuum turning boxes that bring the beams from the NE and NW telescope pairs and directs them onto parallel paths of long fixed components of optical delay that are still in vacuum. (Author's photo.)

Figure 8.23. (Left) The continuously adjustable path-length compensating rail system is viewed from the west end of the long delay line lab. The vacuum components of delay are underneath each rail line. The optical tables in the foreground support the beam sampling systems that separate visible from near-infrared light, reduce the beams to a smaller diameter and send them off to the right to the beam combining lab. (Right) The path-length compensating, retro-reflector carts are in their parked positions at the east end of the delay line rail system. (Courtesy of Steve Golden.)

Beam Synthesis Lab (see Figure 8.21). Upon arrival at the central facility, photons from all six telescopes are made parallel by mirrors inside "turning boxes" (Figure 8.22) that send light into long path-delay sections—still in vacuum. The amount of delay in each path is set by retro-reflectors that pop into the light paths to provide an appropriate fixed delay length. After this down and back trip through the vacuum delay segments, periscopic mirrors take our photon out of vacuum through a second optical window and send it back east into the continuously variable sections of the delay system, the length of which can be envisioned from Figure 8.23.

This last segment of path-length compensation is varied to match the change in delay arising from Earth rotation using delay-line carts (Figure 8.23) tracked by a laser-interferometer metrology system. Each cart is a catseye system that automatically returns the photon in a horizontal plane and parallel to the direction it had traveled from. (See Appendix B.5.1 for more on path length compensation.) The carts, which were designed for us by the JPL interferometry group, feature a nested servo system with the final, most-precise adjustment being a piezo-electric transducer moving the small secondary mirror of the catseye to provide positioning at the ±10 nm level.

Leaving the catseye system (see Figure 8.24), the photon is sent westward again to be intercepted by a telescope that reduces the beam diameter set at the light-collecting telescope by a factor of five for entry into the beam combiners located in a room that forms the small part of the L-shaped Beam Synthesis Laboratory. First, a dielectric beam splitter separates the visible portion from the near-infrared portion of each beam before they make a left turn and head to a beam combiner consistent with the science goal for that observation. There, our photon instantaneously recalls its wave-like split-personality and interferes with those portions of its wave function that have passed through the other telescopes, all arriving at the same time thanks to the respective delay lines. It is detected in combination with all of the other lucky photons that have arrived together to form an image in a final detector. For those

Figure 8.24. (Above) At the west end of the Delay Line Lab are the Beam Directing tables where optics reduce the diameters of the collimated beams emerging from the delay by a factor of five, separate visible light from near-infrared light, turn the beam by 90° and send them into the Beam Combining Lab seen in the left background. (Below) Each of eight large optical tables in the Beam Combining Lab serves a particular beam combiner or other support subsystem. For example, the table in the foreground houses the Michigan Infrared Combiner (MIRC) which has been responsible for producing images of complex objects such as the close binary system β Lyrae and the eclipse of ε Aurigae. (Courtesy of Steve Golden.)

interested in learning about one process through which visibility is extracted from an interferometer, please refer to Appendix B.5.5. For now, let's turn to science from the CHARA Array.

CHARA's first scientific results appeared in *The Astrophysical Journal* in 2005 July. As of this writing—almost exactly 15 years after that leadoff paper—the number of refereed scientific publications from the Array just reached the 200 mark. That respectable level of productivity results from CHARA partnering with several groups from around the world who have brought added value in several ways. This concept was first proposed to CHARA by Steve Ridgway and has paid off enormously. First, in lieu of charging members of the "CHARA Consortium" for access to the GSU facilities, they have generally contributed substantial new beam-combining instrumentation that increases the breadth of accessible science through the addition of capabilities such as imaging and spectroscopy. Second, they expand the user community and hence the publication rate. And, third, they bring new ideas —a healthy additive to any endeavor. The consortium and each member's senior representative presently consists of: the National Optical Astronomy Observatories (Steve Ridgway); University of Michigan (John Monnier); University of Exeter (Stefan Kraus): Observatoire de Paris (Vincent du Foresto); Observatoire de la Cote d'Azur (Denis Mourard), Université de Limoges (Francois Reynaud), Sydney University (Peter Tuthill), Australian National Observatory (Michael Ireland); and Kyoto University (Makato Kishimoto).

The user community has been further expanded by special NSF support to CHARA in return for which up to 60 nights per year are made available to external users who submit proposals through the review process of NSF's National Optical-Infrared Astronomy Research Laboratory (NOIRLab). As a result of this, the CHARA Array is the closest thing to a national optical interferometer facility that now exists in the U.S., something that never entered my mind when Clyde Faulkner authorized CHARA in 1984.

The happy problem with 200 scientific papers now out there is that there is too much to talk about! While they may not comprise the most important science to come out of the Array, I've chosen to emphasize some of the images that have been produced of objects at wondrously small angular scales. These images relate directly to the title of this book—*Seeing the Unseen*.

1. *Rapidly Rotating Stars*—Our first publication dealt with spectroscopic and interferometric modeling of the anticipated oblateness of α Leonis (Regulus; McAlister et al. 2005), which rotates with a period of only 16 hr compared with the Sun's sedate 25 days. It was obvious from the visibility data, which, in their simplest interpretation, provide diameters across sections of the stellar disk. As expected, Regulus was far from round. The first stellar image from the Array was that of another rapid rotator—α Aquilae (Altair; Monnier et al. 2007). This first-ever image of a main sequence star was produced by John Monnier's University of Michigan group using their Michigan Infrared Combiner (MIRC), shown on its optical table in Figure 8.24. Altair appears in Figure 8.25 along with four other oblate stars; all imaged using CHARA/MIRC. This collection includes a subsequent image of Regulus published along with that of β Cassiopeiae (Che et al. 2011), and another jointly-published pair, α Cephei (Alderamin) and α Ophiuchi (Rasalhague; Zhao 2009). Presenting nearly pole-on, β Cas shows only a marginally ellipsoidal distortion, but modeling of the interferometry indicates that the star is rotating at about 92% of its breakup velocity at

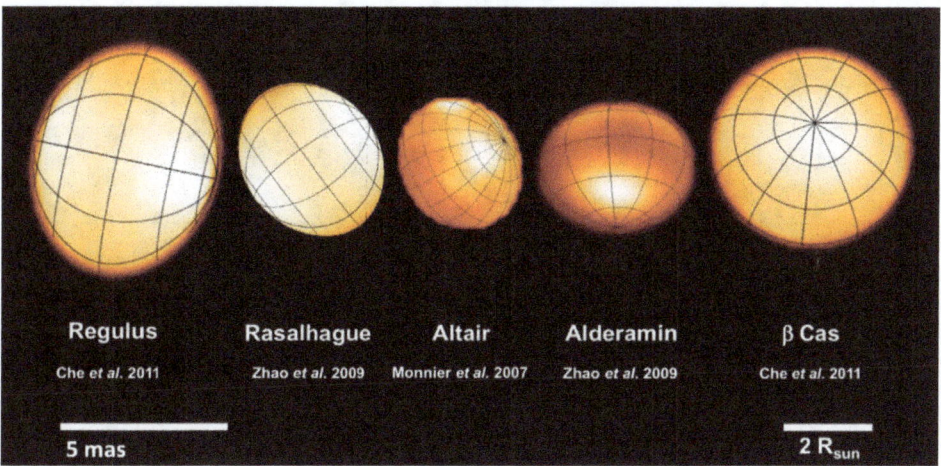

Figure 8.25. Images of five stars showing equatorial oblateness as a result of their rapid rotation rates. They all show gravity darkening at their equators and brightening at their poles. These images are scaled as if they are seen from the same distance. (Adapted from their original publication in *The Astrophysical Journal* and used here courtesy of Ming Zhao and John Monnier. © IOP Publishing. Reproduced with permission. All rights reserved.)

which point the centrifugal force at the star's equator equals the gravitational force. All of these stars demonstrate the latitude dependency of temperature that results in hotter, and radiatively-driven bright poles along with convective and gravity-darkened equatorial regions. Superimposed on these effects is limb darkening, which is responsible for the fading toward the pole of Alderamin and Altair.

2. *Spotted Stars*—The RS Canum Venaticorum (RS CVn) stars are close binaries with evolved and chromospherically-active primary components. Brightness changes superimposed on eclipse-induced variations apparently arise from large spots on the primary associated with chromospheric activity. Techniques have been developed for Doppler tomographic imaging of the stars using high-precision spectroscopic data. These stars are also amenable to so-called light-curve inversion techniques to produce images of the primary from photometric data. The 19.6 day spectroscopic binary σ Geminorum is at a distance that makes it resolvable by the CHARA Array. Rachel Roettenbacher used CHARA/MIRC data obtained during 2011 and 2012 to produce the images shown in the top row of Figure 8.26 (Roettenbacher et al. 2017). She also obtained contemporaneous spectroscopy and photometry enabling a three-way comparison of the imaging of a stellar surface. All three methods show general agreement about the longitude of spots but some disagreement on their latitude. In this first imaging comparison of its type, she concludes that "*the direct imaging of stellar surfaces available with optical interferometry is unmatched in its ability to capture the surface as it appears on the sky.*"

3. *The Expanding Fireball of Nova Delphini 2013*—Nova V339 Del was discovered on 2013 August 14, and within 15 hr of the discovery

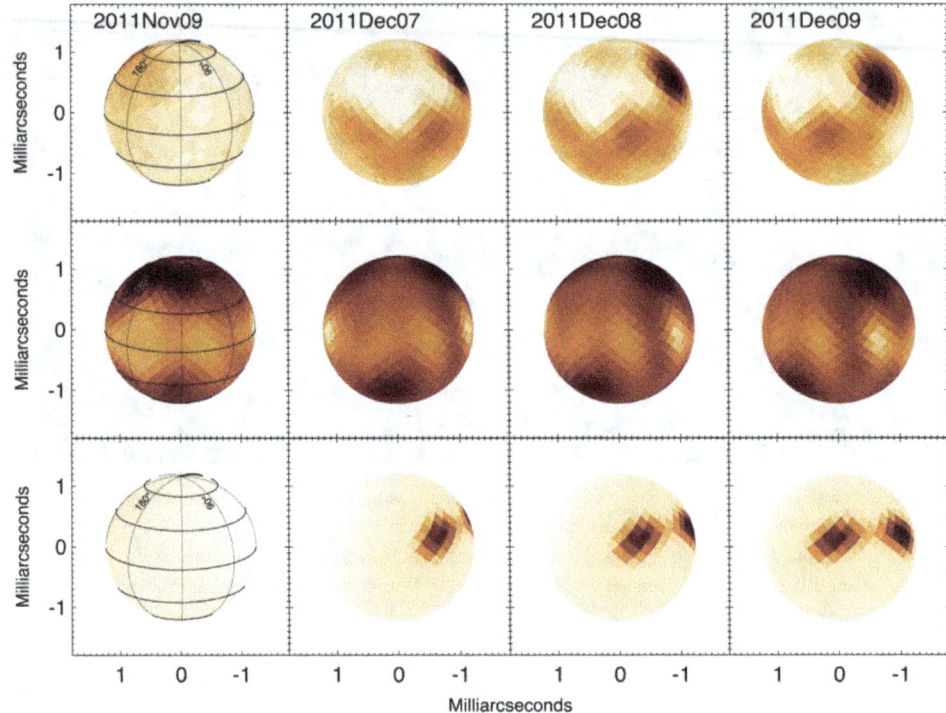

Figure 8.26. Images of the spotted giant star σ Geminorum produced from (top to bottom rows) CHARA/ MIRC interferometry, Doppler imaging from very high-resolution spectroscopy, and light curve inversion of photometric data. (Figure 5 from Roettenbacher et al. 2017. © IOP Publishing. Reproduced with permission. All rights reserved.)

announcement CHARA astronomer Gail Schaefer mounted an intensive observing campaign to detect the expansion of the fireball. The resulting angular diameter curve (Schaefer et al. 2014), obtained from more than a month of observing until the star became too large to measure, is shown in Figure 8.27. The nova fireball initially expanded with an optically thick surface surrounded by a diffuse envelope. The surface showed asymmetries that might be attributed to a bipolar outburst. This event was being followed from a number of observatories worldwide, and infrared photometry showed an increase in brightness attributed to the formation of heated dust in the outer layers of the envelope. By combining the observed angular expansion rate of 0.156 ± 0.003 mas per day with the radial velocity of the expanding surface, a geometrically-determined distance of 4.54 ± 0.59 kiloparsecs was derived. Images of the fireball from the interferometry are inset at the upper left of Figure 8.27. Note that the optically thick component was only 0.5 mas in diameter when it was first imaged. This is an example of how a university-managed observing facility can react with agility and persistence to an opportunity of this type.

4. *A Resolved Spectroscopic Binary*—GSU graduate student Kathryn Lester has carried out a spectroscopic binaries program with the CHARA Array,

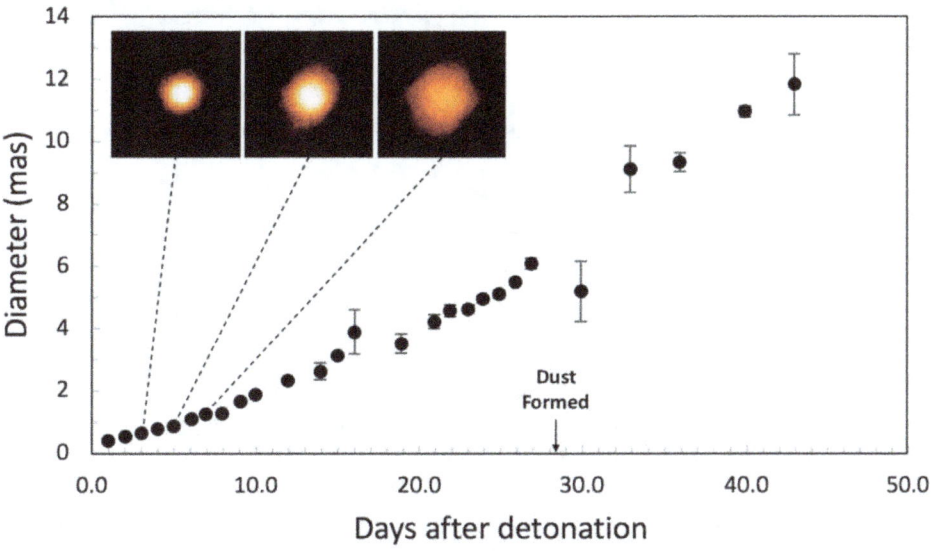

Figure 8.27. Measurements of the angular diameter of Nova Delphini began within one day of its eruption on 2013 August 14. (Author's diagram.)

and I can't resist featuring one of her results here. In this case, it is the double-lined spectroscopic pair comprising HD 185912. The star has a spectroscopically-determined orbital period of 7.640735 ± 0.000004 days, putting it on the far side of the selection effects gorge separating classical visual and spectroscopic binaries. Figure 8.28 shows that the star is easily resolved and accurately measured (Lester et al. 2019). The orbital inclination is so high that the star undergoes eclipses with sharp primary and secondary minima some 0.5 and 0.2 magnitudes below the brightness outside eclipse. The orbital elements are so accurately known as to produce masses with accuracies at the 0.3% level of $M_1 = 1.361 \pm 0.004$ M$_{sun}$ and $M_2 = 1.332 \pm 0.004$ M$_{sun}$. The distance derived from the orbital parallax of 41.02 ± 0.22 parsecs is in reasonable agreement with the Gaia spacecraft distance of 40.47 ± 0.08 parsecs. This is a lovely example of CHARA's construction of that superhighway across the oft-mentioned gorge. And yet this is not the shortest-period nor smallest-scale system resolved to date by the Array. That status goes to the 1.1 day spectroscopic binary σ Corona Borealis resolved by Deepak Raghavan in 2007 for which he found an orbital semimajor axis of 1.225 ± 0.013 mas, twice as small as the 2.57 ± 0.03 mas of HD 185912 (Raghavan et al. 2009).

5. *β Lyrae*—My candidate for the most dramatic result from the CHARA Array is the imaging by Ming Zhao of the famous interacting binary β Lyr (Zhao et al. 2008). CHARA/MIRC images were produced from data acquired on six nights in 2006 and 2007. Four of those images are shown in Figure 8.29. They clearly show the elongation of the brighter Roche lobe-filling donor star as well as the elongated disk of material spilling onto the

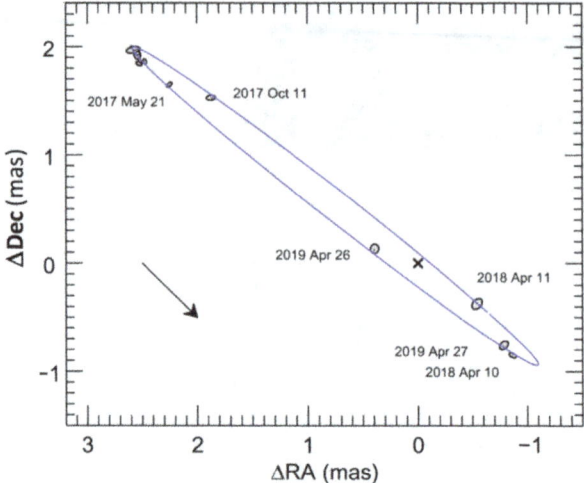

Figure 8.28. Interferometric orbit of the eclipsing binary star HD 185912. (Figure 3 of Lester et al. 2019. © IOP Publishing. Reproduced with permission. All rights reserved.)

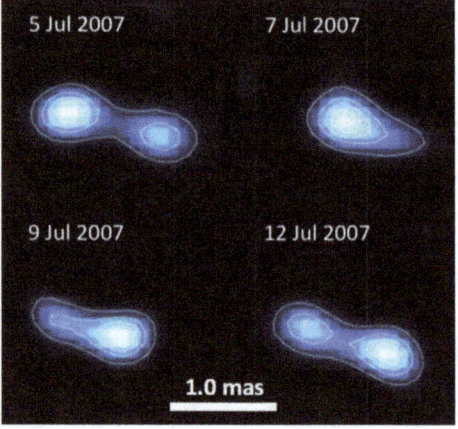

Figure 8.29. Images of the famous eclipsing and interacting binary star β Lyrae. The brighter donor star has filled its Roche lobe and spilling matter onto the thick disk surrounding the fainter mass gainer. (Courtesy of Ming Zhao and John Monnier.)

mass gainer. These are the first such images showing these phenomena occurring in close interacting binaries.

6. _β Persei (Algol)_—Another famous binary with imaging from the Array is the triple system Algol. The 2.87 day eclipsing pair, consisting of components A and B with spectral types B8V and K2IV, is gravitationally bound to an outer component C in an orbit having a period of 680 days. The AB,C system was first resolved with speckle interferometry by Labeyrie and his colleagues in 1973 (Labeyrie et al. 1974) and has continued to be followed by the subsequent users of that technique.

From CHARA/MIRC data obtained between 2006 and 2010, a team led by Fabien Baron—then a University of Michigan graduate student and now a Georgia State astronomy professor—imaged the full triple system of the "Demon Star" (Baron et al. 2012). They refined the orbit of the eclipsing pair with a 15% change in the mass of Algol A and confirmed that the orbital planes of the AB and AB,C systems are essentially perpendicular. While these improvements are important, the images, some of which are seen in Figure 8.30, are extraordinary.

7. *Imaging the 2009–2011 eclipse of ε Aurigae*—Every 27.1 years, this remarkable star system featuring a F0 supergiant undergoes an eclipse by a most unusual secondary. The eclipse lasts approximately two years, but the duration of dimming varies from cycle to cycle. A year or so prior to the

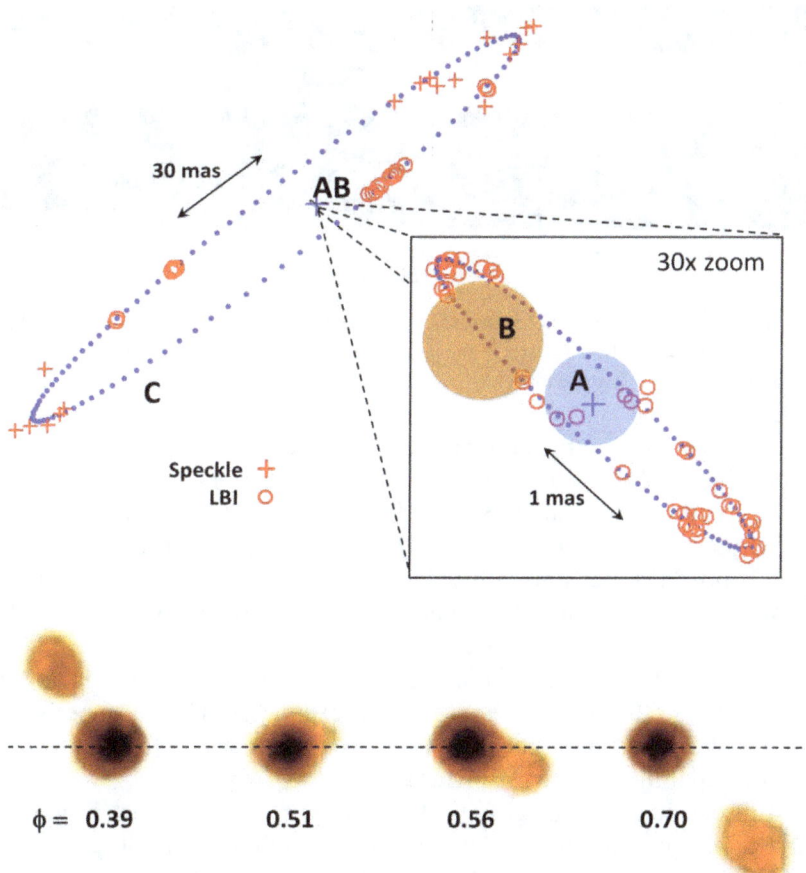

Figure 8.30. (Upper) The geometry of the Algol triple system showing the angular scales of the inner and outer systems and the perpendicularity of their orbits. Measurements from the CHARA/MIRC data are shown as open circles while speckle observations of the AB,C systems are indicated by + signs. (Lower) Images of Algol AB at four orbital phases produced from CHARA/MIRC data by a team led by Fabien Baron. (Author's diagram with images courtesy of Fabien Baron.)

last eclipse, Robert E. Stencel of the University of Denver contacted me to ask if CHARA might observe the upcoming event. I was only somewhat aware of ε Aur, but after a few minutes of listening to Bob, I knew this was something we had to do. Theo and Steve agreed, so we allocated a significant amount of observing time. Figure 8.31 shows a superposition of images produced by Bob's graduate student Brian K. Kloppenborg (Kloppenborg et al. 2010). Those images show that the eclipsing object is a very thick and highly elongated cloud of dark material that presumably enshrouds a star at its center. Rather than elaborate on this system, about which much was learned during the international observing campaign at that last eclipse, I will invoke the old cliché about a picture's worth in words and invite you to simply look at the remarkable imaging results for this perplexing system.

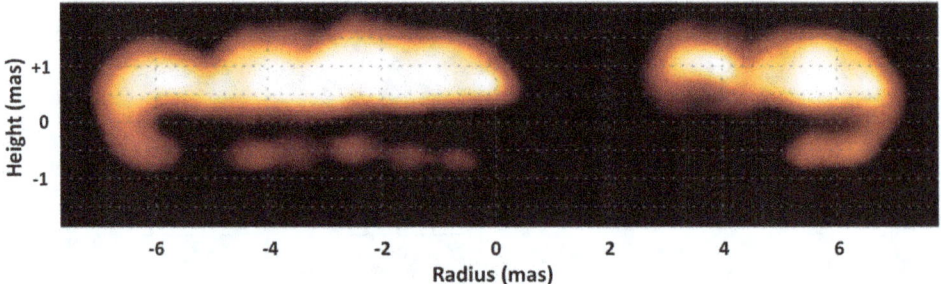

Figure 8.31. Multiple images of the 2009 eclipse of ε Aurigae are aligned as if the visible star is sliding behind the thick disk thereby showing the silhouette of the eclipsing material. (Adapted from Figure 2 of Kloppenborg et al. 2015, courtesy of Brian Kloppenborg).

I reiterate that these seven examples were selected from any number of the results to date from the CHARA Array. Others, such as the work by my former graduate student and present Louisiana State University astronomy professor Tabetha Boyajian, whose paper on the calibration of color–temperature relations for main sequence K and M type stars (Boyajian et al. 2012) has nearly 400 literature citations, may have more scientific impact. But, as visual beings, there is nothing like seeing images of things never before seen.

The diversity of the science emanating from the CHARA Array is apparent in a list of the Arrays "firsts," some of which exemplify Dan Popper's statement regarding unanticipated outcomes as being the hallmarks of an instrument with a new realm of capability. That list includes:

- First direct detection of gravity darkening on a single star (Regulus).
- First direct measurement of the Baade–Wesselink P-factor (δ Cep).
- First detection of hot exozodiacal dust around a main sequence star (Vega).
- First model-independent measurement of an exoplanet diameter (HD 189733b).
- First angular diameter for a halo population star (μ Cas).
- First image of a single, main sequence star (Altair).
- First direct image of an interacting binary (β Lyr).

- Shortest-period (1.14 days) binary star system yet resolved (σ^2 CrB).
- First image of a binary star system in eclipse (ε Aur).
- Earliest measurement of an expanding nova fireball (Nova Delphinus 2013).
- First images of starspots (λ And and ζ And).

This list will likely grow, especially if the vision of CHARA's leadership is fulfilled. Following my retirement in 2015, my longtime GSU colleague Douglas R. Gies, renowned for his expertise in the area of the hottest, most-massive stars, was appointed CHARA Director. Theo ten Brummelaar, who played a key role in making the Array work so well, was made Director of the CHARA Array. I was thrilled by these appointments that put CHARA in the best possible hands. Theo's immediate responsibility was to ensure that the NSF-funded adaptive optics upgrades to the CHARA telescopes were successfully implemented, while overseeing the routine operations of the Array. The talented and dedicated CHARA staff, who rightfully feel a personal ownership stake in the Array, would ensure, as they've done for years, that all would go well. Doug, not satisfied with the status quo, immediately began looking toward the future, asking what should be the next big thing for CHARA. A conservative answer would be to find the money for a seventh telescope.

But, Doug, recognizing that the 2020–2030 Decadal Review was at hand, had bigger things in mind. He asked the CHARA staff and a few of the core collaborators to assemble on Mount Wilson in 2017 September. He kindly invited me to be there as well. Figure 8.32 is a group photograph of the participants in that gathering.

The vision arising from this "CHARA Futures" meeting, and many discussions thereafter, is nothing short of inspiring. A new CHARA Michelson Array (CMA) would replace the current facility. It would grow from the present six 1 m telescopes

Figure 8.32. At the CHARA Futures meeting on 2017 September 22: Staff and collaborators gather to plan for the future of CHARA. From left are: Steve Ridgway, Gail Schaefer, Doug Gies, Judit Sturmann, John Monnier, Craig Woods, Theo ten Brummelaar, Nils Turner, Alicia Rice, Chris Farrington, Larry Webster, Laszlo Sturmann, Matt Anderson, Robert Klement, the author, and Norm Vargas. (Author's photo.)

to twelve 2 m telescopes equipped with higher-order adaptive optics. Five of the 2 m light collectors would form an outer array that provides baselines as long as 1.2 km yielding a factor of 3.6 gain in resolution—or a practical limit of about 25 micro-arcseconds (µas).

Light from the outer array would be brought to the Beam Synthesis Facility by optical fiber cables rather than through evacuated pipes. The six existing telescope locations would each be upgraded to 2 m apertures and a seventh 2 m added at the center of the array. Altogether, the CMA would have 66 baseline pairs compared with the current 15—a fourfold gain in (u,v)-plane sampling (see Appendix B.5.6).

This is not a mere, incremental upgrade to the CHARA Array. CMA is _the_ next-generation long-baseline, optical/near-infrared successor to the CHARA Array. Its ambitious gains in resolution, sensitivity, and aperture synthesis would open new windows on the universe. It would resolve and image exoplanets in silhouette as they transit their Suns, image the protoplanetary disks around the youngest stars, resolve the accretion features in the black hole binary Cygnus X-1, probe the SS 433 micro-quasar jet, and map the inner regions of galaxies with active nuclei powered by super-massive black holes.

All this comes at a price tag estimated at $100M to be spent over an eight-year project span. It will require support of the astronomical community and will undergo extensive peer review. How likely is CMA to become a reality? Only time will tell. In the meantime, the CMA has already gained momentum through a $2.5M NSF grant to fund CMAP—the CHARA Michelson Array Pathfinder. CMAP involves the addition of the longed-for seventh 1 m telescope. But, it won't be located at a fixed position. Instead, the new telescope will be portable and movable to multiple stations selected for optimal baseline coverage for specific targets. Optical fiber cables will transport light to the delay lines—a key step toward demonstrating the practicality of the CHARA Michelson Array while substantially enhancing CHARA's near-term imaging capabilities.

While the best may be yet to come for CHARA—if funding is forthcoming—what happens beyond that? Here, I can only once again invoke Dan Popper:

"History has indeed taught us that whenever a new technique enters a new realm of observational phase space, the most striking and productive results tend to be those not anticipated by even the most prescient thinkers."

8.3 Afterword

This book traces the descendants of the Hale–Michelson–Anderson–Pease progen-itors of the Mount Wilson branch of the interferometry family tree—the CHARA Array being the youngest of the progeny. Some of the descendants were inspired by the progenitors but fulfilled their ambitions elsewhere. Others were drawn to build on the family homestead. This is admittedly a fanciful picture I'm attempting to create, but, like a real family tree, there are traceable connections through the descendant timeline shown in Figure 8.33.

As an important reminder, there are other major branches in the interferometry genealogy—in particular the French branch—but our attention has been

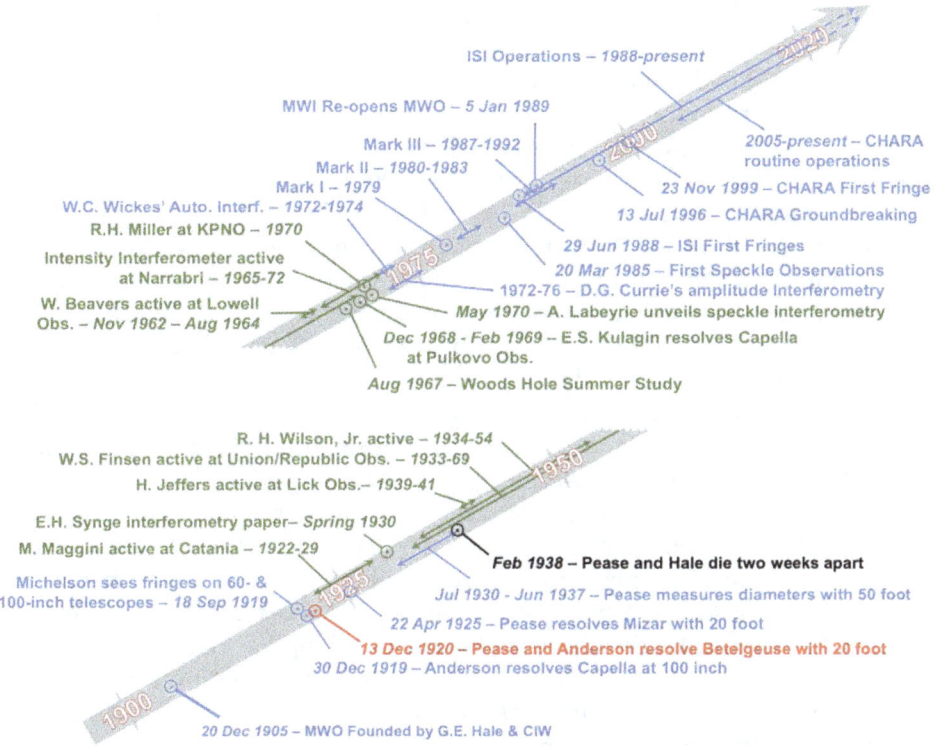

Figure 8.33. The Mount Wilson interferometry timeline for 1900–2020. Events taking place on the mountain are in blue, off the mountain in green, and milestone moments in red and black. (Author's diagram.)

purposefully narrowed to include those individuals or groups with logical or overt ties to Mount Wilson Observatory. A particular example of this focus is that while there were other practitioners of Anderson's double-star interferometry methodology who are given attention in Daniel Bonneau's book (see Table B.1), I've chosen a subset who represent the most and least fastidious of those who followed Anderson.

The successor generation to Pease kicked off in 1930, well before Pease's early death, with the unknown Irish genius "Hutchie" Synge laying out in an obscure journal a plan for a long-baseline successor to the 50 foot. Pease seems to have never heard of him, and one has to wonder how history might have been changed had Synge instead published in *Nature*. You can bet that Russell and Eddington would have been all over it.

In 1933, Bill Finsen launched a highly-productive program of Anderson-type visual interferometry of double stars that would span a third of the century. His handheld interferometer simply plugged into the eyepiece holder of the Johannesburg 26½ in refractor putting 0.1 arcsec resolution at his fingertips. Finsen's success derived from his understanding of the rigors of visual interferometry and scrupulous attention to its demands. The decade of the Great Depression closed with Ray Wilson briefly trying to resurrect the unwieldy 50 foot just a few months after Pease's unexpected death. Had Pease lived, perhaps Wilson's assistance would have averted the abandonment of the interferometer telescope. In any event, Wilson took up double-star interferometry in the northern hemisphere to complement Finsen's work in the south.

The 40s and 50s were left to Finsen and Wilson until the latter quit interferometry in 1954. Finsen was the lone wolf from 1954–1962, and his final series of measurements were made in early 1969 prior to the closure of Johannesburg Observatory in a reorganization of South African astronomy. From 1962–1964, Willet Beavers, working almost entirely on his own, set out to build only the second Michelson–Pease interferometer—getting and photographing fringes but not quite achieving his goal of quantifying visibilities. Although Beavers did not observe at Mount Wilson, he most certainly was a descendant of the line.

After decades of hiatus, interferometry once again returned to Mount Wilson with, not just one, but two independent projects being undertaken at the same time. Bill Wickes' achromatic interferometer succeeded beautifully in automating Anderson's visual interferometer for accurate double-star observations. During 1972–1974, Wicke's obtained accurate measurements of selected binaries, and his technique might have become the standard for doubles had he not been spirited off to the Hewlett–Packard Corporation where he played a major role in developing HP's line of handheld scientific calculators. Overlapping Wicke's work on the mountain was Doug Currie, who was employing his amplitude interferometer concept on observing runs during 1972–1976. Not since the early 1920s was there such interferometry action at the homestead!

Wickes and Currie made periodic visits to use their specialized instruments on the Observatory's telescopes. Henceforth, the progeny would see Mount Wilson not for its telescopes but instead for its real estate under night skies offering fine astronomical seeing. And, thus came Mike Shao's Mark I, II, and III interferometers that cleared a technical path for their successors. Charlie Townes brought an entirely new approach to interferometry and opened up a new spectral window with the ISI's first fringes in 1988. Eleven years later, CHARA celebrated its first-fringe milestone, but got back to work finishing off its full complement of telescopes before commencing routine observing at the end of 2004. The descendancy stops there for now.

Whenever I'm on Mount Wilson, I like to walk around the Observatory grounds late at night—the domes and towers silhouetted against a sky softly glowing from the illumination of the LA basin. On those excursions, I see specters, mostly imagined but sometimes not. In the fall of 2009, Susan and I went out to Mount Wilson during the extreme threat of the Station Fire (McAlister & McAlister 2019). Firefighters and their equipment had been staged on the mountain to defeat the fire should it take a run up the north slope as it appeared to be contemplating. On one of those nights around midnight, Susan and I were out roaming after a long and tense day. Flames were lighting up a distant slope as if the fire were signaling what it might do to our mountain. As we headed back to the Kapteyn Cottage, we crossed over the same footbridge taken by Anderson and Pease on 1920 December 13 for their night-lunch discussion of the "whopper" they had just resolved. We didn't encounter those gentlemen that night of the fire, but we did see ghostly apparitions scattered in the arroyo beneath the footbridge. It took but an instant to realize that is was the Hotshot crew from Helena, Montana, tucked in their sleeping bags recharging for another day of hacking fire line along a ridge the fire might be fancying for its assault on the Observatory. I mention this episode at some length as the Station Fire came close to terminating the Mount Wilson interferometry branch then and there.[3]

I hope to take that midnight walk around the grounds yet again. This time I'll look for real spirits through the gloom. Perhaps I'll hear the footsteps of Betelgeuse's resolvers as they cross the footbridge, their "heads full of sums." Or on some other night, as I walk by the 50 foot building, Francis Pease will be standing just outside its entrance, hands in his pockets while slowly shaking his head as he talks himself out of yet another night's struggle to make the interferometer telescope do his bidding. Maybe, I'll quietly enter the closed 100 inch telescope dome and slowly climb the long metal stairway up to the observing floor to see if Pease is at work at the eyepiece of the 20 foot beam with Mr. Kimple perched high on the telescope's upper end awaiting instructions. Or could that figure standing near the CHARA S2 telescope enclosure be Hutchie smiling in the darkness at this embodiment of his overlooked 1930 scheme for a long-baseline interferometer?

Now, I don't expect to ever see these spirits, mind you, but I do often contemplate my interferometry ancestors with a sense of awe and gratitude for what they created and left to those of us who came after them.

References

Armstrong, J. T., Mozurkewich, D., Rickard, L. J., et al. 1998, ApJ, 496, 550

Avasthi, A. 2005, Charles Townes Wins Templeton Prize Science, www.sciencemag.org/news/2005/03/charles-townes-wins-templeton-prize

Baron, F., Monnier, J. D., Pedretti, E., et al. 2012, ApJ, 752, 20B

Bester, M., Danchi, W. C., Hale, D., et al. 1996, ApJ, 463, 336

Boyajian, T. S., von Braun, K., van Belle, G., et al. 2012, ApJ, 757, 112

Boyd, R. 2015, Natur, 519, 292

Breckinridge, J. B., McAlister, H. A., & Robinson, W. G. 1979, ApOpt, 18, 1034

Che, X., Monnier, J. D., Zhao, M., et al. 2011, ApJ, 732, 68

Danchi, W. C., Bester, M., Degiacomi, C. G., Greenhill, L. J., & Townes, C. H. 1994, AJ, 107, 1469

Danchi, W. C., Bester, M., & Townes, C. H. 1988, in NOAO-ESO Conf. High-Resolution Imaging by Interferometry, ed. F. Merkle (Garching: ESO), 867

Fry, M. C. 1973, The Cat's Whisker, 3, 1

Glindemann, A., & Käufl, H. U. 2006, in Visions for Infrared Astronomy (Paris: Lavoisier), 347

Haskins, C. P. 1965, Report of the President, Carnegie Institution of Washington Year Book 64 (Washington, DC: Carnegie Institution of Washington), 62

Ireland, M. J., & Monnier, J. D. 2014, Proc. SPIE, 9146, 914612

Johnson, M. A., Betz, A. L., & Townes, C. H. 1974a, BAAS, 6, 450

---

[3] In 2020 September, the Bobcat Fire took another shot at the mountain. It was thwarted by the efforts of hundreds of firefighters, and the Observatory was again unscathed.

Johnson, M. A., Betz, A. L., & Townes, C. H. 1974b, PhRvL, 33, 1617

Kloppenborg, B. K., Stencel, R. E., Monnier, J. D., et al. 2015, ApJS, 220, 14

Kloppenborg, B. K., Stencel, R., Monnier, J. D., et al. 2010, Natur, 464, 870

Labeyrie, A., Bonneau, D., Stachnik, R. V., & Gezari, D. Y. 1974, ApJ, 194, L147

Lester, K. V., Gies, D. R., Schaefer, G. H., et al. 2019, AJ, 158, 218

Lockwood, S., Ravi, V., Wishnow, E. H., et al. 2014, in ASP Conf. Ser. 487, Resolving the Future of Astronomy with Long-Baseline Interferometry, ed. M. J. Creech-Eakman, J. A. Guzik, & R. E. Stencel (San Francisco, CA: ASP), 337

McAlister, H. A., ten Brummelaar, T. A., Gies, D. R., et al. 2005, ApJ, 628, 439

McAlister, H. A., ten Brummelaar, T. A., & Ridgway, S. T. 2014., Proc. SPIE, 9146, 91460D

McAlister, H. A., & McAlister, S. J. 2019, The 2009 Station Fire Threat to Mount Wilson Observatory, 88

Monnier, J. D., Zhao, M., Pedretti, E., et al. 2007, Sci, 317, 342

Moravveji, E., Guinan, E. F., Khosroshahi, H., & Wasatonic, R. 2013, AJ, 146, 148

Overby, D. 2020, New York Times,

Raghavan, D., McAlister, H. A., Torres, G., et al. 2009, ApJ, 690, 394

Ravi, V., Wishnow, E., Lockwood, S., & Townes, C. H. 2011, in ASP Conf. Ser. 448, 16th Cambridge Workshop on Cool Stars, ed. C. M. Johns-Krull, M. K. Browning, & A. A. West (San Francisco, CA: ASP), 1025

Ridgway, S. T., McAlister, H. A., & ten Brummelaar, T. A. 2014, Proc. SPIE, 9146, 91460E

Roettenbacher, R. M., Monnier, J, D., Korhonen, H., et al. 2017, ApJ, 849, 120

Sandage, A. R. 1997, GriO, 61, 1

Schaefer, G. H., ten Brummelaar, T. A., Gies, D. R., et al. 2014, Natur, 515, 234

Schliesser, A., Picque, N., & Hänsch, T. W. 2012, NaPho, 6, 440

Scholz, M., & Takeda, Y. 1987, A&A, 186, 200

Sutton, E. C., Storey, J. W. V., Betz, A. L., Townes, C. H., & Spears, D. L. 1977, ApJ, 217, L97

Sutton, E. C., Storey, J. W. V., Townes, C. H., & Spears, D. L. 1978, ApJ, 224, L123

Sutton, E. C., Subramanian, S., & Townes, C. H. 1982, A&A, 110, 324

Tatebe, K., Hale, D. D. S., Wishnow, E. H., & Townes, C. H. 2007, ApJ, 658, L103

ten Brummelaar, T. A., McAlister, H. A., & Ridgway, S. T. 2014, Proc. SPIE, 9146, 91460F

Townes, C. H. 1966, THINK (IBM magazine), 32, 2

Townes, C. H. 1995, Making Waves (Woodbury, NY: AIP Press), 157

Townes, C. H., Wishnow, E. H., Hales, D. D. S., & Walp, B. 2009, ApJ, 697, L127

Townes, C. H., & Schawlow, A. 1955, Microwave Spectroscopy (New York: McGraw-Hill)

Tsay, W.-S., Bagnuolo, W. G., McAlister, H. A., White, N. M., & Forbes, F. F. 1990, PASP, 102, 1339

Tuthill, P. G., Monnier, J. D., & Danchi, W. C. 1999, Natur, 398, 487

Weiner, J., Danchi, W. C., Hale, D. D. S., et al. 2000, ApJ, 544, 1097

Weiner, J. 2004, ApJ, 611, L37

Weiner, J., Hale, D. D. S., & Townes, C. H. 2003a, ApJ, 588, 1064

Weiner, J., Hale, D. D. S., & Townes, C. H. 2003b, ApJ, 589, 976

Zhao, M., Gies, D. R., Monnier, J. D., et al. 2008, ApJ, 684, L95

Zhao, M., Monnier, J. D., Pedretti, E., et al. 2009, ApJ, 701, 209

Appendix A

Why Mount Wilson?

Why has Mount Wilson attracted interferometrists to this very day. The blithe answer is—*because it's there*. At several levels, though, that's not such a thoughtless response. The *it* aspect of the answer is that here is a privately-owned facility with large-aperture telescopes possessing superb optics made available for use by outsiders. That was particularly true for the 60 inch after the 100 inch came on line. The *there* is a location that has had transcontinental rail connections to most anywhere in the US since well before Hale latched onto the site. As someone who has flown the ATL–LAX–ATL round trip some 250 times, I can testify to the continued accessibility of the Observatory! A logistical bonus of the location is that if something relatively minor breaks in your instrument, you can likely run down off the mountain to LA in the morning and be back that afternoon with whatever widget you needed. Even more importantly, the site enjoys better weather and clearer skies than do the home institutions of most of the astronomers who benefited from the kindness of the Observatory's directors for the past century.

As we shall see, it is the *there* aspect that brought late 20th century interferometers to the mountain as the observatory ecompasses considerable real estate set aside for astronomy. That property originally belonged to the Mount Wilson Toll Road Company, not to the Carnegie Institution. Through a process of transfers, the land was given to the U.S. Department of Agriculture in the 1960s for inclusion in the Angeles National Forest. These transfers all preserved the original 1904 leasehold agreement for the Observatory, and that agreement was renewed in 2002 for another 100 years. Such longevity lends an aura of permanence to anyone wishing to locate a new facility on the Observatory grounds.

Also important to the planning and budgeting for a new venture are existing roads, power, and water as well as experience with the implications of the National Environmental Policy Act. One apparent shortcoming is the absence of flat and vacant land as would ideally be sought for an interferometric telescope array. While the leasehold is large, "flat and vacant" it is not. The mountaintop is topographically convoluted and populated with dozens of buildings, but three multi-telescope, long-baseline interferometers have been shoe horned in among those buildings since the 1980s.

Although the brightening night skies of Southern California blinded the telescopes to very distant galaxies decades ago, this phenomenon has no effect on interferometry's targets as they are far brighter than the background light pollution. The Observatory itself is resistant to local incursion of lighting—or any development at all—due to its presence in the 650,000 acre Angeles National Forest, established by President T. Roosevelt in 1908 through the consolidation of earlier federal land reserves. Another layer of protection came about in 2014 when President Obama set aside a portion of the ANF as the San Gabriel Mountains National Monument. This action encompassed most of the prominent peaks in the mountain range, including Mount Wilson. Dark skies do occasionally glorify the night over the mountain when the marine layer rolls in off the Pacific to cast the Los Angeles basin in a "June gloom." The mountain's 5700 ft elevation more often than not makes it an island above the clouds with blue skies during the day and the Milky Way at night. On such nights, we see the stars as they presented themselves to the native Americans who summered on those mountain tops long before the first Europeans entered what would become known as California.

Weather, accessibility, logistics, infrastructure, acreage, and long-term endurance are all important attributes for any observatory. But, for interferometry in particular, atmospheric seeing trumps all else. As noted in Section 7.2, a paper presented at the 1985 January Tucson meeting of the American Astronomical Society by Pete Nisenson (1941–2004), Bob Noyes, and Mike Shao of the Harvard Smithsonian Center for Astrophysics asked in its title "Is Mt. Wilson the Best Interferometric Site in the World? (Nisenson 1984)" That's a tall order.

The significance of seeing to certain of astronomy's needs was apparent from the earliest planning by the Carnegie Institution's Advisory Committee on Astronomy, whose activities are described in Section 2.1. In the Committee's "General Plan for Furthering Special Researches in Astronomy," the third paragraph begins with: *"The need for special undertakings in astronomy, such as are contemplated by us, arises in part from the desire to secure special conditions of atmosphere. If the atmosphere were everywhere in a perfect state of calm, with no difference of temperature except those due to increasing altitude, the telescope images of celestial objects seen through it would be perfectly steady and distinct. Under the actual conditions of observation as we experience them, the telescope image is in a state of rapid and incessant vibration."* The quest for sites with the best seeing topped their list of selection criteria even then.

Spoiler alert—Mount Wilson is inherently suited to good seeing as described in Appendix B.6, which outlines the parameters developed to characterize the seeing of astronomical sites.

The most frequently used parameter for characterizing seeing is Fried's parameter r_0 (Fried 1965), which is is related to the size of the seeing disk of an unresolved star imaged by a telescope by:

$$\text{Seeing Disk FWHM (in radians)} \sim 0.97\lambda/r_0,$$

where the wavelength λ scales as $\lambda^{-0.2}$. This relation yields 1 arcsec seeing when r_o equals 10 cm and $\lambda = 500$ nm. Another representation of seeing is the atmospheric redistribution time τ_o, essentially the time it take the atmosphere to move the distance corresponding to the size of the r_o seeing cell. That parameter is obviously related to wind speed and is of interest to designers of adaptive optics system that must keep up temporally with τ_o, but it does not necessarily correlate with r_o.

In addition to where you locate an observatory, it has been realized for some time that the design of the telescope dome plays a major role in the achievable seeing. Telescope enclosures built with inattention to thermal management can turn an intrinsically good site into a mediocre one. Without naming names, the worst example of dome design I have personally seen involved a telescope at a high-elevation, dark-sky site atop a small, heated building whose furnace was exhausted by a pipe opening to the air immediately to the south of the dome. No kidding—what were they thinking?

As a counter example, it is worth noting that Hale's nighttime telescope domes have inner and outer walls between which ambient air can enter from the ground level and rise up to exit at the top of the domes. This prevents heat from building up excessively during the day that would otherwise boil out of the dome as its interior cooled upon opening.

Although seeing conditions were rated at the 60 inch telescope prior to 1917, it was that year when a scale, shown in Table A.1, was established and thereafter recorded nightly. It originally was on a 1–10 scale, but practice showed that seeing was rarely rated above 8. So, the scale was modified accordingly as described by H. W. Babcock in a 1963 paper that showed how a photoelectric seeing monitor that stared at Polaris could provide reliable measurements of seeing that were well correlated with the Mount Wilson scale (Babcock 1963). This device incorporated a crossed Ronchi lattice pattern that rotated to modulate the image and produced a signal from which seeing-induced excursions from a stationary position could be detected. Ten years earlier, Babcock had set forth the underlying principles of adaptive optics (Babcock 1953). Understanding and measuring seeing was clearly a

Table A.1. The Mount Wilson Seeing Scale

Seeing Value	Image Size (arcsec)	Criteria
<1	5+	Image extremely soft and extended
1	3–5	Image soft and blurred
2	2	Image soft and blurred to 2 arcsec
3	1	Image about 1 arcsec with condensation
4	<1	Image sharp with slight motion
5	0.6–0.7	Image sharp and nearly motionless
6	0.5–0.6	Image sharp, round and motionless
7	0.4–0.5	...
8	0.3	...

priority of his. This was brought home to me not long after our 1995 selection of Mount Wilson as the site for the CHARA Array. Dr Babcock called me one afternoon to explain that seeing was highly dependent upon height above ground level and suffered from tree effects in the vicinity of domes. He lamented that this should not have been allowed to happen. He noted in particular that the 100 inch telescope building and pier placed that instrument essentially at the tree tops, and we should not expect the best seeing conditions unless we elevated our telescopes to similar heights. That was a condition that we would be unable to fulfill, but we have subsequently collected an enormous amount of seeing data ourselves. As will be seen below, we now have millions of r_o measurements of seeing from CHARA's telescopes.

But first, let's take a look at what had been learned about Mount Wilson seeing in prior years. As early as 1924, Pease described visual inspection of stars outside of focus to see the turbulent streams of air passing through the entrance pupil with the speed and degree of "flickering" being representative of seeing conditions (Pease 1924). It would take 70 years before the first quantitative measurements of Mount Wilson seeing were available. In 1994 and 1995, David Buscher and his colleagues published measurements of τ_o made during observations with the Mark III stellar interferometer described Section 7.2 (Buscher 1995; Buscher et al. 1995). In reviewing this appendix, CHARA Array Director Theo ten Brummelaar told me that he believes "*that τ_o is quite large on Mount Wilson compared to most other sites, including Mauna Kea, and this means one can have much longer exposure times. This plays as much into interferometry as it does into adaptive optics, especially for fringe tracking, which I suppose is a form of AO. It is anecdotal but generally acknowledged that while Mauna Kea has better r_o values, Mount Wilson has slower seeing giving it an advantage for both AO and interferometry. One could also argue that AO systems will be better behaved in large τ_o seeing, making r_o less important.*"

In 1997, Donald Walters and William Bradford of the Naval Postgraduate School presented results they accumulated over a twenty-year period from 18 observatory sites around the world from which they measured r_o and the isoplanatic angle α_o, the angular distance of over which the seeing perturbations are more or less instantaneously uniform—see Appendix B.6 for more. Their Mount Wilson samples, taken between 1986 and 1990, gave $<r_o> = 12.9$ cm and $<\alpha_o> = 2.15$ arcsec. Mount Wilson and the Air Force space reconnaissance site on Mount Haleakala, Maui, were tied for number one in terms of large values of r_o among the 18 sites tested (which did not include Mauna Kea), and Mount Wilson excelled over all other continental sites (Walters & Bradford 1997).

In a particularly fascinating study, Scott Teare and Laird Thompson of the University of Illinois, who were collaborating on a laser guide star adaptive optics program at the 100 inch telescope, along with telescope operators Colleen Gino and Kirk Palmer, undertook an analysis of 80 years of Mount Wilson observing records (Teare et al. 2000). These included hours of operation published in the Carnegie Yearbooks during 1917–1919, data from the "Mount Wilson Weather Bulletin" published between 1920 and 1973, and seeing estimates using the scale from Table A.1 recorded in the observing logs of the 60 inch and 100 inch telescopes during 1974–1999. They found average seeing disk diameters of about 1.3 arcsec during this 80 year period with sub-arcsecond seeing on nearly one-third of all nights. They also found a trend over time toward slight degradation in seeing, possibly due to regrowth of trees and the construction of new facilities in the vicinity of the telescopes. Teare et al. pointed out that the period during which measurements were made by Walters and Bradford spanned the poorest years in the entire data sample.

The most recent Mount Wilson seeing measurements are those that are routinely recorded at all six telescopes of the CHARA Array. These data result from the logging of tip/tilt corrections required to stabilize images before they are interferometrically combined. Figure A.1 shows quarterly averages of r_0 calculated from a total of several million of these measurements from the 11 year period 2009–2019. These data are a subset of those that have been collected from the CHARA telescopes since 2006 and were kindly provided to me for this analysis by their curator, CHARA Research Scientist Nils Turner.

A glance at Figure A.1 reveals several aspects of observing conditions on the mountain. The absence of data during the first quarter of most years is the result of CHARA using, until recently, this period of clouds and poor seeing for maintenance, upgrades, and engineering development. The first generalization from the remainder of these years is that the best seeing occurs in summer, which is not news. Indeed, it was known from early on that July and August were the primo periods to take advantage of sub-arcsecond seeing. Next, we see that not all years are created equal, a fact also previously known. Likewise, and lastly, seeing varies with location on the mountain. In this case, telescope W1, located on the northwest tip of CHARA's Y-shaped array, enjoys better seeing than does telescope E1 located at the end of the northeast arm of the Array on the mountain's escarpment. This degradation is perhaps due to air currents convecting up the steep slope immediately adjacent to E1 while W1 is in a relatively level area that is locally heavily forested with mountain oaks. Those surroundings are counter to Dr Babcock's advice about trees, but removing trees is, of course, orthogonal to the purpose of a national forest. As can be seen in Figure A.1, 2019 is more or less in the mid-range among these 11 years.

The numerical entries incorporated into Figure A.1 are given in Table A.2 from which we see that the grand average seeing over 2009–2019 at telescope E1 was 9.85 ± 1.01 cm in r_0 corresponding to 1.02 ± 0.10 arcsec in terms of seeing disk diameter. W1 produced the better seeing as represented by the values 11.26 ± 1.05 cm and 0.89 ± 0.08 arcsec. These results are more favorable than what Teare et al. found, probably as a result of the different proxies for "seeing" among the various methods utilized. All in all, though, there is consistency among all the pertinent data.

In Figure A.2, the nightly normalized occurrence distributions of r_0 for the two telescopes are shown averaged over the 2019 year and for that summer quarter. The distributions, for this particular year at least, show favorable tendencies for a higher frequency of very good seeing nights over very poor nights.

Getting back to the question—is Mount Wilson the best interferometric site in the world? In terms of seeing—No. Mauna Kea on Hawaii, which once hosted the

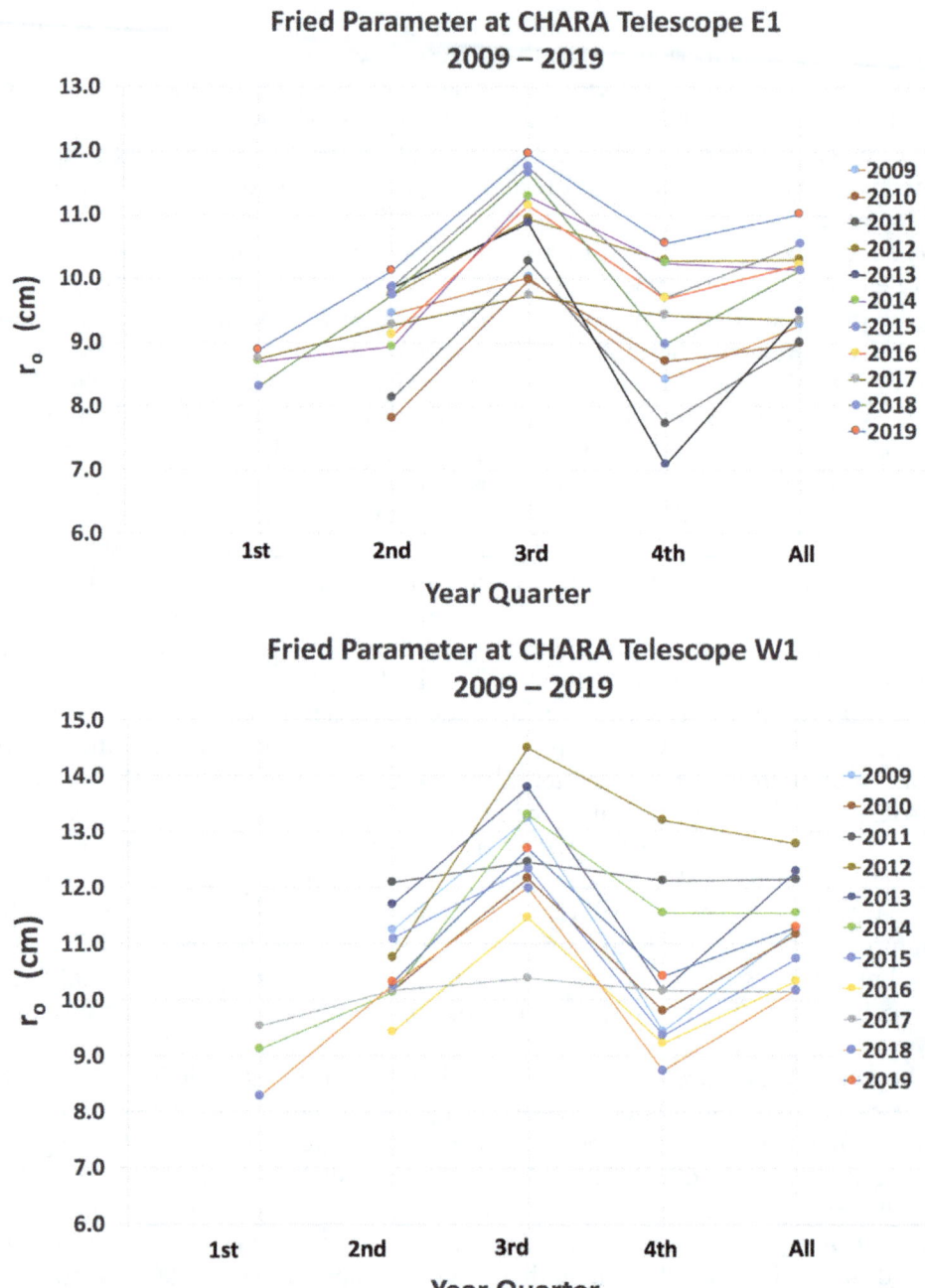

Figure A.1. Quarterly averages of seeing at CHARA telescopes E1 and W1 from 2009 through 2019. The average dispersion in the mean r_o values is ±0.8 cm. (Author's diagram.)

Table A.2. Values of r_o (cm) from CHARA Telescopes E1 and W1

	1st Qtr		2nd Qtr		3rd Qtr		4th Qtr		Full Year	
	E1	W1	E1	W1	E1	W1	E1	W1	E1	W1
2009	–	–	9.4 ± 0.9	11.2 ± 0.7	10.0 ± 0.8	13.2 ± 2.0	8.4 ± 1.2	9.4 ± 1.1	9.3 ± 0.9	11.3 ± 1.5
2010	–	–	7.8 ± 1.0	10.2 ± 1.2	10.0 ± 1.4	12.2 ± 0.7	8.7 ± 1.2	9.8 ± 1.2	9.0 ± 1.2	11.2 ± 1.0
2011	–	–	8.1 ± 0.8	12.1 ± 1.0	10.3 ± 0.8	12.5 ± 0.7	7.7 ± 0.9	12.1 ± 0.9	9.0 ± 0.8	12.2 ± 0.9
2012	–	–	9.7 ± 1.0	10.7 ± 1.2	10.9 ± 0.9	14.5 ± 1.2	10.3 ± 0.8	13.2 ± 1.0	10.3 ± 1.0	12.8 ± 1.2
2013	–	–	9.9 ± 1.1	11.7 ± 1.2	10.9 ± 1.9	13.8 ± 1.0	7.1 ± 0.8	10.2 ± 1.1	9.5 ± 1.5	12.3 ± 1.1
2014	8.7 ± 0.9	9.1 ± 1.2	8.9 ± 1.1	10.1 ± 1.1	11.3 ± 1.0	13.3 ± 1.0	10.2 ± 1.0	11.5 ± 1.1	10.1 ± 1.0	11.6 ± 1.1
2015	–	–	9.9 ± 0.9	11.1 ± 0.7	11.7 ± 0.8	12.3 ± 0.9	9.7 ± 0.8	9.4 ± 0.7	10.5 ± 0.9	10.7 ± 0.9
2016	–	–	9.1 ± 1.1	9.4 ± 1.2	11.1 ± 0.8	11.5 ± 1.0	9.7 ± 1.0	9.2 ± 1.0	10.2 ± 1.0	10.3 ± 1.0
2017	8.7 ± 0.7	9.5 ± 1.3	9.3 ± 0.8	10.2 ± 0.8	9.7 ± 0.7	10.4 ± 0.8	9.4 ± 1.0	10.2 ± 0.8	9.3 ± 0.8	10.1 ± 0.9
2017	8.3 ± 0.7	8.3 ± 0.9	9.7 ± 1.1	10.3 ± 1.1	11.7 ± 0.8	12.0 ± 0.8	9.0 ± 0.7	8.7 ± 0.8	10.1 ± 0.9	10.2 ± 0.9
2019	8.9 ± 1.7	–	10.1 ± 0.8	10.3 ± 1.0	12.0 ± 1.1	12.7 ± 0.9	10.6 ± 0.8	10.4 ± 1.0	11.0 ± 0.9	11.3 ± 1.0
r_o—all years (cm)	8.65 ± 1.02	8.97 ± 1.12	9.27 ± 0.96	10.66 ± 1.06	10.87 ± 1.01	12.57 ± 1.00	9.16 ± 0.93	10.38 ± 0.97	9.85 ± 1.01	11.26 ± 1.05
Seeing Disk FWHM—all years (arcsec)	1.16 ± 0.14	1.11 ± 0.13	1.08 ± 0.11	0.94 ± 0.09	0.92 ± 0.09	0.80 ± 0.06	1.09 ± 0.11	0.96 ± 0.09	1.02 ± 0.10	0.89 ± 0.08

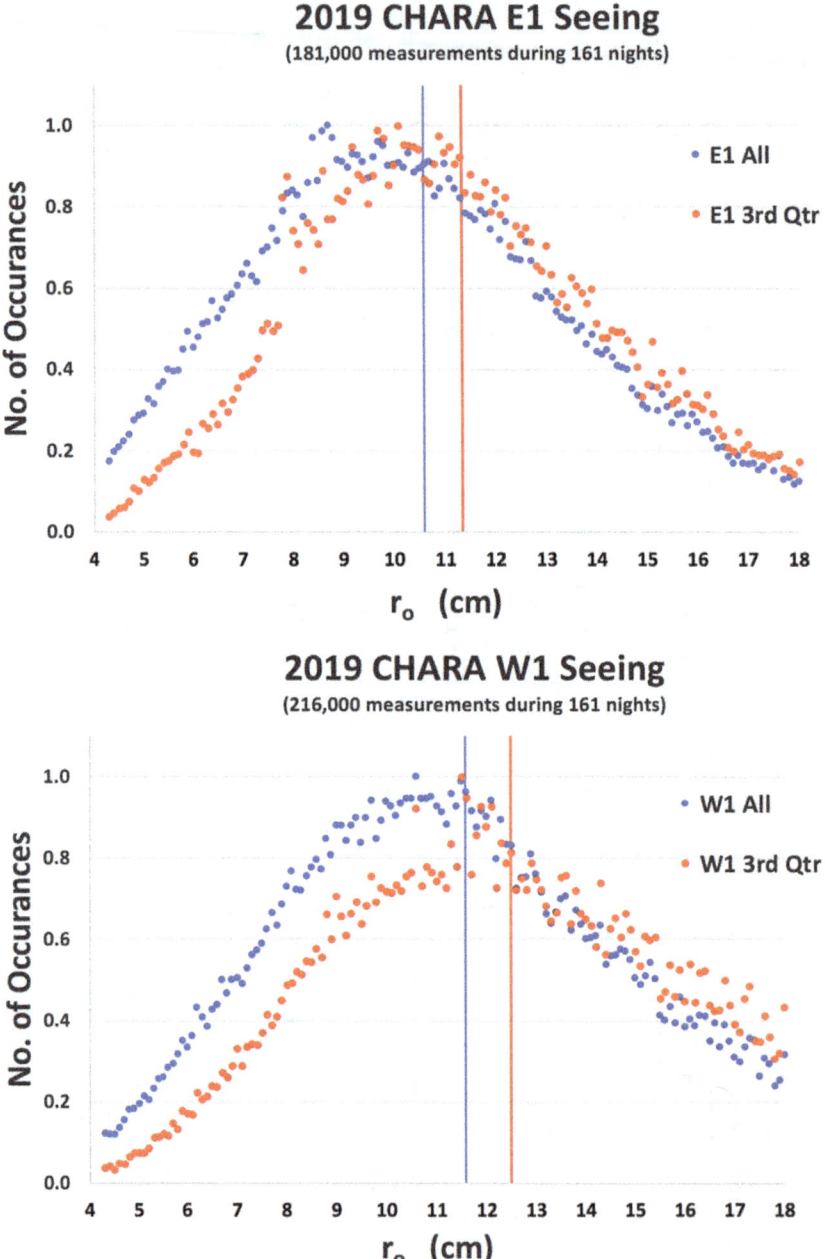

Figure A.2. Normalized frequency distributions of ro measurements made during 2019 at CHARA telescopes E1 and W1. The vertical lines represent the means of the full-year (blue) and third-quarter (red) data. (Author's diagram.)

now-canceled Keck Interferometer, and Cerro Paranal, home to the European Very Large Telescope Interferometer, most likely deserve that title. Once while doing speckle interferometry of binary stars at the 3.6 m Canada–France–Hawaii Telescope on Mauna Kea, I saw 0.25 arcsec seeing for a substantial fraction of a night. Sure, that's purely anecdotal—there's plenty of real data for that superb site—but I've never personally experienced such exquisite seeing anywhere else. Even if one has the budget for such remote sites, good luck in getting permission to build on them nowadays.

What might be a fair statement about Mount Wilson is that it is an outstanding choice when all considerations of seeing, logistics, and costs (both construction and operating) are folded into one's site selection equation. If limited to the continental U.S. and ignoring Mauna Kea's sometimes very fast seeing, then the answer to the question of Nisenson, Noyes, and Shao is, in my opinion, yes.

References

Babcock, H. W. 1953, PASP, 65, 229

Babcock, H. W. 1963, PASP, 75, 1

Buscher, D. F. 1995, Proc. SPIE, 2200, 260

Buscher, D. F., Armstrong, J. T., Hummel, C. A., et al. 1995, ApOpt, 34, 1081

Fried, D. L. 1965, JOSA, 55, 1427

Nisenson, P., Noyes, R., & Shao, M. 1984, BAAS, 16, 908

Pease, F. G. 1924, PASP, 36, 191

Teare, S. W., Thompson, L. A., Gino, M. C., & Palmer, K. A. 2000, PASP, 112, 1496

Walters, D. L., & Bradford, L. W. 1997, ApOpt, 36, 7876

Seeing the Unseen
Mount Wilson's role in high angular resolution astronomy
Harold A McAlister

Appendix B

An Interferometry Primer

B.1 Purpose

The intention of this appendix is to provide an elementary, top-level introduction to assist in better understanding this book. I could have inserted it at various places, but it would have distracted from the theme of the book. Instead, I've put parenthetical references to sections of Appendix B at appropriate spots. More thorough presentations of interferometry are listed in Table B.1. For those with knowledge and experience in the field as well as those who simply don't wish to see such details, feel free to ignore this material.

B.2 Interference Fringes

B.2.1 Young's Double Slit Experiment

Being the starting point for astronomical interferometry, *Young's Double-Slit Experiment* is shown schematically in Figure B.1 in which two aperture masks illuminated by plane-parallel wavefronts of light from a bright, distant source produce a pattern of light and dark bands, or *interference fringes*, at right. More detail is found in Figure B.2, which shows the additive nature of the maxima and minima of the two interfering light sources upon arrival at the screen with varying path-length differences.

The dashed red lines indicate propagation of the waves perfectly in phase so as to produce successive fringe maxima at intervals in x. The mathematical relationship between the observed wavelength λ, interference order m, slit spacing d, and the angular distance α of the first-order maximum and the central fringe as seen from the slit center point is known as *Young's Double Slit Equation*.

We can easily derive that equation from inspection of Figure B.3. The red right triangle is formed by lines from the slit-pair center through the overlapping maxima of the two wavefronts illuminating the screen at right at the locations of the $m = 0$ and 1 fringe maxima. Note that the two wavefronts at $m = 0$ are perfectly in phase while for that at $m = 1$ the wavefront from the lower slit is lags by 1λ. This *path*

doi:10.1088/2514-3433/abb4dech10

Table B.1. A Partial List of Stellar Interferometry Resources

Principles of Long Baseline Stellar Interferometry: Course Notes from the 1999 Michelson Summer School, edited by P. R. Lawson, Jet Propulsion Laboratory Publication 00-009 07/000, 353 pp., 1999.

Optical Interferometry in Astronomy, by John D. Monnier, 2003, *Reports on Progress in Physics*, **66**, 789–857.

Optical Interferometry, 2nd Edition, by P. Hariharan 2003, Elsevier, 351 pp.

An Introduction to Optical Stellar Interferometry, by A. Labeyrie, S. G. Lipson & P. Nisenson, 2006, Cambridge: Cambridge Univ. Press, 360 pp.

All You Ever Wanted to Know About Optical Long Baseline Stellar Interferometry, But Were Too Shy to Ask Your Advisor, by F. Millour, 2008, HAL Archives-Ouvertes, https://hal.archives-ouvertes.fr/hal-00273465.

Astronomical Optical Interferometry I. Methods and Instrumentation, by S. Jankov, 2010, *Serbian Astron. Journ.* No. 181, 1–17, and *II. Astrophysical Results*, by S. Jankov, 2011, *Serbian Astron. Journ.* No. 183, 1–35.

Practical Optical Interferometry: Imaging at Visible and Infrared Wavelengths, by D. F. Buscher, 2015, Cambridge: Cambridge Univ. Press, 286 pp.

Astronomy at High Angular Resolution: A Compendium of Techniques in the Visible and Near-Infrared, edited by H. M. Boffin, G. Hussain, J.-P. Berger & L. Schmidtobreick, 2016, Berlin: Springer, 274 pp.

Mieux Voir les Etoiles: 1er Siecle de l'Interferometrie Optique, by Daniel Bonneau, 2019, EDP Sciences, 209 pp.

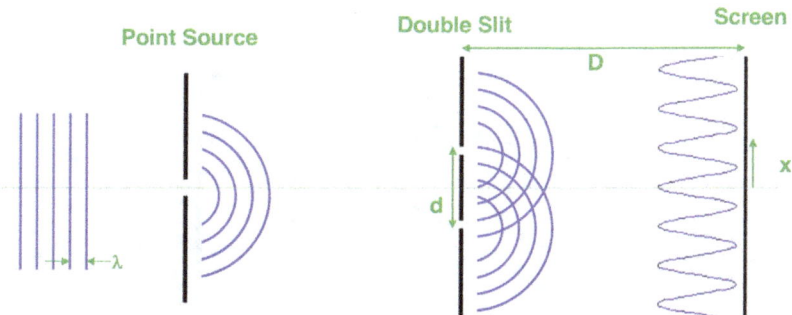

Figure B.1. Young's Double-Slit Experiment. Left to right: successive crests of incoming waves of wavelength λ from a distant source—Young used sunlight—arrive at a pinhole, creating a new point source for illumination of the double-slit mask. The slit pair produces two mutually coherent point sources from which the wave crests and troughs interfere to produce the "fringe" pattern of alternating bright and dark bands, of sinusoidal intensity, on the screen at right. (Author's diagram.)

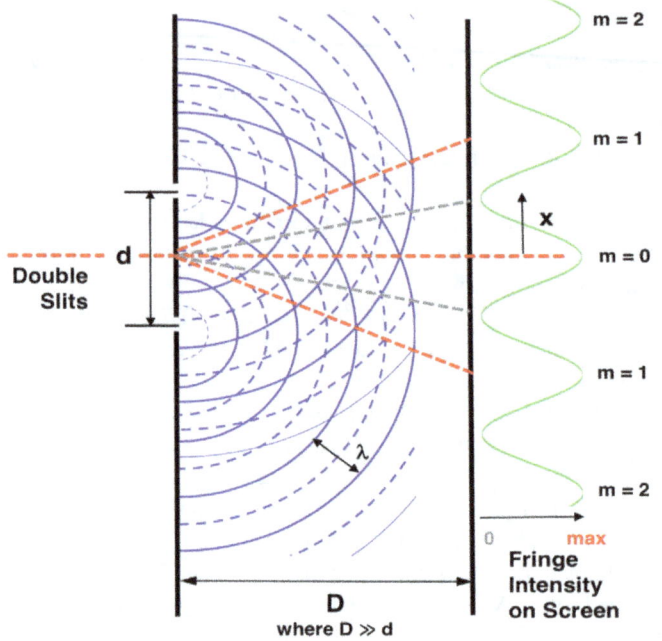

Figure B.2. Young's Fringes of alternating light and dark bands appear on the screen at right with spacing dependent on the slit spacing d, projection distance D, and the wavelength λ. The solid and dotted blue semicircles represent the maximum and minimum amplitudes of the wave fronts spreading from each of the two slits as they move toward their addition at the screen. The green sine curve shows the variable brightness of the resulting fringe pattern on the screen along with the order m of the successive fringe maxima on either side of the central fringe at $m = 0$. (Author's diagram.)

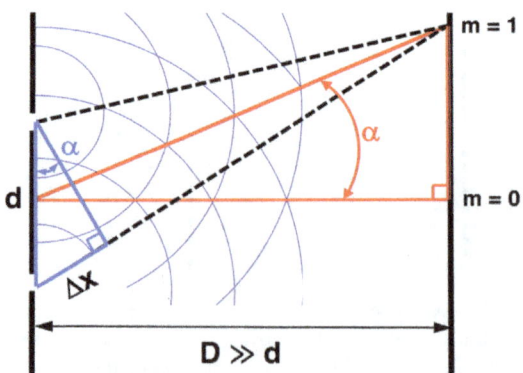

Figure B.3. Double slit equation setup results from the blue and red right triangles being similar triangles in the case of $D \gg d$. (Author's diagram.)

length difference Δx increments by λ with each step in m producing a bright fringe. Thus,

$$\Delta x = m \cdot \lambda$$

where m is the *order of interference* and is given by

$$m = 0, 1, 2, 3, \ldots$$

and the $m = 0$ fringe is called the *central fringe*. Thus, bright fringes occur when the path length difference of the paths from each slit opening to the screen are integer increments of the wavelength, or

$$0, \lambda, 2\lambda, 3\lambda, \ldots = m \cdot \lambda$$

In the limit of $D \gg d$, it is apparent that the blue triangle, whose short side ΔX equals the path length difference traveled by the two interfering beams at the $m = 1$ point, is a similar triangle with the red one. Thus, the smaller of the acute angles of the blue triangle is also equal to α.

It then follows that

$$\sin \alpha = \frac{\Delta x}{d} \quad \text{or}$$
$$\Delta x = d \sin \alpha$$
$$= m \cdot \lambda$$

from which we arrive at Young's Double Slit Equation

$$m \cdot \lambda = d \sin \alpha.$$

The fringe intensity at any point x is given by

$$I(x) = 4A^2 \cos^2\left(\frac{2\pi}{\lambda} \frac{xd}{D}\right)$$

where A is the amplitude of the light radiating from each of the double slits and $D \gg d$.

B.2.2 Effects of Finite Spectral Bandwidth

We have so far treated λ as being monochromatic when in reality we must use some finite range of wavelengths to have sufficient light to analyze. Figure B.4 shows the effect on the observed fringe in the practical case of observing over a specific spectral bandwidth of radiation to give a *polychromatic fringe*. Instead of having fringe maxima at arbitrarily large orders, the fringe dampens out from the central fringe in a manner that, in the absence of noise, is symmetric about $m = 0$. This leaves us with a *fringe packet* rather than an infinitely extend fringe. Keeping in mind that the path length difference between orders is λ, the practical challenge of actually finding the fringe packet of an interferometer is not a small one!

The resulting fringe intensity for polychromatic fringes is given by:

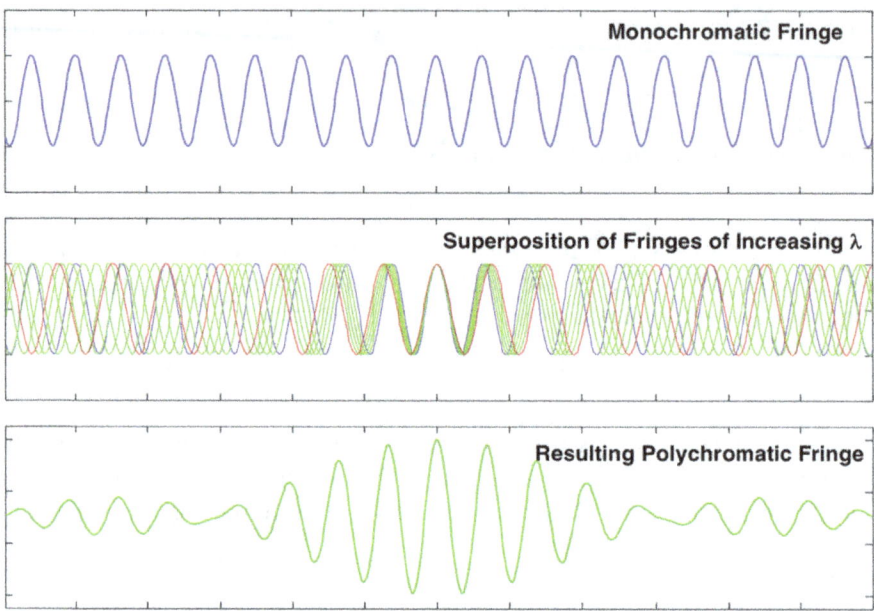

Figure B.4. We see here that the practical case of having a finite bandwidth as opposed to a spectral delta-function results in a fringe that dampens at increasing orders of m, producing a fringe packet. (Author's diagram.)

$$I(x) \approx \frac{4A^2 \sin^2 \beta}{\beta^2}$$

$$= 4A^2 \mathrm{sinc}^2 \beta \quad \text{where}$$

$$\beta = \pi \cdot \frac{\Delta\lambda}{\lambda_0^2} \cdot S$$

and S is the path length difference or *delay*. The quantity $l_c = \lambda_0^2/\Delta\lambda$ is defined as the *coherence length* and is related to the width of the fringe packet in delay space. The longer the coherence length, i.e., the narrower the wavelength bandpass, the broader will be the fringe packet as is shown in Figure B.5. The observational tradeoff is the desire for a broader fringe packet, as provided in the narrow bandwidth case, in order to find the fringe in delay space against the need for a higher flux of light in order to more accurately determine the fringe amplitude resulting from a broader bandwidth.

B.2.3 Fringe Visibility

Michelson defined the quantity

$$V = \frac{I_{\max} - I_{\min}}{I_{\max} + I_{\min}}$$

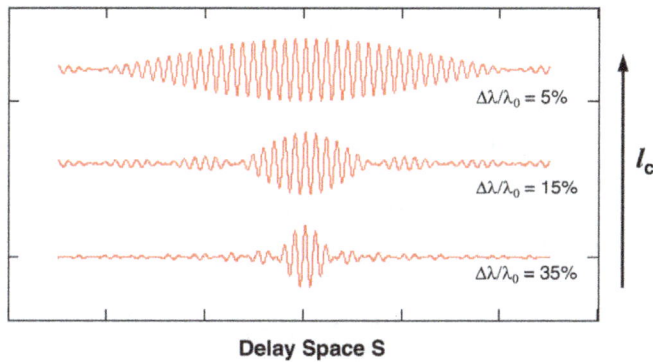

Delay Space S

Figure B.5. The broadening of the fringe packet as the spectral bandwidth narrows is demonstrated above. (Author's diagram.)

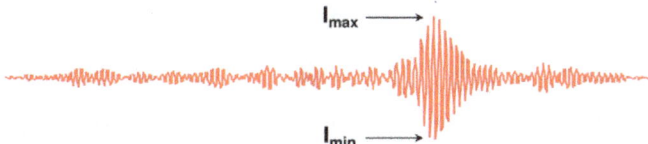

Figure B.6. The maximum and minimum of a fringe packet are indicated here. (Author's diagram.)

as the *visibility* of a fringe. This value ranges between 0 and 1 and is the basic observable of astronomical interferometry. As seen in Figure B.6, V is a normalized version of the amplitude of the central fringe of a packet.

B.3 Binary Stars

B.3.1 Basic Considerations

During the 1920s, John Anderson, Paul Merrill, and Francis Pease measured the angular separations and relative orientations of the components of binary star systems with the 20 foot and 50 foot interferometers. Although Michelson had already come up with the concept of fringe visibility and fully understood how it related to binary star geometries, there was no accurate way to quantify V other than when it went to a minimum. In the case of equal brightnesses, such as the components of Capella very nearly are, the fringe contrast would go to zero, i.e., $V = 0$. The eye and brain are sensitive to subtle contrast variations, and a careful and patient observer can vary the orientation and separation of the slits of an interferometer to home in on minimum visibility. This method was perfected and used for many years after the Mount Wilson era by W. S. Finsen in Johannesburg, South Africa.

Before we consider how binaries are observed interferometrically, I'll give a brief overview of the relevant basics of binary star astronomy. A binary star system consists of two stars that are gravitationally bound and orbit about a common center of mass—the *barycenter*—as shown in Figure B.7. At every instant, the stars are opposite the barycenter from each other by amounts obeying the relation

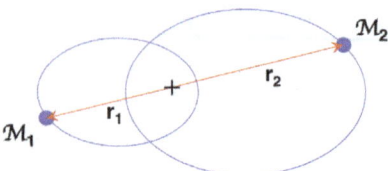

Figure B.7. Orbital geometry of the two unequal mass components of a binary with the + sign indicating the position of the barycentric. (Author's diagram.)

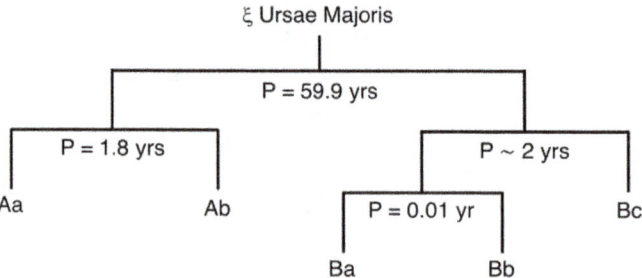

Figure B.8. Noting that the distance between components in a bound binary star is proportional to the period P to the two-thirds power (from Kepler's Third Law), this "mobile diagram" of a quintuple system shows the gravitational nesting of the five stars to achieve long-term dynamical stability. (Author's diagram.)

$$\frac{\mathcal{M}_1}{\mathcal{M}_2} = \frac{r_2}{r_1}$$

where $\mathcal{M}_1 > \mathcal{M}_2$ and $\mathcal{M}_1$ is referred to as the mass of the *primary* or *A component* while $\mathcal{M}_2$ is the mass of the *secondary* or *B component*. Note that a script $\mathcal{M}$ is typically used for stellar mass to distinguish it from a star's absolute magnitude M. Gravitationally-bound star systems can have more than two stars, and the highest order of multiplicity known is seven. In all cases, *multiple stars* are hierarchical in nature in order to be dynamically stable. For example, a distant third component can be in a stable orbit with a close binary pair, or two close pairs can orbit each other at a considerable distance, etc. In Figure B.8 we show a "mobile diagram" for the quintuple system comprising the star ξ Ursae Majoris.

The terms *binary star* and *double star* are casually used as synonymous, but chance near alignments along the line of sight make some stars at vastly different distances masquerade as binaries. Such objects are called *optical double stars*. Thus, we may generalize that all "binary stars" are physical while some "double stars" are imposters. They are a nuisance to binary star statistical studies and can be unmasked by spectroscopic and photometric observations that show they have discordant distances.

The next level of nomenclature for a particular binary star depends upon the technique used to discover it. For example, those seen to be two separate stars through a telescope are called *visual binaries* while others discovered from having variable radial velocities indicative of orbital motion are *spectroscopic binaries*, the great majority of which are unresolvable by the eye. Similarly there are *photometric (or eclipsing) binaries, composite spectrum binaries, astrometric binaries, etc.*

Another important descriptive parameter for binaries is their brightness difference, which is estimated by visual observers and can be determined quantitatively by other means, including interferometry. If the *magnitude difference* Δm is known along with the total brightness of both components $m_{(1+2)}$ (or equivalently $m_{(A+B)}$), then a handy equation for calculating the individual magnitudes of either component is given by:

$$m_{1,2} = m_{1+2} + 2.5 \log(1 + 10^{-,+\Delta m})$$

where $\Delta m = m_2 - m_1 > 0$.

B.3.2 Observational Aspects

Visual binary stars are systems for which both components can be seen and measured in the polar coordinates *angular separation* ρ (in seconds of arc) and *position angle* θ (in degrees). The *primary* or brighter star of the pair is fixed at the origin of the coordinate system with the *secondary* (fainter) star imagined to be orbiting the primary. In reality, of course, both are orbiting the barycenter under the circumstances shown in Figure B.7. A complete measure of a visual binary consists of a dataset containing (ρ, θ, t, Δm) where t is the epoch of the observation and Δm is an estimate of the brightness difference in magnitudes. Figure B.9 shows the polar geometry for visual binary stars, and Figure B.10 is a schematic of a *filar micrometer*, the instrument mounted at the focus of a telescope and used to obtain measurements of ρ and θ.

However, interferometers employing a slit pair—like Anderson's device—have resolution only in the direction of the line adjoining the slits. Interferometers do not measure ρ and θ explicitly. Instead they yield a value of visibility that is a response to the projected separation of the pair of stars onto the line adjoining the slits, which is referred to as the *baseline* for a long-baseline interferometer with separate telescopes acting as a slit pair. This projection is shown in Figure B.11.

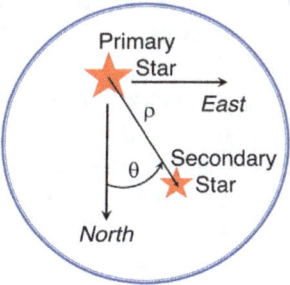

Figure B.9. Observational data for resolved double star systems consists of the angular separation ρ between the two stars and the position angle θ of the line from the brighter or primary star to the fainter secondary star, measured eastward from north for a particular epoch of time. By convention, the orientation is shown inverted as viewed through a telescope, i.e., north is down, east is to the right. (Author's diagram.)

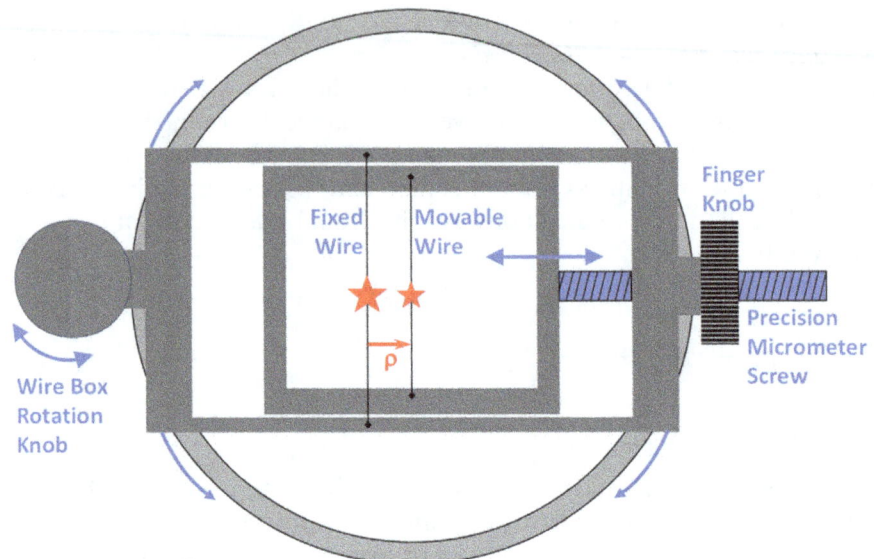

Figure B.10. Schematic diagram of a filar micrometer mounted in the focal plane of a telescope. It superimposes two wires onto the images of the resolved binary star as observed through a high-power eyepiece. The position angle is measured by rotating the micrometer so that the fixed wire is parallel to line joining the two stars. The "wire box" is then rotated 90° with the "fixed wire" bisecting the primary star. The position of the "movable wire" is then set to bisect the primary by the "micrometer screw." The values of ρ and θ are recorded from and engraved scale for position angle and the position of the micrometer screw for older micrometers or from encoders that register the values for the last models of micrometers built ca. 1960–1980. (Author's diagram.)

In Figure B.12, we see how the angular separation of the stars in a binary system results in a tilt in the wavefronts from the two stars, color coded in red and blue, each of which will produce fringes displaced on a Young's screen by an amount related to the stars' angular separation ρ. In this example, five specific slit-separation settings are made, as shown below the incoming wavefronts. The corresponding responses of the interferometer, each of which shows the relative displacement of the red and blue fringes along with their summation shown by the resultant black net fringe. We see in this example the diminishment of contrast as d increases to near 0 at step 4 and then climbs out of the null at step 5.

So, if we decide that obs_4 is our best bet at a null, we select slit separation d_4 as the nulling value. What then? The answer to this is simple. The net fringe null occurs when the individual fringes are 180° out of phase or half a fringe apart. Thus, we can define an interferometer's limiting resolution for binary stars as

$$\rho_{\min} = \frac{\lambda}{2d}$$

from which we see that increased resolution is achieved by decreasing λ and/or by increasing d.

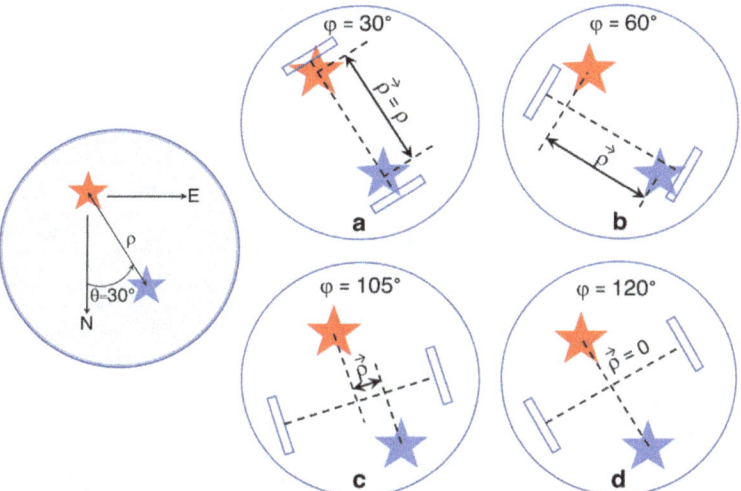

Figure B.11. (Left) In this example, we see how an interferometer responds to the resolution of a binary star with $\theta = 30°$. (Right) Four projections of ρ onto slit baseline in an Anderson-type interferometer show how that projection depends upon φ, the position angle of interferometer slit pair or baseline projected onto the plane of the sky. The projection is maximum as in frame a when $\varphi = \theta$ and 0 when φ and θ are perpendicular as in frame d. Note that from the symmetry of the slits, the visibility curve as a function of φ repeats itself after 180° of rotation. (Author's diagram.)

Noting that resolution only exists along the line adjoining the slits, when we find a null, we are finding the component of the true separation vector $\vec{\rho}$ described by the combination of angular separation ρ and position angle θ projected onto the slit-separation vector. Thus, the angular separation we calculate from d_4 is actually the dot product of the slit vector and the binary's separation vector, or $\rho_d = \vec{d_4} \cdot \vec{\rho}$. Measurements made at multiple baselines and slit position angles can produce the values of ρ and θ as inputs to the determination of orbital elements that may then provide stellar masses if complementary data are available as we shall see is the case of a resolved double-lined spectroscopic binary.

B.3.3 Orbital Elements

An observation of a binary star consists of (ρ, θ, t) where t is the time of the observation. A series of such data triplets can be fit by a variety of techniques to yield the orbital elements of the system. These seven elements uniquely describe the orbit in its true plane that is inclined to the plane of the sky onto which ρ and θ are projected as shown in Figure B.13. Keep in mind that the orbit determined from interferometric measurements is considered a "visual" orbit because interferometers measure ρ and θ in angular units as do classical visual micrometers. Some distinctions between visual orbits and spectroscopic orbits are pointed out in the definitions of the orbital elements.

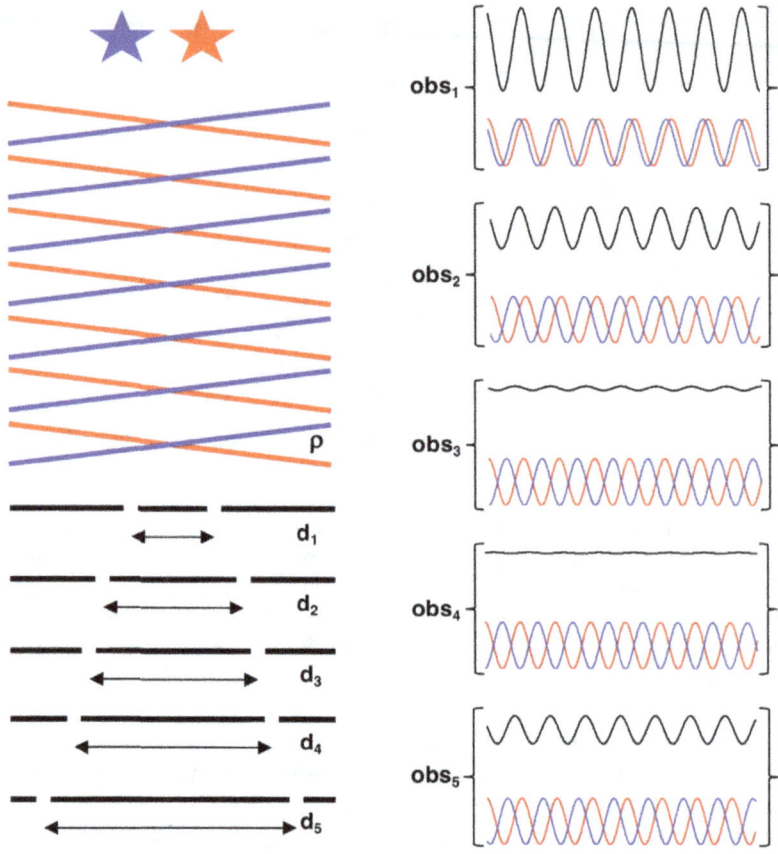

Figure B.12. The effect of increasing double slit spacing d (left) is shown (at right) on the summation of the two fringe packets arising from the components of a double star. The observer sees the summation—in black—of the two fringes and notes the fringe null or disappearance at spacing d_4 by seeing no fringe modulation in obs_4. (Author's diagram.)

The orbital elements are:

P = *period of orbital revolution*—A frequently used related quantity is the "mean motion" defined as $\mu = 360/P$. P is typically measured in days for spectroscopic binaries and years for visual binaries, which includes what might otherwise be called interferometric binaries. A convenient link between days and years is provided by the "Besselian Year" (based upon one complete circuit in right ascension of the "mean Sun") and the "Julian Date" (the interval of time in days since 4713 BC Jan 1). The two are related by $JD = 2,433,282.423 + 365.2422(BY - 1950.0)$.

a = *semimajor axis of the orbit*—This is measured in the plane of the orbit, not the sky. For visual binaries, a is in angular units as the sum of the semimajor axes of the two components while spectroscopic binaries measure the projection of a—in linear not angular units—in the line of sight for the individual components relative to the barycenter.

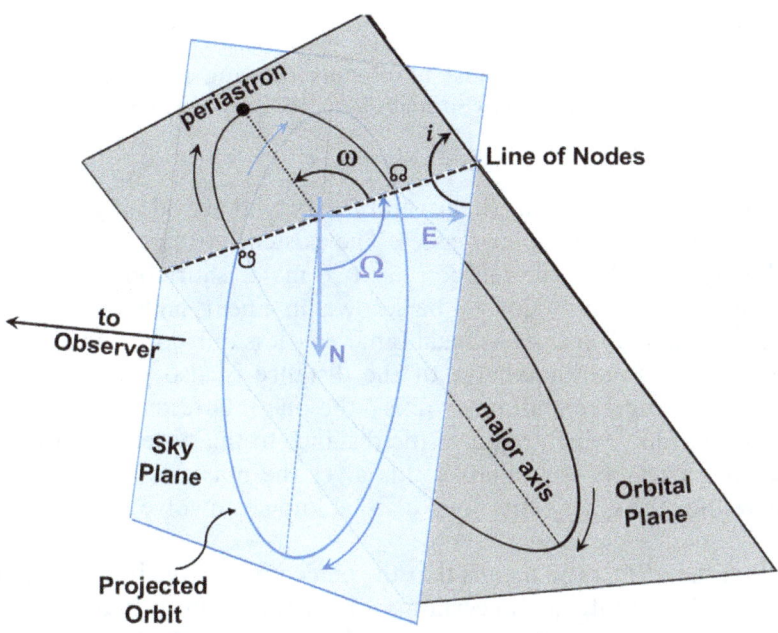

Figure B.13. The projection of the true orbit of a binary star system onto the sky and the geometric elements specifying that projection are shown here. (Author's diagram.)

i = *inclination of the orbital plane with respect to the plane of the sky*—For "direct" motion in the orbital (θ increasing with time), $0° < i < 90°$ whereas for "retrograde" (θ decreasing with time), $90° < i < 180°$.

T = *epoch of periastron passage*—For circular orbits, T is often taken as a passage through a "node" or point of intersection between the planes of the orbit and the sky.

e = *orbital eccentricity*—For closed orbits, 0 (circular) $< e < 1$ (parabola), and the periastron (point of closest approach to barycenter) and apastron (farthest from barycenter) distances are given by $\rho_{P/A} = a(1 \pm e)$.

Ω = *position angle of the line of nodes*—This is measured in the plane of the sky. There are two nodes differing in θ by 180° and we adopt the "ascending node" (radial velocity > 0) if known; otherwise take $\Omega < 180°$. This is subject to precession and hence must be quoted for a specific equinox. The symbols for the ascending and descending nodes are an upright and inverted Ω, respectively.

ω = *longitude of periastron*—This is measured in the plane of the orbit from the node specified by Ω to the periastron point measured in the direction of orbital motion. In the case of $e = 0$, T is usually taken as the time of periastron passage setting $\omega = 0$. For visual binaries, ω is measured with respect to the secondary while for spectroscopic binaries it is measured with respect to the primary, causing the two values to differ by 180°.

B.3.4 Stellar Masses

Kepler's Third Law applies not just to planets orbiting the Sun but to any two gravitationally-bound, mutually-orbiting bodies. We write it here as

$$\mathcal{M}_1 + \mathcal{M}_2 = a^3/P^2$$

where a = orbital semimajor axis (in au) and P = orbital period (in years). The values of M will then be in units of solar masses. The existence of this law is basically why we study binary stars. But, the relation has two major shortcomings:

- It requires the semimajor to be known in linear units, and for the most common class of binaries—visual binaries—it is only known in angular units. Thus, an accurate knowledge of the distance is also generally required to convert the angular semimajor axis—the one interferometry determines—to the linear value, which requires the distance to the system to also be known.
- Even if a parallax is available, this gives the mass sum, not the individual masses. So, the mass ratio must also be known to solve for them.

Note also that the a^3/P^2 ratio means that the percentage errors in the determinations of a and P propagate into the uncertainty of the mass sum increased by factors of 3 and 2 respectively. So, we need really good data to get good masses.

We are left with the problem of the missing mass ratio in order to solve for the individual masses. Unfortunately, the discovery regimes of the various techniques have little overlap, so the majority of binary stars can only be studied by one technique. Historically, binaries that can be observed spectroscopically and have orbital planes in the line of sight have angular separations too small to be resolved by visual methods. Similarly, the latter have long periods and small variations in radial velocities undetectable by traditional radial velocity methods. so as to produce eclipses can also be studied photometrically. Combining orbital parameters from the two methods reveals the masses and even the relative diameters of the components.

Since interferometers provide very-high angular resolution and modern radial velocity techniques have exquisite precision, the overlap of the two methods is now quite broad. And therein lies a wonderful outcome—we get the masses of each component star <u>and</u> their distance from us. This is what John Anderson and Paul Merrill did with Capella. The current generation of long-baseline interferometers, like Georgia State's CHARA Array on Mount Wilson, can also measure the diameters of the components. Let's dig a little deeper here.

Spectroscopically determined radial velocities show the variation in V_r with time, which look like distorted sinusoids with velocity amplitudes of k (in km s^{-1}). If these variations can be measured for both stars in a system, we then have $V_r(t)$ for each component and hence values of k_1 and k_2. Such binaries are called "double-lined spectroscopic binaries" (DSBs). We can then take advantage of the fact that $\mathcal{M}_1/\mathcal{M}_2 = a_2/a_1 = k_2/k_1$, and there we have it—the mass ratio and with that, we solve for $\mathcal{M}_1$ and $\mathcal{M}_2$.

But, there's more. The velocity amplitudes are related to their respective values of a by

$$a_{1,2} \sin i = 13{,}751(1 - e^2)^{0.5} k_{1,2} P$$

in km if P is in days and k is in km s^{-1}. Since for DSBs we have both k_1 and k_2, and we have i from the interferometry (and e from both methods), we can solve for the two a values. We then use the fact that $a = a_1 + a_2$ to get the linear value that links to the angular value a of the semimajor axis by the distance to the system as $\pi = 1/d = a/a$. And, voila, we have the *orbital parallax* of the resolved, double-lined spectroscopic binary. A measurement of the magnitude difference provided by the interferometry leaves us with the luminosities as well as the masses of the two stars. This elegant process of combining two complementary methods of observation provides two of the most important parameters we can measure for stars that can then be used to constrains models of stellar formation and evolution.

B.4 Visibility Relations

Interferometers measure parameters of stars, such as the *angular diameter* Θ of a single star and θ and ρ for a binary star, by measuring values of visibility at a variety of baselines and then fitting data points to the appropriate visibility relations. The reduction and calibration of instrumental visibilities requires careful processing as will be described in Section B.5. First, let's see how those value relate to measurements of star parameters.

B.4.1 Response to Stellar Angular Diameters

Visibility as a function of baseline d, wavelength λ, and the uniform disk angular diameter Θ_{UD} are related by the expression

$$V(d) = \frac{2[J_1(\pi \Theta_{\mathrm{UD}} d / \lambda)]}{(\pi \Theta_{\mathrm{UD}} d / \lambda)}$$

where J_1 is the first-order Bessel function and Θ is measured in radians. Having only a single unknown, the fitting of this relation to a set of observed visibilities is a simple process.

You will note from the two curves in Figure B.14 that larger stars are, as expected, resolved at shorter baselines than are smaller stars. Similarly, if the observations had been made at shorter wavelengths, both curves would shift to the left.

The correction from a uniform disk to a radially-symmetric, limb-darkened disk involves a more generalized Hankel transform relation than that of the previous equation:

$$V(d) = \frac{|\int I(r) J_1(xr) r \, dr|}{\int I(r) r \, dr}$$

where $x = \frac{\pi \Theta d}{\lambda}$. Rather than fitting to this more complicated equation, which involves a model-dependent surface brightness distribution $I(r)$, various approximations have been calculated. Particularly useful are the limb- and gravity-

B-14

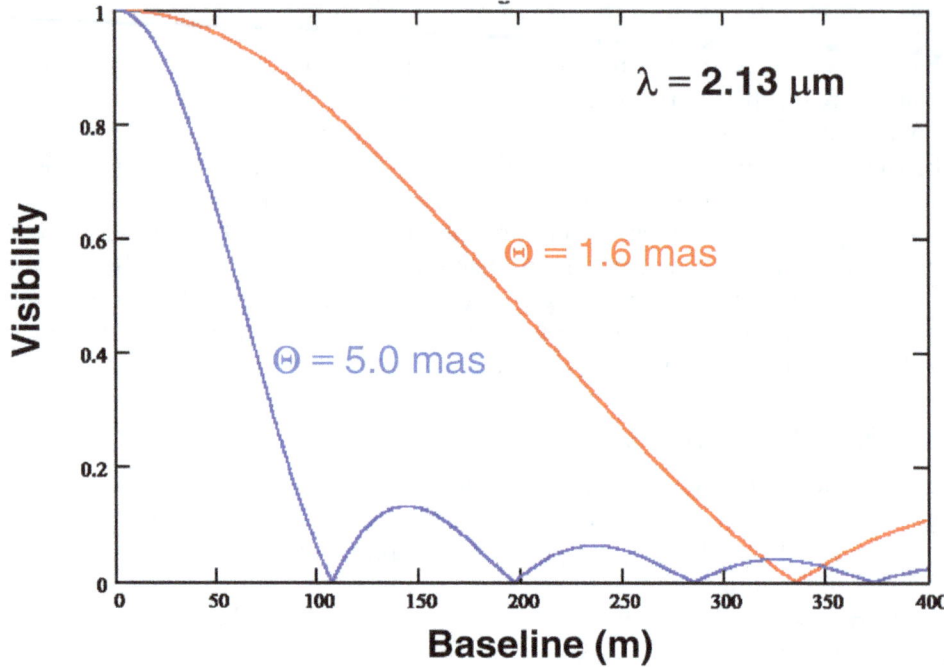

Figure B.14. Visibility curves are shown for two values of Θ measured at $\lambda = 2.2$ mm for two values of Θ. (Author's diagram.)

darkening (which relates to equatorial darkening that arises for rapidly rotating and oblately distorted stars) corrections modeled over the years by Antonio Claret, who has a series of publications presenting tables of coefficients within grids of metallicity, surface gravity, and effective temperature (e.g., Claret 2020).

B.4.2 Response to Binary Stars

The visibility relation for binary stars is naturally more complicated that for single stars as it must accommodate the presence of two stellar disks, generally of unequal brightness, and their relative positions. In visibility squared form, the relation is:

$$V^2(d) = (1 + \beta)^{-2}\{\beta^2 V_1^2(d) + V_2^2(d) + 2\beta V_1 V_2(d)\cos[2\pi\beta\lambda^{-1}\rho\cos\psi]\}$$

where for a magnitude difference Δm, $\beta = 10^{0.4\Delta m}$, $\psi = |\theta - \varphi|$, and with ρ, θ, and φ as already defined. The visibilities arising from the angular diameters of each component star are given by:

$$V_{1,2}(d) = \frac{[J_{1,2}(\pi\Theta_{1,2}d/\lambda)]}{(\pi\Theta_{1,2}d/\lambda)}$$

where the subscripts 1 and 2 refer to the primary and secondary stars.

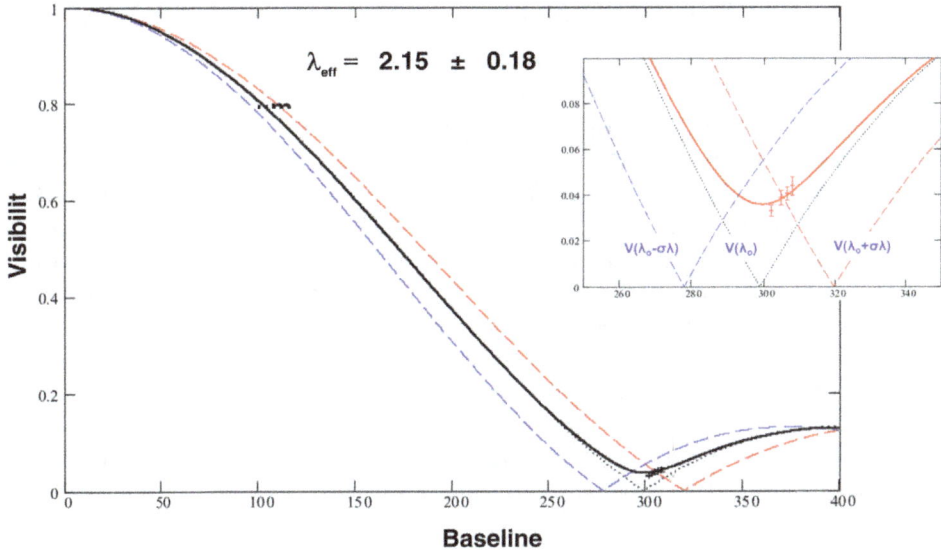

Figure B.15. Taking fringe data over a finite bandwidth results in a smearing of baselines that will yield the wrong value for the target star's diameter if not considered. The inset shows actual visibilities from the CHARA Array measured at values of 0.04 using a filter with a bandwidth of ~8% of its effective wavelength in the K-band near infrared. (Author's diagram.)

B.4.3 Baseline Smearing

Because of the finite width of the observed bandpass $\lambda_o \pm \sigma_\lambda$, one must correct angular diameter fits for a smearing effect that is especially pronounced near nulls. The effect of a finite bandwidth results in *baseline smearing* at the wavelength center. As can be seen in the inset to Figure B.15, the assumption of a monochromatic wavelength will result in an incorrect diameter fit. Observations near the null in visibility are sensitive to stellar limb darkening and can provide an understanding of the observed star's radial surface brightness distribution.

While on the subject of observational spectral bandpass, if one's goal is to measure stellar effective temperatures to an accuracy of $\pm 1.0\%$ or better, it is also necessary to correct the wavelength specified on the manufacturer's interference filter for the spectral responses of the atmosphere, optical surfaces transmissions and reflections, and the detector spectral response as well as for the flux distribution of the star being observed.

B.5 Reducing Interferometry Data

Let's walk through the various steps involved in taking data with an interferometer and reducing it to calibrated visibilities from which can be extracted, for example, stellar diameters and effective temperatures, and binary star measurements for mass determinations. Although the contemporary optical interferometers are all multiple telescope arrays, we'll consider a simple interferometer of two telescopes. After all, if you can do this with two, you can do it with 20. All it takes is money!

B.5.1 Optical Path Length Compensation

Let's say we wish to measure the diameter of a star 100 lt-yr from the Sun. Picture a wave front of starlight having left our target a century ago, and now a portion of it is intercepted by one of the telescopes of our long-baseline interferometer. Let's say the pair is oriented east–west, and the star is well east of the celestial meridian. As is shown in Figure B.16, that tiny disk of starlight is gathered by the eastern telescope —#1 in the diagram. A ten-millionth of a second later, that same wave front arrives at Tel. #2. Multiple optical surfaces relay the two pieces of wave front to a point of intersection in a beam splitter where fringes are formed that will lead to your knowing the star's diameter and temperature. Sounds easy—but, here's the rub. In order for the two patches of starlight to produce fringes rather than just pool their light together, the two patches must have traveled the same distance—matched to about the length of a typical bacterium—from their point of first interception by Tel. #1 down the entire optical train until they reunite in the interfering beam splitter. Inspection of Figure B.16 shows that to achieve mutual coherence of the two beams,

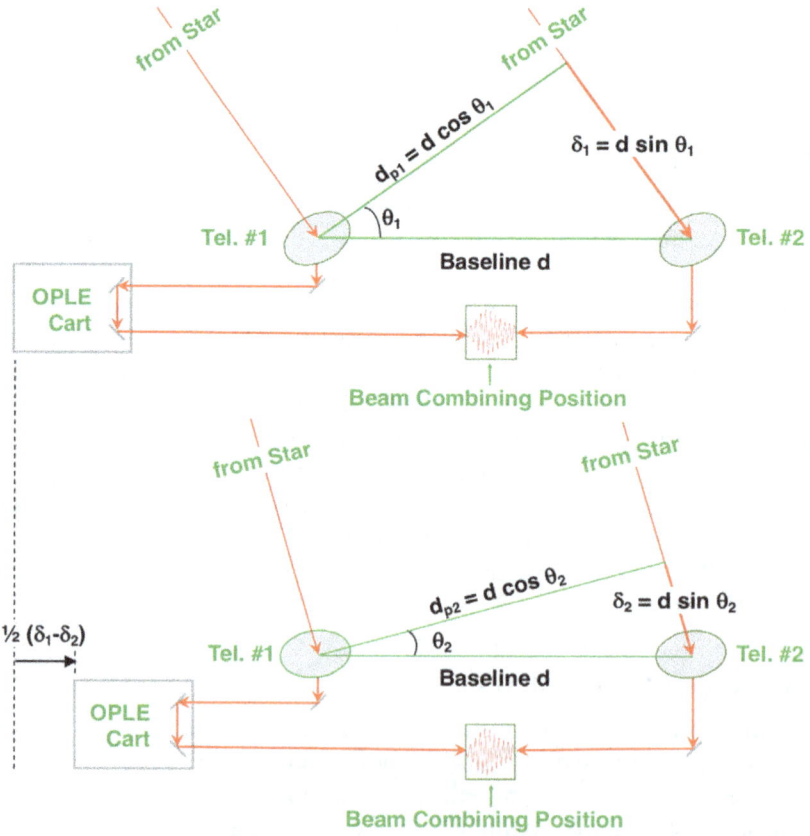

Figure B.16. The correction of the optical path length variation arising from Earth rotation is made with an "optical path length equalizer" moving on precision rails as schematically shown above. (Author's diagram.)

and *optical delay* δ must be inserted into the light path from Tel. #1 so that it doesn't arrive early and fringes are nowhere to be seen.

And this leads us to the second rub. Because of Earth rotation, the optical delay is constantly changing by an amount in this idealized figure equal to

$$\frac{d\delta}{dt} = d\,\cos\theta\frac{d\theta}{dt}\text{or}$$

$$\dot{\delta} \simeq 0.5\,d\,\cos\theta$$

$$\simeq 88\cos\theta\ mh^{-1}\text{ for the longest baseline of the CHARA Array}$$

$$= 2.4\cos\theta\ \text{cms}^{-1}\text{, so that}$$

$$\dot{\delta}_{max} = 2.4\ \text{cms}^{-1}.$$

This is a very large overhead burden for a long-baseline interferometer. For example, the CHARA Arrays longest baseline between a pair of its six telescopes is 330 m, which gives rise to a maximum speed of 2.4 cm s^{-1} for the "OPLE Cart" shown in Figure B.16. OPLE stands for Optical Path Length Equalizer, and the catseye mirror double-pass retroreflector—not the two turning mirror arrangement shown for simplicity in the figure—travels on high-precision, stainless-steel rails in an "optical trombone" arrangement.[1]

B.5.2 Extracting Visibility

The "Interferometry 101" reduction and analysis process is not particularly complicated, and some readers may find it interesting to go through the details to see how one gets from starlight entering a pair of interferometer telescopes to knowing the angular diameter of that star. So here goes. We will adopt the procedure described in detail by Jim Benson, Mel Dyck, and Bob Howell (Benson et al. 1995), all then at the University of Wyoming, and show its application to a single scan from the CHARA Array. In order to get a personal feel for the then new interferometer we had built and were commissioning in the early 2000s, I wrote my own reduction code using Mathcad. The real reduction pipeline for CHARA was developed by Theo ten Brummelaar, now the Director of the CHARA Array on Mount Wilson.

In the previous subsection, we visited the requirement for optical path length equalization, so let's assume we've done that (and a few other things) correctly and we are ready to interfere light from a pair of telescopes that are both pointing at and tracking a star whose diameter we wish to know. In Figure B.17 you will find a schematic view of the "CHARA Classic" (we're based in Atlanta, home to Coca Cola—get it?) beam combiner. This system, which operated in the K-band near infrared spectral region, was CHARA's primary pair-wise beam combiner during much of its science operation. The components of Classic are described in the

---

[1] For a detailed overview of the various subsystems of the CHARA Array, this is the go-to paper: ten Brummelaar et al. (2005), First Results from the CHARA Array. II. Description of the Instrument.

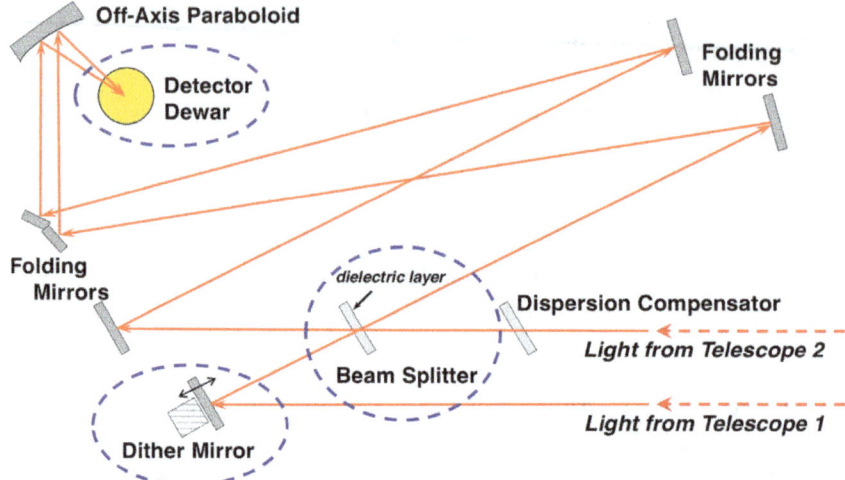

Figure B.17. The CHARA Classic beam combiner for the K-band near infrared receives collimated light beams from two telescopes for which the optical path delay has been compensated to ensure the path difference is very close to zero at the dielectric layer of the beam splitter where interference then occurs. The "Dither Mirror" is driven back and forth using a piezoelectric crystal stack to scan beam #1 back and forth with respect to beam #2. This is equivalent to the dielectric layer scanning the point of interference across the fringe envelope with the resulting intensity variations recorded by the detector in the nitrogen cooled dewar. Note that, because beam #1 enters the beam splitter from the glass side, an identical flat window is placed in beam #2 so that it also passes through the same thickness of glass upon interference. This ensures that both beams suffer identical dispersions upon interference. (Author's diagram.)

figure's caption. The beam splitter is the key to the rest of this narrative as it is where the interference occurs.

So, let's start from there by looking at a closeup in Figure B.18. Entering the beam splitter to be interfered are the two input beams from telescopes #1 and #2. At the dielectric, components of each beam are transmitted and reflected, and their averaged intensities emerge from sides A and B where

$$\langle I_A \rangle = G_A \langle I_{1R} + I_{2T} \rangle \text{ and}$$
$$\langle I_B \rangle = G_B \langle I_{1T} + I_{2R} \rangle$$

with the subscripts T and R refer to the reflected and transmitted components of I_1 and I_2. The dither mirror scans in x along the beam direction by sawtooth varying the delay of I_1 with respect to I_2 by an amount on either side of the predicted fringe position to ensure detection of the fringe. The intensities of the separately detected A- and B-side signals are given by:

$$I_A(x) = 1 + \left\{ \frac{[2V(I_1 I_2)^{0.5}|r||t|]}{[I_1 |r|^2 + I_2 |t|^2]} \right\} \text{sinc}(\pi \Delta \sigma x)\cos(2\pi \sigma_o x + v) \text{ and}$$

$$I_A(x) = 1 - \left\{ \frac{[2V(I_1 I_2)^{0.5}|r||t|]}{[I_1 |r|^2 + I_2 |t|^2]} \right\} \text{sinc}(\pi \Delta \sigma x)\cos(2\pi \sigma_o x + v)$$

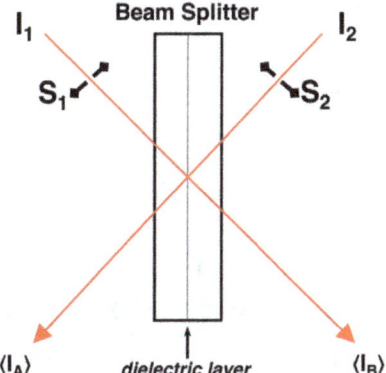

Figure B.18. Upper: beams of intensity I_1 and I_2 arrive at the beam splitter from telescopes 1 and 2. The shutters S_1 and S_2 are used to separately measure signal levels as described in the text. Lower: The emerging beams I_A and I_B each have reflected and transmitted components of interference, and I_A and I_B are averages of those fringe scans. (Author's diagram.)

where G is the detector gain, r and t are the complex reflection and transmission coefficients of the beam splitter, σ is the wavenumber $1/\lambda$, x is the path difference between the two beams, v is a phase term caused by instrumental and atmospheric effects, and V is the visibility of the object.

At the zero path difference—the fringe center—the previous two equations simplify to:

$$I_A(x) = 1 + \frac{[2V(I_1I_2)^{0.5}|r||t|]}{[I_1\,|r|^2 + I_2\,|t|^2]} \text{ and}$$

$$I_B(x) = 1 - \frac{[2V(I_1I_2)^{0.5}|r||t|]}{[I_1\,|r|^2 + I_2\,|t|^2]}.$$

These can be further simplified by introducing the following defined quantities:

$$\alpha = I_2/I_1 \text{ and}$$
$$\beta = |r|^2/|t|^2 = R/T,$$

which leave us with:

$$I_A(x) = 1 + \frac{[2V(\alpha\beta)^{0.5}]}{[\alpha + \beta]} \text{ and}$$

$$I_B(x) = 1 - \frac{2V(\alpha\beta)^{0.5}}{(1 + \alpha\beta)}.$$

Taking the difference of these two equations gives:

$$I_A(x) - I_A(x) = 2V(\alpha\beta)^{0.5}[1/(\alpha + \beta) + 1/(1 + \alpha\beta)],$$

which can be solved for the visibility and rewritten as:

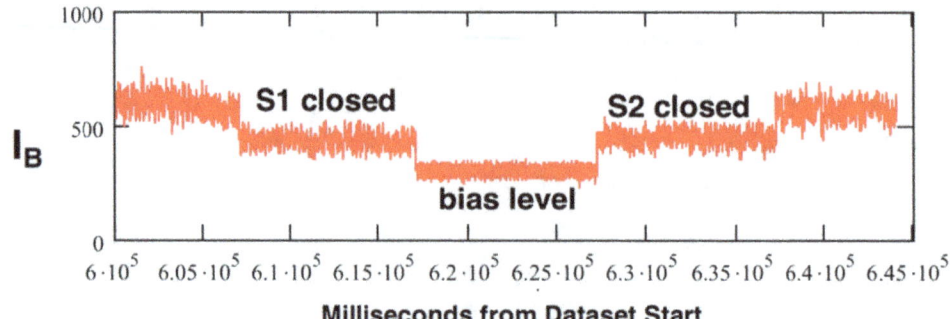

Figure B.19. The effects of shuttering from either and both sides of the beam splitter is shown here. The "bias level" is the detector noise. (Author's diagram.)

$$V = 0.5[I_A(x) - I_B(x)](\alpha\beta)^{-0.5}[1/(\alpha + \beta) + 1/(1 + \alpha\beta)]^{-1} \text{ or}$$
$$V = 0.5\Gamma[I_A(x) - I_B(x)] \text{ where}$$
$$\Gamma = (\alpha\beta)^{-0.5}[1/(\alpha + \beta) + 1/(1 + \alpha\beta)]^{-1}.$$

We go to all this trouble because the parameters we've defined above can be measured using shutters S_1 and S_2 on each of beams I_1 and I_2, as shown in Figure B.18, and taking data with them opened and closed. Figure B.19 is an example of how the shutters affect the signal level. When both are closed, the "bias" level and noise of the detector are measured. When shutter S_1 is closed, the signal reaching the A-side detector is I_2T while that reaching detector B is I_2R. Similarly, when S_2 is closed the A detector sees I_1R and B sees I_1T. In this instance, the detector was a relatively early InSb device with considerable noise. These procedures allow us to use the shuttered signal levels to calculate the following quantities:

$$\alpha = I_2/I_1 = I_{AS1}/I_{BS2} = I_{BS1}/I_{AS2}$$
$$\beta = R/T = I_{AS2}/I_{BS2} = I_{BS1}/I_{AS1}$$
$$\alpha\beta = (I_2/I_1)(R/T) = I_{BS1}/I_{BS2} \text{ and}$$
$$\alpha/\beta = (I_2/I_1)(T/R) = I_{AS1}/I_{AS2}.$$

We now have what we need to extract estimates of visibility from a series of hundreds of dither mirror scans. The first step is to slice up a dataset into the individual scans and prepare those for Fourier filtering by packing the ends of each scan with means before fast Fourier transforming them as is shown in Figure B.20.

Next, to correct for low-frequency seeing fluctuations, a 15 Hz filter is applied to remove those components, which otherwise distort the fringe packet. The straightened signal is then normalized by dividing each data point by the scan mean giving the result in Figure B.21.

Because the A and B scans have reversed signs, the subtraction of B from A (Figure B.22) gives the combined signal that is ready for the final step in fringe processing. Here, the Fourier transform is inspected to locate the central spatial frequency of the fringe, and a 50 Hz-wide bandpass filter is applied to remove the low-frequency seeing

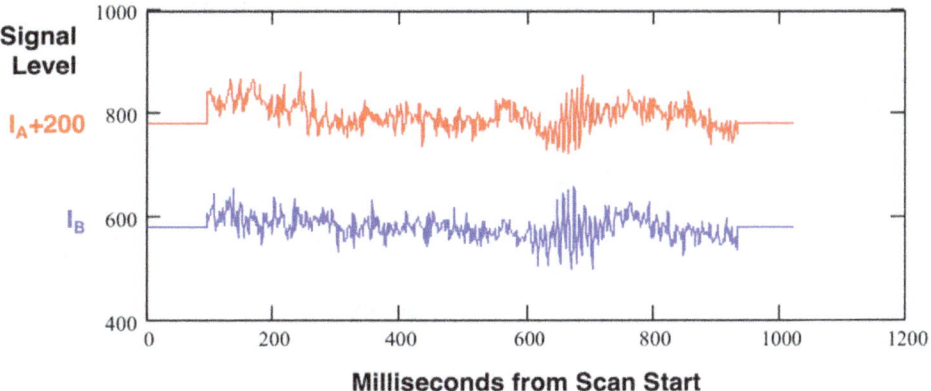

Figure B.20. Individual A- and B-side fringe scans after slicing. (Author's diagram.)

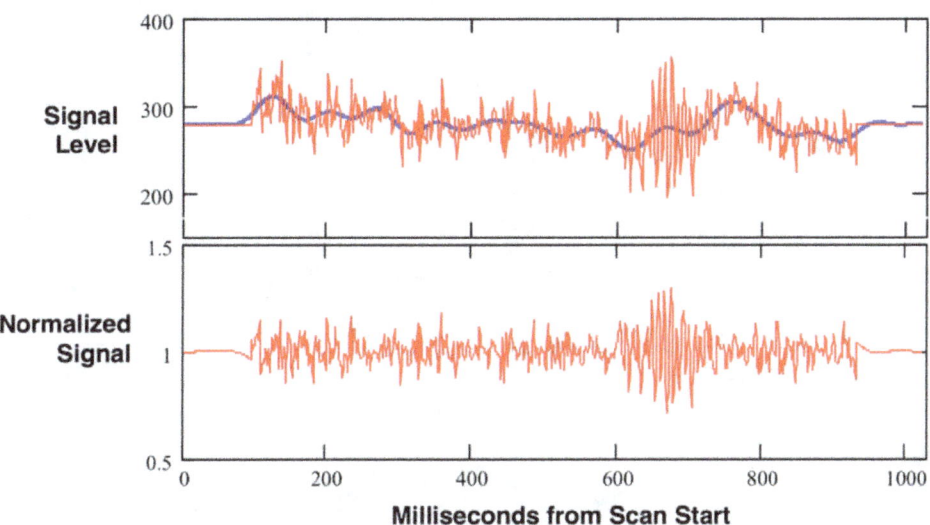

Figure B.21. A fringe scan before and after the low-frequency seeing correction is applied and the fringe normalized. (Author's diagram.)

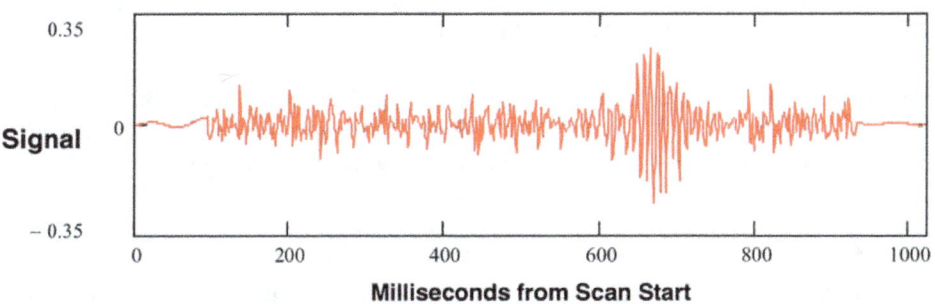

Figure B.22. The net scan obtained by subtracting scan B from A. (Author's diagram.)

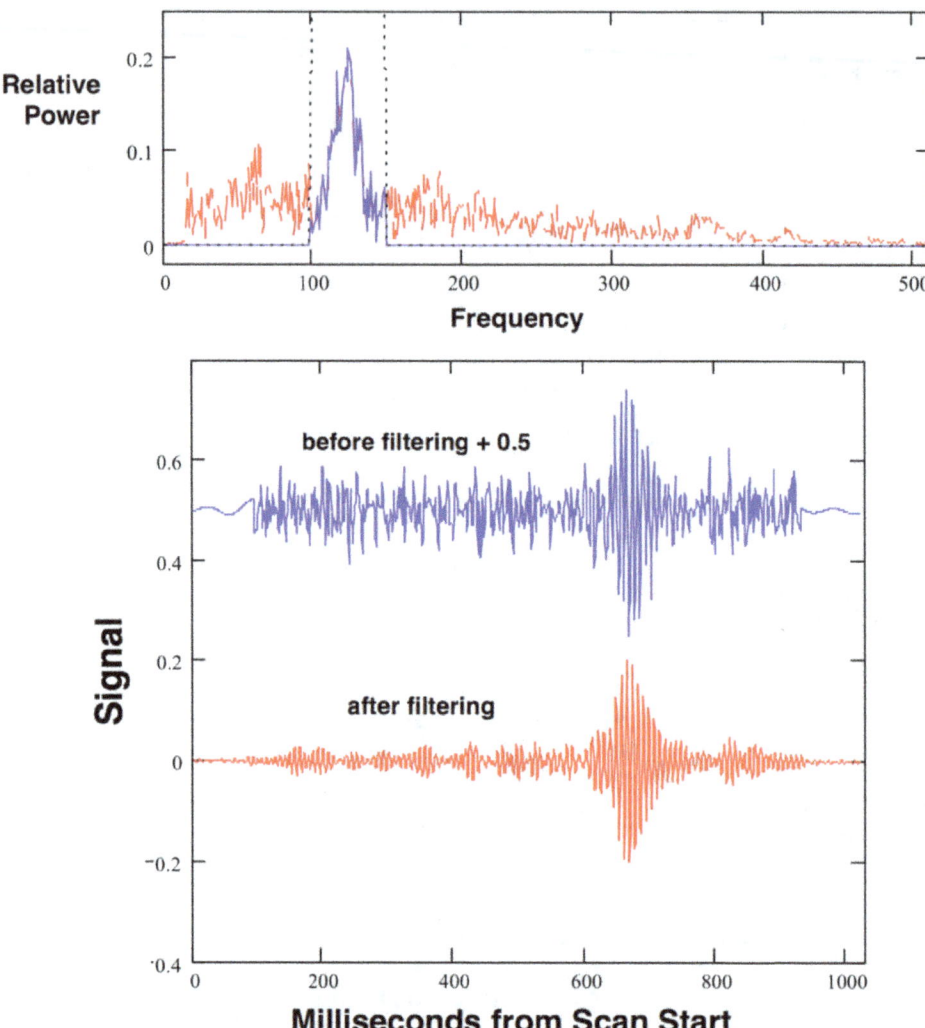

Figure B.23. (Above) The location of the selected 50 Hz wide, bandpass filter is shown for a fringe scan. The spatial frequencies outside the filter will be omitted, and the resulting inverse transform (middle) are the subtracted, net fringe. Last is the smoothed fringe after application of the bandpass filter (bottom)—ready for visibility fitting by the $V(x)$ relation. (Author's diagram.)

remnants and the high-frequency detector noise fluctuations. Figure B.23 contains an example of this bandpass filter, and its effect on the subtracted net fringe. The final product to this point is the smoothed scan at the bottom of the figure.

B.5.3 Calibrating Visibilities

This is important and worth bearing down on. An interferometer measures an "instrumental visibility" V_i—sometimes referred to as "raw visibility" or simply the "correlation." This value is always diminished in comparison with the true visibility presented by the object due to atmospheric effects, instrumental vibrations, optical aberrations, and things we are probably not thinking of.

To convert V_i to a "calibrated visibility" V_c—aka the "true visibility" or just "visibility"—we compare the instrumental visibility of the target/science star with that of a "calibrator star" selected on the basis of having a known or true visibility assuming that:

$$\frac{V_{c/\text{target}}}{V_{i/\text{target}}} = \frac{V_{\text{true/calibrator}}}{V_{i/\text{calibrator}}}.$$

If valid, we simply rearrange to get

$$V_{c/\text{target}} = \frac{V_{\text{true/calibrator}}}{V_{i/\text{calibrator}}} V_{i/\text{target}}.$$

So, the trick is to estimate $V_{\text{true/calibrator}}$ in the best possible manner, which amounts to estimating the angular diameter of your calibrator. Here are a few ways of going about selecting calibrator stars for this important process, recognizing that whatever errors you have in the calibrator's estimated visibility propagate into the angular diameter you deduce for the target star.

- Spectral type and luminosity class or spectral type and trigonometric parallax (if a good one exists)—For example: HD 83362 was used to calibrate CHARA's 2004 Regulus observations and has the following observationally determined properties: Sp.Type = G8; Lum. Class = III. According to *Allen's Astrophysical Quantities*, a star with these properties should have a radius of 13 solar radii and an absolute V magnitude of +0.80. The relationship $(m-M) = 5\log(D/10)$ gives a "spectroscopic parallax" of 0.095 arcsec corresponding to a distance of 105 pc. At that distance, the star would have an angular diameter of 0.79 mas. This would result in visibilities of 0.92 and 0.69 at baselines of 150 m and 300 m, respectively, in the k-band infrared.

  Adopting the Hipparcos trig parallax of 6.19 ± 0.89 mas instead, one gets an angular diameter of 0.75 mas leading to visibilities at the two baselines of 0.92 and 0.72.

  This looks pretty good, but there are considerable risks in basing calibration upon data that represents mean values from a large sample of stars. What if the star is slightly evolved? What if the spectral type is inaccurate? Likewise for the parallax—is it accurate? What if the star is a rapid rotator and not spherical? And—here's the big risk—what if it's a binary rather than a single star? There is a non-trivial probability of that being the case, and your values of $V_{i/\text{calibrator}}$ will likely be affected by the presence of the second star.

  This is a good way to screen potential calibrators, but final selection needs to be more refined. This leads us to the next bullet...

- From radiative transfer theory, it can be shown that at spectral frequency ν there is a relationship between the integrated flux f_ν, determined from photometry, and the surface flux F_ν, which measures the net rate of energy flow across a unit area in the atmosphere per unit frequency (see, for example, Mihalas' Stellar Atmospheres equation 1–7). The two quantities are related by

$$f_\nu = 2\left(\frac{R}{D}\right)^2 \pi \int I_\nu(0, \mu)\mu d\mu = \left(\frac{R}{D}\right)^2 F_\nu$$

for a star of radius R and distance D and where $F_\nu = \sigma T_{\text{eff}}^4$ and σ is the Stefan–Boltzmann constant. This leaves us with

$$f_\nu = \frac{1}{4}\Theta^2 F_\nu.$$

This lets you estimate the angular diameter Θ by fitting the photometrically observed spectral energy distribution f_ν, sampled by multiple brightness measurements at a variety of photometric bandpasses, to a model stellar atmosphere. The model will give you the effective temperature T_{eff}, the integrated flux over all frequencies F_{bol}, and the best-fit angular diameter for the calibrator from which you can then calculate $V_{\text{true/calibrator}}$ and you are all set to calibrate your measurements of the instrumental visibilities for the star of interest.

A lot of trouble, but you arrive at a good result!

- Two final points on calibration:
  - Pick calibrators as unresolved by the interferometer as possible, which is a challenge for very long baselines. This is because the effect of the uncertainty of the calibrator's diameter increases with decreasing visibility, i.e., the more resolved is the calibrator, the larger error you will get for the target star.
  - When taking data, each observing sequence of the target star should be sandwiched between observations of the calibrator, i.e., *cal–target–cal–target–cal* or even better *cal1–cal2–target–cal1–cal2–target ... target–cal1*. And, yes, you will spend more time observing calibrators than science targets.

Now that we have calibrated visibilities, we simply go back to the visibility equations for single and binary stars in Section B.4 and calculate a fine grid of values that we can use to best fit a diameter and estimate the error of that fit.

B.5.4 Basic Considerations

The resolution of an interferometer derives from the fact that a baseline produced by a pair of telescopes possesses sensitivity to the object structure at an angular scale inversely related to the size of the baseline. Thus, longer baselines resolve finer structure.

Thus, the simplest interferometer is a pair of telescopes that can recover the amplitude of the complex visibility function:

$$V(k, B) = \int d\alpha \, d\beta A(\alpha, \beta) \, F(\alpha, \beta) e^{-ik(\alpha Bx + \beta By)},$$

where $k = 2\pi/\lambda$, B is the baseline vector, α and β are orthogonal angular coordinates (most conveniently taken as RA and Dec), and B_x and B_y are the projected components of the baseline in the coordinate frame. A describes the effective collecting apertures (normally a circular entrance pupil) while F is the wavelength dependent brightness distribution across an object. The van Cittert–Zernike theorem states that an interferometer's response is related to the Fourier transform of the brightness distribution F.

Because visibility is related to a Fourier transform, it is standard to define spatial frequencies u and v such that:

$$u = B_x/\lambda = kB_x/2\pi$$

and

$$v = B_y/\lambda = kB_y/2\pi$$

for which the visibility becomes

$$V(k, B) = \int d\alpha \, d\beta \, A(\alpha, \beta) \, F(\alpha, \beta) e^{-2\pi i(\alpha u + \beta v)},$$

which has dimensions of power.

The spatial frequency coordinates u and v, in units of cycles/arcsec (sometimes also expressed as $M\lambda$ i.e., mega wavelengths), are given by:

$$u = (B_E \cos h - B_N \sin b \sin h + B_L \cos b \sin h)/206{,}265\lambda$$

and

$$v = (B_E \sin \delta \sin h + B_N(\sin b \sin \delta \cos h + \cos b \cos \delta)$$
$$- B_L(\cos b \sin \delta \cos h - \sin b \cos \delta))/206{,}265\lambda,$$

where h = object hour angle, δ = object declination, b = interferometer latitude, and E, N, and L represent the east, north and vertical components of the baseline. The optical delay, which must be compensated in an interferometer to equalize path lengths, is given by

$$\Delta = B_E \cos \delta \sin h - B_N \sin b \cos \delta \cos h - \cos b \sin \delta)$$
$$+ B_L(\cos b \cos \delta \cos h + \sin b \sin \delta).$$

B.5.5 Closure Phase

In order to produce images of an object from interferometric observations, one must have both the amplitude and phase at all sampled points in the incoming wave front. Standard two-telescope interferometry yields visibilities, which only give the amplitude at the sampled point. But, if one can find a method for recovering the visibility phase in addition to its amplitude, then images can be reconstructed of the object.

Each telescope samples an incoming wave front at a particular phase, but even if the telescopes of an array are perfectly "phased up," the atmosphere introduces differential phase displacements because the light is sampled by each telescope through a unique column of air. With three or more telescopes, a method first described in 1958 by the British radio astronomer Roger C. Jennison (1922–2006) results in the cancellation of the atmospheric components. Jennison called it "self-calibration." A minimum of three telescopes is required in this process, which proceeds as follows. For each telescope in a triplet and at each ith instant, a telescope measures a visibility having the phase:

$$\Phi_{obsi} = \Phi_{truei} + \varphi_{atmi}.$$

Consider the three possible phase differences in a triplet:

$$\psi_1 = \Phi_{obs(1-2)} = \Phi_{true(1-2)} + (\varphi_{atm1} - \varphi_{atm2}),$$
$$\psi_2 = \Phi_{obs(2-3)} = \Phi_{true(2-3)} + (\varphi_{atm2} - \varphi_{atm3}), \text{ and}$$
$$\psi_3 = \Phi_{obs(3-1)} = \Phi_{true(3-1)} + (\varphi_{atm3} - \varphi_{atm1}).$$

Defining *closure phase* as:

$$\psi_{closure} = \psi_1 + \psi_2 + \psi_3,$$

you will see from above that this has the happy effect of canceling out the atmospherically-induced phase errors, each of which appear twice with opposite signs. For a circularly symmetric object, like a normal star, the closure phase will be zero. Non-zero closure phases indicate asymmetric structure in the object. For N telescopes, there are $(N-1)(N-2)/2$ closure phases and $N(N-1)/2$ baseline pairs. As a result of the diurnal rotation of the Earth, the projected baselines from a ground-based multiple-telescope interferometer will change, enabling the interferometer to sample different point in the object's (u, v) plane as described below. This permits for a large body of closure phases and visibility amplitudes to be accumulated from which an image can be reconstructed.

B.5.6 (u, v) Coverage and Array Considerations

As a result of Earth rotation, h changes with time, causing the (u, v) points to change from moment to moment. This has the great advantage of providing visibility measurements over a large ensemble of spatial frequencies sampled by regions within the array's effective aperture. *Earth rotation aperture synthesis* is an important tool for interferometric imaging of complex objects. Figure B.24 shows the (u, v) coverage from the originally funded CHARA Array with five telescopes and how the placement of a sixth telescope was chosen so as to fill in a region of the (u, v) plane that would have been inadequately sampled without that addition.

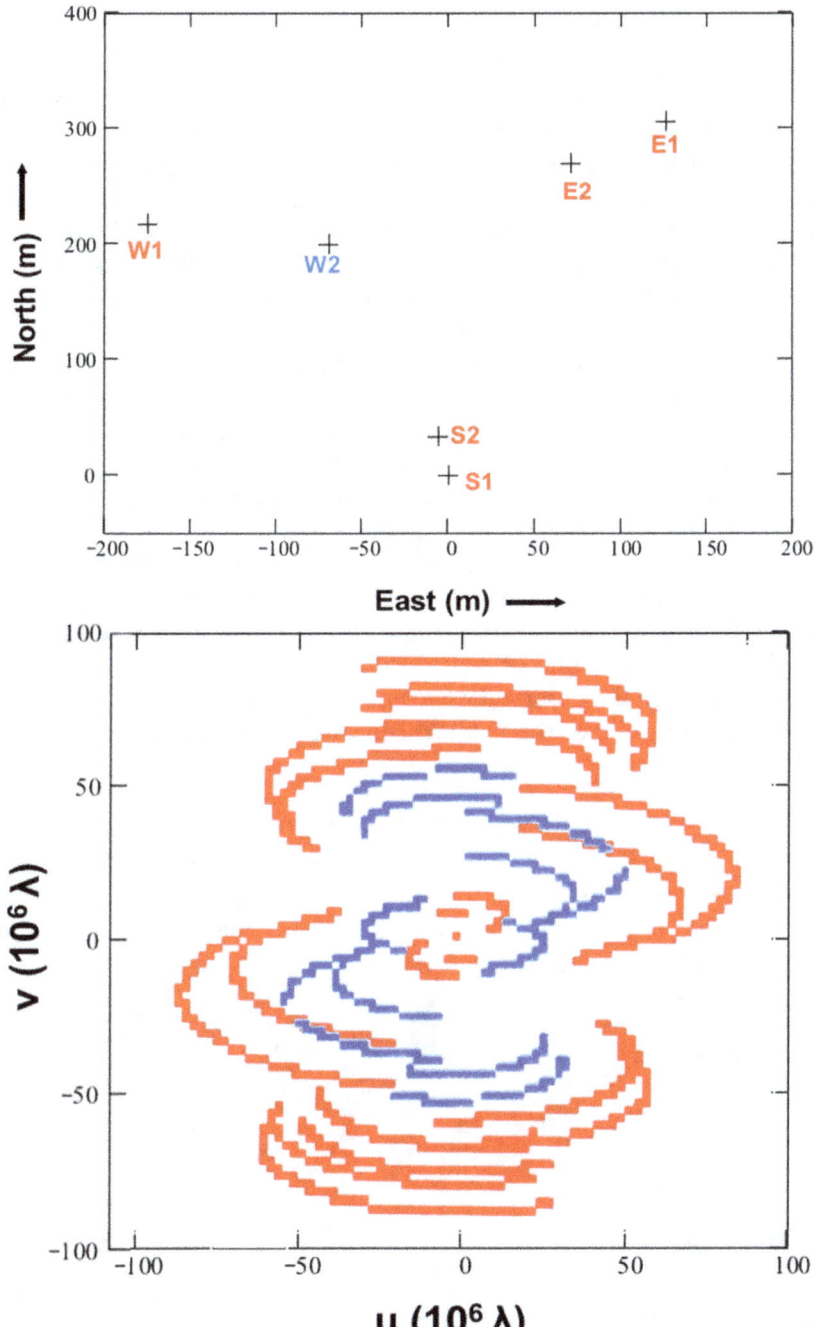

Figure B.24. (Above) The five originally funded telescopes in red give rise to the red (u, v)-plane (below), which was supplemented by the blue coverage from the sixth telescope later funded at position W2. (Author's diagram.)

B.6 Astronomical Seeing

Mount Wilson is a natural location for good seeing. As shown schematically in Figure B.25, the cold Pacific Ocean along the Southern California coast acts like a heat sink in calm weather and supports a prevailing laminar air flow onto the Los Angeles basin, which climbs gradually to the foothills of the San Gabriel Mountains. That weirdly east–west range rises abruptly and forms an obstacle that causes the air flow to become turbulent as it crests the San Gabriel ridges. The dissipation of this turbulent energy through a cascading effect of ever small eddies results in the production of a shortest or "inner scale" having a characteristic dimension of l_o that originated from the beginning longest or "outer scale" L_o eddies.

This approximation to the complexities of atmospheric turbulence was originally developed around 1940 by Andrey Kolmogorov (1903–1987). His approach was expanded by David L. Fried in 1965 to a full description of wave front distortion in which Fried specified what amounts to the effective aperture that the atmosphere imposes on a telescope looking through Kolmogorov turbulence at a given time (Fried 1965). *Fried's parameter r_o* is a measure of the characteristic scale of turbulence and might be thought of a bit simplistically as a patch of air of uniform density, temperature, and index of refraction. If r_o were the size of your telescope aperture, you'd have, at least for an instant until the wind blows another turbulent eddy, or eddies, into your line of sight, a diffraction-limited telescope. The Fried parameter is a function the observed wavelength λ, the zenith angle z of your viewing line of sight, and a profile of the turbulence power as a function of altitude C_n^2:

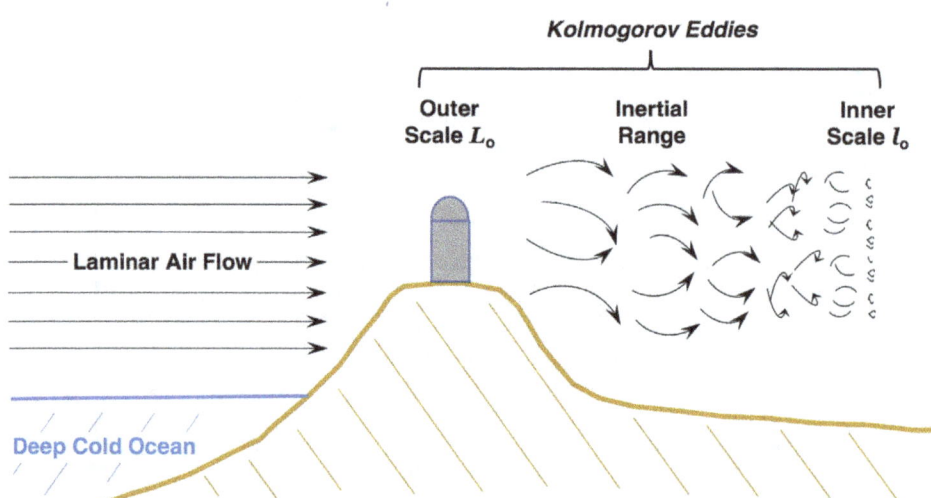

Figure B.25. A schematic of how laminar air flow off the cold Pacific Ocean is made turbulent when it hits a mountain barrier. The outer scale of turbulence initially set up results in less curvature induced on incoming starlight wave fronts yielding more compact seeing disks. Farther inland, the turbulent eddies have degraded into smaller scales that produce higher spatial frequency curvature and greater blurring. This is why sites like Mount Wilson normally have better seeing than do sites on the interior of the continent. Mauna Kea on Hawaii is an ideal manifestation of these circumstances. (Author's diagram.)

$$r_o \sim \lambda^{6/5} [\sec z \, C_n^2]^{-3/5}.$$

The resulting seeing disk of an unresolved star imaged by a telescope through turbulence characterized by r_o is given by:

Seeing Disk FWHM (in radians) $\sim 0.98\lambda/r_o$,

where the wavelength λ scales as $\lambda^{-0.2}$. This relation yields 1 arcsec seeing when r_o equals 10 cm and $\lambda = 500$ nm.

Another representation of seeing is the *atmospheric redistribution time* τ_o, essentially the time it takes the atmosphere to move the distance corresponding to the size of an r_o seeing cell. It is approximately given by:

$$\tau_o \sim \frac{0.3 r_o}{V}$$

where V is the mean wind speed. It is equally important as r_o and sets an upper limit to integration times for speckle interferometry and for actuator response times for adaptive optics. There is no correlation between τ_o and r_o, and very large values of r_o that would yield great seeing conditions might be totally undone by a high wind speed V giving a τ_o that one's instrument cannot keep up with.

Sites with excellent astronomical seeing have local ground or "boundary layer" turbulence and tropospheric conditions that give rise to large r_o and long τ_o.

B.7 Speckle Interferometry

Labeyrie's invention of speckle interferometry was the result of a brilliant insight into the nature of astronomical seeing and the power of the Fourier transform in pulling out scientific results from what appear to be chaotic images. We will here give a very brief overview of speckle interferometry or speckle imaging as it is now more commonly referred to as there is no such thing as a "speckle interferometer." All that is needed to see astronomical speckle is a high-speed camera with detector pixels that resolve structure at the diffraction limit of a given telescope. Because speckle techniques utilize the full aperture of a telescope rather than slits or sub-apertures as do true interferometers, the formal resolution limit of the method is given by the Rayleigh criterion:

$$\theta_{\text{lim}} = 1.22\lambda/A \text{(in radians)}$$

where λ is the wavelength of the observation and A is the aperture of the telescope's objective mirror or lens. Plugging in values for the 100 inch telescope gives $\theta_{\text{min}} = 50$ mas. This would be the smallest angular separation that a binary star could have and still be fully resolved by this 2.54 m telescope.

We also focus on binary star speckle imaging as this is what has been done at Mount Wilson and dominates the scientific product of the field. Two speckle images of binaries are shown in Figure B.26 demonstrating the pairing of speckle that is the telltale sign of a binary in a speckle frame. In surveys for new doubles, the observer often knows immediately if the star being observed is a binary just from glancing at the display of incoming speckle images.

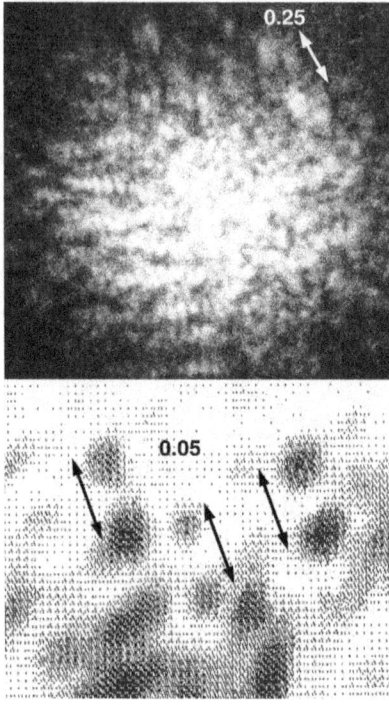

Figure B.26. Speckle images taken at the Kitt Peak 4 m telescope of the binary stars κ Ursae Majoris (above) and Capella (below). Capella is quite bright, and the identified speckle pairs in this small section of a speckle frame are each diffraction-limited images of the binary star. (Author's images and diagram.)

The namesake speckle, a term from laser light scattering from surfaces that Labeyrie brought into his discovery, arise from the wave front entering a telescope having suffered delays and tilts from the r_o-size turbulence patches that rearrange themselves, dissolving and reforming on τ_o timescales. At any instant, there will be many random samples of cells across the telescope pupil that are of the same phase and will constructively interfere to produce a speckle. The larger the aperture, the more of these common-phase cell groupings there will be, and so one detects hundreds of speckles in a snapshot to freeze them in place. Since each set of common-phase patches has components sampling the full aperture of the telescope, each resulting speckle will be a noisy rendition of an Airy disk subject to the limiting resolution of the Rayleigh criterion. This also means that the recording of sharply imaged speckles requires the use of a relatively narrow bandwidth filter because of the dependence on λ of their sizes.

Not only does speckle possess higher resolution than classical methods, it also yields those data with significantly enhanced accuracy. This is seen in Figure B.27. While it requires large aperture telescopes of limited accessibility, speckle observing is capable of measuring 200+ stars on a single night. As a result of these favorable attributes, speckle imaging has largely replaced visual micrometry as the standard method of following the orbital motions within resolved binary star systems. This is

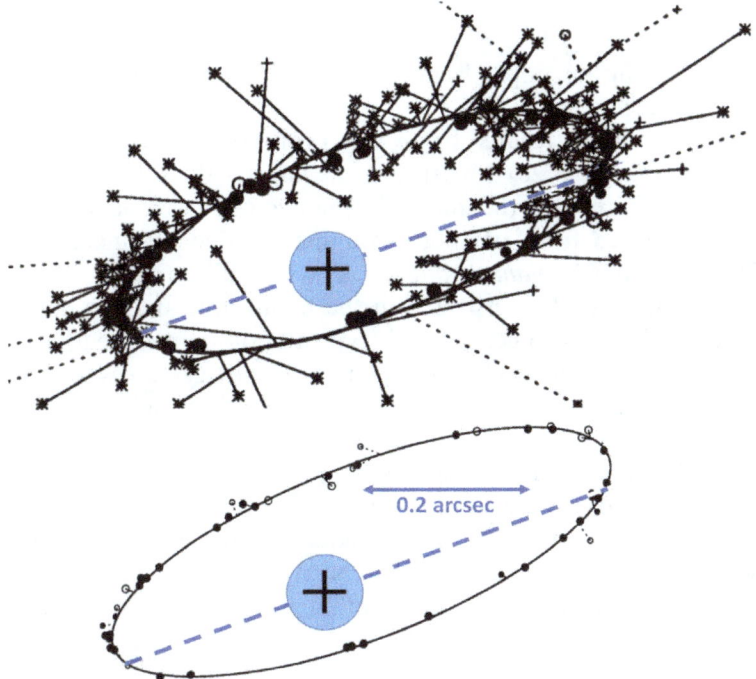

Figure B.27. These measurements of the binary system κ Pegasi obtained by visual micrometry (above) and speckle imaging (below) plotted along with the speckle orbit clearly demonstrate the enhanced accuracy obtained using Labeyrie's technique. The blue disks at the origin represent the limiting resolution in these observations. (Author's images and diagram.)

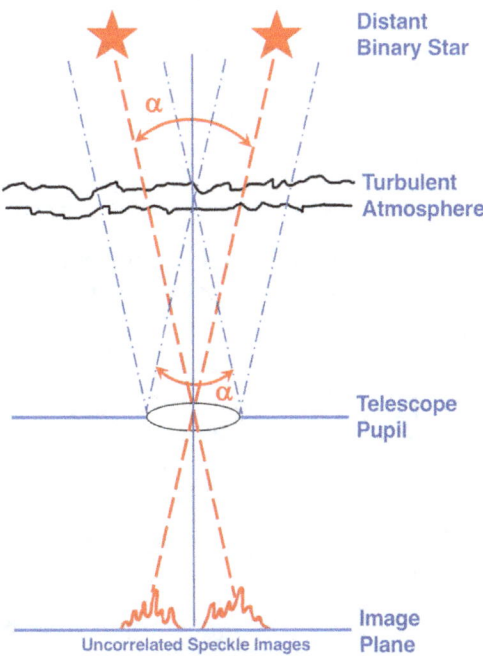

Figure B.28. The two stars have an angular separation at which no parts of their wave fronts collected by the telescope pass through the same turbulence region. They therefore have no mutual correlation and are not amenable to measurement by speckle techniques. (Author's diagram.)

in spite of there being an upper limit to the angular separation that the modern technique can handle.

This wide-angle limit results from the property of *isoplanatism* as shown schematically in Figure B.28. Measuring the (θ, ρ) values for a binary star via speckle methods requires that the images of the two components have a high degree of correlation obtained from passing through the same turbulence patch screen. Figure B.28 defines the *isoplanatic angle* α at which all correlation is lost and the speckle analysis fails. As demonstrated in the figure, this angle is dependent upon altitude of the dominant turbulence and is typically thought to be a few arcseconds. That is more than adequate to give speckle interferometry access to the majority of binary stars with periods amenable to orbit studies ranging from months to hundreds of years.

The origin of speckle and Labeyrie's original Fourier processing scheme for binary star observations are shown in Figure B.29. We begin on the top line of the figure with a pair of unresolved stars of equal brightness represented by two δ functions separated by angular resolution ρ. Next, upon passing in common through a layer of atmospheric turbulence, the object intensities $O(x)$ are convoluted against

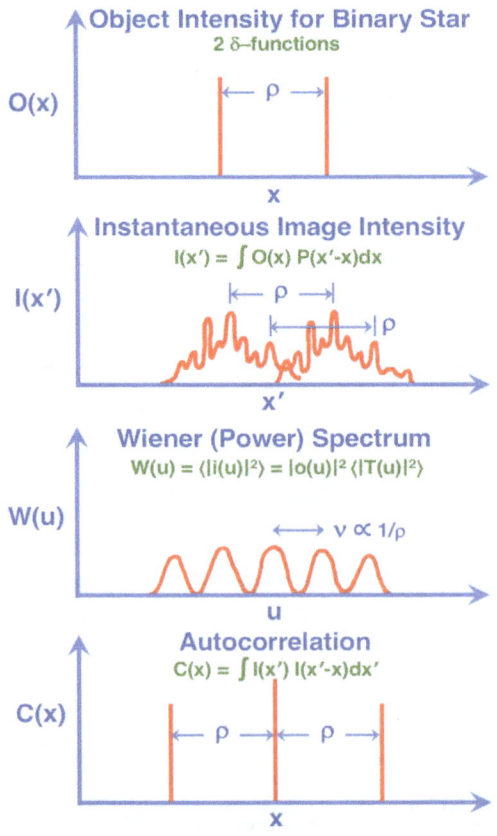

Figure B.29. Outline of Labeyrie's original speckle processing. (Author's diagram.)

the point-spread function of the shared patch of atmosphere $P(x')$ to produce the highly correlated speckle images at $I(x')$. Direct integration of these frames will quickly blur out the speckles to give a standard Gaussian-like seeing disk possessing no information at the diffraction limit.

An ensemble of speckle images are thus recorded in quantity with an appropriately high-speed camera for post hoc or real-time processing, which can take the form of the final two lines in the diagram. The first step is to take the squared modulus of the two-dimensional Fourier transform of each frame and add them in Fourier space. The result is the mean spatial-frequency power spectrum of the ensemble, integrated to provide a high signal-to-noise ratio for accurate measurement. In this process, all phase information is lost, with the result that an inverse transformation does not return an image and instead yields the autocorrelation of the binary. This faithfully provides the (θ,ρ) measurements that, along with the time of the observation, are the observables needed for orbit determination. However, the brightness ratio of the two stars is lost in this process and a 180° ambiguity in θ arises, which can sometimes lead to adoption of the wrong orbital period and eccentricity. Subsequent evolution of the processing of binary star speckle data has eliminated this ambiguity and recovers the magnitude difference Δm.

References

Benson, J. A., Dyck, H. M., & Howell, R. R. 1995, ApOpt, 34, 51

Claret, A. 2020, A&A, 634, 93

Fried, D. L. 1965, JOSA, 55, 1427

ten Brummelaar, T. A., McAlister, H. A., & Ridgway, S. T. 2005, ApJ, 628, 453

www.ingramcontent.com/pod-product-compliance
Lightning Source LLC
Chambersburg PA
CBHW080403190526
45161CB00003B/115